METHODS IN MOLECULAR BIOLOGY™

Series Editor
John M. Walker
School of Life Sciences
University of Hertfordshire
Hatfield, Hertfordshire, AL10 9AB, UK

For further volumes:
http://www.springer.com/series/7651

Prenatal Gene Therapy

Concepts, Methods, and Protocols

Edited by

Charles Coutelle

National Heart and Lung Institute, Molecular and Cellular Medicine Section,
Imperial College London, London, UK

Simon N. Waddington

Institute for Women's Health, Gene Transfer Technology Group,
University College London, London, UK

☀ Humana Press

Editors
Charles Coutelle
National Heart and Lung Institute
Molecular and Cellular Medicine Section
Imperial College London
London, UK

Simon N. Waddington
Institute for Women's Health
Gene Transfer Technology Group
University College London
London, UK

ISSN 1064-3745   e-ISSN 1940-6029
ISBN 978-1-61779-872-6 e-ISBN 978-1-61779-873-3
DOI 10.1007/978-1-61779-873-3
Springer New York Dordrecht Heidelberg London

Library of Congress Control Number: 2012937797

Printed on acid-free paper

Humana Press is part of Springer Science+Business Media (www.springer.com)

Preface

The emerging new field of prenatal gene therapy is based on the rapid scientific and technical advances in fetal medicine, molecular biology, and gene therapy over the last two decades. This novel and still preclinical research subject aims at applying gene therapy during pregnancy for the prevention of human diseases caused by early onset congenital or gestation-related conditions. The present volume summarizes the accumulated scientific knowledge and practical experience over more that 15 years of research by leading scientist in the fields of gene therapy, fetal medicine and medical ethics. It provides a unique and comprehensive overview of the concept of prenatal gene therapy, its potential target diseases, its advantages and possible adverse effects and of the ethical and societal implications of this approach. This book contains detailed protocols for vector production, for breeding and husbandry of the animal models, for the surgical procedures of gene delivery in large and small animals, and for the methods of gene transfer analysis. The various chapters are introduced by overviews covering the different vector systems, animal models and analysis methods used in basic research on prenatal gene therapy, and in preparation for human application. Although prenatal disease is the main target of application in this volume, the chapters on vector generation, production, and testing compiled here provide detailed state of the art knowledge useful for other gene therapy projects beyond the scope of fetal medicine.

Written for: Gene therapists, obstetricians, specialists in perinatal medicine, human geneticists, molecular biologists, medical ethicists.

London, UK   *Charles Coutelle*
Simon N. Waddington

Contents

Contributors

KHALIL N. ABI-NADER • *Prenatal Cell and Gene Therapy Group, EGA Institute for Women's Health, University College London, London, UK*

RAUL ALBA • *Institute of Cardiovascular and Medical Sciences, University of Glasgow, Glasgow, UK*

ORESTIS ARGYROS, PHD • *Faculty of Medicine, Molecular and Cellular Medicine Section, National Heart and Lung Institute, Imperial College London, London, UK*

RICHARD ASHCROFT, PHD • *School of Law, Queen Mary, University of London, London, UK*

ANDREW H. BAKER, PHD • *Institute of Cardiovascular and Medical Sciences, University of Glasgow, Glasgow, UK*

CHRISTOPHER J. BINNY • *Department of Haematology, UCL Cancer Institute, London, UK*

ARIJIT BISWAS • *Experimental Fetal Medicine Group, Department of Obstetrics and Gynaecology, Yong Loo Lin School of Medicine, Singapore, Singapore*

MICHAEL BOYD • *Biological Services Unit, Royal Veterinary College, London, UK*

SUZANNE M.K. BUCKLEY, PHD • *Gene Transfer Technology Group, Institute for Women's Health, University College London, London, UK*

DAVID CARR • *Prenatal Cell and Gene Therapy Group, EGA Institute for Women's Health, University College London, London, UK; The Rowett Institute of Nutrition and Health, University of Aberdeen, Aberdeen, UK*

ANIL CHANDRASHEKRAN, PHD • *Department of Surgery and Cancer, Institute of Reproduction and Development Biology, Hammersmith Hospital, Imperial College London, London, UK*

JERRY K.Y. CHAN, MD, PHD • *Experimental Fetal Medicine Group, Department of Obstetrics and Gynaecology, Yong Loo Lin School of Medicine, Singapore, Singapore; Department of Reproductive Medicine, KK Women's and Children's Hospital, Singapore, Singapore; Cancer and Stem Cell Biology Program, Duke-NUS Graduate Medical School, Singapore, Singapore*

MAHESH CHOOLANI • *Experimental Fetal Medicine Group, Department of Obstetrics and Gynaecology, Yong Loo Lin School of Medicine, Singapore, Singapore*

CHARLES COUTELLE, MD, DSC • *National Heart and Lung Institute, Molecular and Cellular Medicine Section, Imperial College London, London, UK*

ANNA L. DAVID, PHD, MRCOG • *Prenatal Cell and Gene Therapy Group, EGA Institute for Women's Health, University College London, London, UK*

JULIETTE M.K.M. DELHOVE • *Gene Transfer Technology Group, Institute for Women's Health, University College London, London, UK*

MASAYUKI ENDO • *Department of Surgery, Children's Center for Fetal Research and Center for Fetal Diagnosis and Treatment, Children's Hospital of Philadelphia, Philadelphia, PA, USA*

ALAN W. FLAKE, MD • *Department of Surgery, Children's Hospital of Philadelphia, Philadelphia, PA, USA*

RICHARD P. HARBOTTLE, PhD • *Faculty of Medicine, Molecular and Cellular Medicine Section, National Heart and Lung Institute, Imperial College London, London, UK*

BRONWEN R. HERBERT • *Department of Surgery, Children's Center for Fetal Research and Center for Fetal Diagnosis and Treatment, Children's Hospital of Philadelphia, Philadelphia, PA, USA*

STEVEN J. HOWE, PhD • *Molecular Immunology Unit, Wolfson Centre for Gene Therapy, UCL Institute of Child Health, London, UK*

PABLO LAJE • *Department of Surgery, Children's Center for Fetal Research and Center for Fetal Diagnosis and Treatment, Children's Hospital of Philadelphia, Philadelphia, PA, USA*

CITRA N. MATTAR • *Experimental Fetal Medicine Group, Department of Obstetrics and Gynaecology, Yong Loo Lin School of Medicine, Singapore, Singapore*

TRISTAN R. MCKAY • *Department of Endocrinology, William Harvey Research Institute, Charterhouse Square London EC1M 6BQ*

VEDANTA MEHTA • *Prenatal Cell and Gene Therapy Group, EGA Institute for Women's Health, University College London, London, UK*

AMIT C. NATHWANI, MD, PhD • *Department of Haematology, UCL Cancer Institute, London, UK*

STUART A. NICKLIN, BSc (HONS) PhD • *Institute of Cardiovascular and Medical Sciences, University of Glasgow, Glasgow, UK*

DONALD PEEBLES, MD • *Prenatal Cell and Gene Therapy Group, EGA Institute for Women's Health, University College London, London, UK*

AHAD A. RAHIM, PhD • *Gene Transfer Technology Group, Institute for Women's Health, University College London, London, UK*

JESSICA L. ROYBAL • *Department of Surgery, Children's Center for Fetal Research and Center for Fetal Diagnosis and Treatment, Children's Hospital of Philadelphia, Philadelphia, PA, USA*

MICHAEL THEMIS, PhD • *Gene Therapy and Genotoxicity Research Group, Brunel University, London, UK*

JESSE D. VRECENAK • *Department of Surgery, Children's Center for Fetal Research and Center for Fetal Diagnosis and Treatment, Children's Hospital of Philadelphia, Philadelphia, PA, USA*

SIMON N. WADDINGTON, PhD • *Institute for Women's Health, Gene Transfer Technology Group, University College London, London, UK*

JACQUELINE WALLACE • *The Rowett Institute of Nutrition and Health, University of Aberdeen, Aberdeen, UK*

SUET PING WONG, PhD • *Faculty of Medicine, Molecular and Cellular Medicine Section, National Heart and Lung Institute, Imperial College London, London, UK*

Chapter 1

The Concept of Prenatal Gene Therapy

Charles Coutelle and Simon N. Waddington

Abstract

This introductory chapter provides a short review of the ideas and practical approaches that have led to the present and perceived future development of prenatal gene therapy. It summarizes the advantages and the potential adverse effects of this novel preventive and therapeutic approach to the management of prenatal diseases. It also provides guidance to the range of conditions to which prenatal gene therapy may be applied and to the technical approaches, vectors, and societal/ethical considerations for this newly emerging field of Fetal Medicine.

Key words: Prenatal (in utero, fetal) gene therapy, Fetal Medicine, Prenatal disease, Genetic disease, Prevention, Vectors, Animal models, Ethical considerations

Gestation and intrauterine development are closely interrelated physiological conditions with a high vulnerability to adverse influences from endogenous fetal and maternal or from exogenous stress factors. This has led to the development of Fetal Medicine as a specialty that combines the expertise of many disciplines including obstetrics, neonatology, pediatric surgery, and genetics in order to provide the best scientifically based health care to pregnant women and their future offspring. Particular emphasis of Fetal Medicine is placed on early risk detection and prevention or treatment in order to avoid maternal or fetal disease. This book presents prenatal gene therapy, as a promising concept to achieve this goal. Both, fetal and maternal conditions, which may benefit from this approach, are outlined in Chapter 2 of this volume.

Many maternal factors, such as preexisting maternal disease or preeclampsia, impact on fetal development and neonatal outcome. Advances in maternal and neonatal medical care and better identification of pregnancies at risk of these complications have lead to improvements in intact neonatal survival and long-term outcomes. Nevertheless in many obstetric complications, the best current management lies with elective premature delivery of the fetus.

Charles Coutelle and Simon N. Waddington (eds.), *Prenatal Gene Therapy: Concepts, Methods, and Protocols*, Methods in Molecular Biology, vol. 891, DOI 10.1007/978-1-61779-873-3_1, © Springer Science+Business Media, LLC 2012

With increasing insight into many of these diseases, new therapeutic strategies may emerge, which could also include genetic therapies.

Fetal risk-factors such as congenital malformations and some genetic diseases can be identified by ultrasound scan, amniotic fluid or chorionic villus sampling for karyotyping, and more recently, selected DNA diagnosis in known at-risk-pregnancies.

The rapid progress in human genome analysis has made it possible to pinpoint the mutations of virtually any monogenetic disease and to develop methods for DNA diagnosis. The speed by which DNA sequencing technologies advance gives us confidence to predict that individual genome analysis will soon become part of modern medical diagnostics and disease prevention including Fetal Medicine. However, despite these new opportunities for the detection of many, often early in life manifesting, devastating genetic diseases, little progress has been made so far to turn the emerging knowledge of molecular pathogenesis into new therapeutic approaches.

At present, prenatal diagnosis of a severe genetic disease confronts the family with the often desperate dilemma to choose between preparation for postnatal symptomatic care of an affected child or termination of the pregnancy by abortion. Depending on the disease and its severity, therapy is frequently not available or inefficient, expensive, and a great burden on the affected individual and family as well as on society in terms of financial and health care resources (1). Quite obviously new forms of a causal therapy, which may prevent or cure such devastating diseases, are urgently needed.

Since genetic disease is caused by faults in the DNA sequence, gene therapy appears as the most logical strategy to treat such condition. Ideally, therapy for a genetic disease would aim to be effective for a prolonged, if possible life-long, time span after a single vector application.

The advent of gene cloning in the 1970s and the following international endeavor to map and sequence the human genome in the 1980s found one of its practical applications already at the beginning of the 1990s in experimental and even early clinical attempts to use these newly developed tools and techniques for the treatment of genetic disease by postnatal gene therapy. Most present gene therapy approaches aim at supplementing the cells of the affected patients with normal gene sequences in order to reverse the pathological gene expression caused by the underlying genetic defect. All clinical gene therapy trials so far have been conducted postnatally, mostly on adult patients. However, many genetic diseases manifest very early in life and by the time postnatal treatment, including gene therapy, can be applied; the inflicted damage to the patient is often irreversible. This may prevent cure even if normal gene expression were later restored.

The idea that such early and irreversible damage could be prevented if the therapeutic gene sequences could already be delivered prenatally took shape in the early 1990s after the development of surgical methods for intrauterine intervention on the human fetus for the treatment of congenital malformations. These interventions take advantage of scarless healing during fetal life and were first conducted by open uterine techniques (2), followed by fetoscopic (3) methods. First experiments using ex vivo retroviral gene transfer to mid-gestational cord blood cells followed by reinfusion in sheep and primates were started in 1985 and provided evidence for sustained gene transfer in hematopoietic cells for over 2 years (4–6). Transcutaneous ultrasound guided minimally invasive injection of fetal liver cells into the umbilical vein was first applied in the early 1990s to treat severe immune deficiencies and thalassemia (7). By 1995, several studies had also been conducted to explore direct gene delivery in utero using different routes and vectors to apply marker genes to the rodent and sheep fetus (8–13).

Subsequently, these techniques have been perfected and adapted to target different organ systems in several animal models (14). Ultrasound guided minimally invasive transcutaneous delivery techniques, to reach virtually all major fetal organs, which could be applied to the human fetus, have been developed on the sheep model (15) and are beginning to be applied successfully to nonhuman primates. Proof of principle for therapeutic and even curative in utero gene therapy has been shown in several rodent models for human genetic diseases (16).

The advances of in utero gene delivery to the fetus were followed by gene delivery to extraembryonic and maternal tissues (17–20) with a view to treatment of prenatal diseases. Senut et al. (21) showed already in 1998 that ex vivo genetically modified cells, including autologous placenta-derived cells and primary fibroblasts, transplanted into the placenta, produced a transgenic marker protein that was detectable in the fetal body and suggested that this could be used for transient production of therapeutic proteins. More recently, transient prenatal gene delivery approaches have been extended to target maternal organs such as the uterus (22) in order to prevent fetal or maternal complications during pregnancy or delivery (17, 23).

The envisaged benefits of in utero gene delivery have been reviewed by us and others (16, 24–29).

In summary, this concept offers a potential (a) to prevent early onset disease manifestation and tissue damage caused by inherited diseases or by acquired perinatal maternal and fetal conditions; (b) to correct genetic disease permanently by the use of integrating vectors and by targeting the abundant and expanding stem cell populations present during fetal life or by perhaps repeated transduction of postmitotic tissues with nonintegrating vectors; (c) to avoid immune reactions against the vector and/or transgenic protein by administration of the transgene before maturation of the immune system; and (d) to target organs or tissues, which are difficult to access later in life, but which can be easily reached and have a higher capacity of scarless repair in the fetus. The beneficial ratio of fetal body mass to vector is also a distinctive advantage of gene application in utero.

Although simple in concept, the efficient and safe application of gene sequences to the cells of the organism requiring treatment has turned out to be one of the main challenges of gene therapy since its beginnings. Various modified viruses or nonviral nanoparticles are used as vectors to enter the cell and deliver the therapeutic gene sequences. These vectors need to overcome the natural defense barriers of the recipient organism, including its immune system, against viruses or foreign particles, nucleic acids, and proteins without compromising these defenses to fight off natural pathogens. The vectors need to reach and enter a sufficient number of cells in the tissue or organ that requires correction, and to express the therapeutic protein preferentially only in these cells. One of the main tasks of gene therapy research is therefore the development and generation of effective, nontoxic gene delivery vectors in high purity and quantity. Chapters 3–7 review those vectors and their properties, which have been most successfully applied in prenatal applications.

Chapter 8 of this volume presents ex vivo gene therapy methods to modify isolated autologous stem cell populations in vitro, which are then been readministered prenatally to achieve a therapeutic effect. This alternative approach to direct topical or systemic vector application aims at overcoming some of the described barriers and avoiding possible adverse effects arising from indiscriminate vector spread.

Animal studies have played an indispensible role in preparing for clinical application of in utero gene therapy. Over the last 10 years, many studies on fetal rodents, rabbits, dogs, guinea pigs, sheep, and more recently on nonhuman primates have been conducted by several groups in order to investigate and improve safety and efficiency of prenatal transgene delivery and expression and to develop techniques applicable to the human fetus (for review, see refs. 16, 30).

However, differences in scale and physiology have often been the reason for disappointment when trying to translate successful animal experiments into clinical application. It is therefore important to understand the technical and biological advantages and limitations of the different animal models used in preparation for human gene therapy application for prenatal gene therapy. Chapters 9–12 provide overviews and detailed protocols for the murine, sheep, and primate animal models, respectively.

The correct choice and critical validation of the different available physiological and biochemical endpoints, as described in Chapter 13, is an important factor for the correct assessment of therapeutic success in preclinical and clinical gene therapy studies.

Despite successful proof of principle of the concept of in utero gene therapy in animal models, this approach remains presently still at the stage of preclinical research. Already in 1998, in response to a pre-proposal to conduct human in utero gene therapy trials for hemoglobinopathies and Adenosine Deaminase Deficiency (ADA) (31), the US NIH Recombinant Advisory Committee (RAC) concluded that more research into possible adverse effects was needed before any such trial could be considered (32, 33). Indeed, the potential for adverse effects such as vector-mediated oncogenesis, germ line transduction, developmental aberrations, or other so far unknown long-term alterations are frequently cited as reason against clinical application of in utero gene transfer. The last 10 years have seen the first clinical successes of postnatal gene therapy as well as serious adverse effects (34, 35) of which particularly the observation of vector-related oncogenicity (36, 37) is highly relevant to prenatal gene application. Chapters 14–16 of this book present on-going research to identify the risks of gene transfer, to provide a basis for scientific risk assessment, and to increase the safety of prenatal gene therapy using different animal models.

While the benefits of successful prenatal gene therapy to treat and potentially even cure genetic disease has been clearly shown in animal models, a translation to human application will require rigorous demonstration of high reliability and long-term safety before it becomes a realistic alternative option to either the acceptance of an affected child, abortion, or longer-term family planning for in vitro fertilization and embryo selection. Especially in the beginning of such translation, this will be a very difficult choice for a family confronted with genetic disease. The personal, societal, ethically, and legal considerations involved in this decision are discussed in more detail in the final Chapter 17 of this book.

In conclusion, prenatal gene therapy research has now reached a stage of maturity, to provide the tools, which can achieve efficient transgene delivery and therapeutic levels of transgene expression and could technically be applied to the human fetus. However, more studies on rodents and larger mammalian animal models are required to investigate and avoid possible long-term adverse effects of prenatal vector delivery and to study prenatal gene therapy strategies on a per-disease basis. The scientific background and protocols provided in this book may contribute to the longer-term goal of clinical application of prenatal gene delivery for the prevention of severe human genetic diseases and nongenetic prenatal conditions.

References

1. McCandless SE, Brunger JW, Cassidy SB (2004) The burden of genetic disease on inpatient care in a children's hospital. Am J Hum Genet 74:121–127

2. Flake AW, Harrisson MR (1995) Fetal surgery. Ann Rev Med 46:67–78

3. Harrison MR (2000) Surgically correctable fetal disease. Am J Surg 180:335–342

4. Anderson W, Kantoff P, Eglitis M et al (1986) Gene transfer and expression in nonhuman primates using retroviral vectors. Cold Spring Harb Symp Quant Biol 51(Pt 2):1073–1081

5. Eglitis M, Kantoff PW, McLachlin JR et al (1987) Gene therapy: efforts at developing large animal models for autologous bone marrow transplant and gene transfer with retroviral vectors. Ciba Found Symp 130:229–246

6. Kantoff PW, Flake AW, Eglitis MA, Scharf S et al (1989) In utero gene transfer and expression: a sheep transplantation model. Blood 73:1066–1073

7. Touraine JL (1992) In-utero transplantation of fetal liver stem cells into human fetuses. Hum Reprod 7:44–48

8. Hatzoglou M, Lamers W, Bosch F et al (1990) Hepatic gene transfer in animals using retroviruses containing the promoter from the gene for phosphoenolpyruvate carboxykinase. J Biol Chem 265:17285–17293

9. Hatzoglou M, Moorman A, Lamers W (1995) Persistent expression of genes transferred in the fetal rat liver via retroviruses. Somatic Cell Mol Biol 2:265–278

10. Holzinger A, Trapnell BC, Weaver TE, Whitsett JA, Iwamoto HS (1995) Intraamniotic administration of an adenovirus vector for gene transfer to fetal sheep and mouse tissue. Pediatr Res 38:844–850

11. McCray PB, Armstrong K, Zabner J, Miller DW et al (1995) Adenoviral-mediated gene transfer to fetal pulmonary epithelia in vitro and in vivo. J Clin Invest 95:2620–2632

12. Pitt BR, Schwarz MA, Pilewski JM et al (1995) Retrovirus-mediated gene transfer in lungs of living fetal sheep. Gene Ther 2:344–350

13. Tsukamoto M, Ochiya T, Yoshida S et al (1995) Gene transfer and expression in progeny after intravenous DNA injection into pregnant mice. Nat Genet 9:243–248

14. Waddington SN, Kramer MG, Hernandez-Alcoceba R et al (2005) In utero gene therapy: current challenges and perspectives. Mol Ther 11:661–676

15. David A, Peebles D (2007) Gene therapy for the fetus: is there a future? Best Pract Res Clin Obstet Gynaecol 22:203–218

16. Waddington S, Buckley SMK, David AL et al (2007) Fetal gene transfer. Curr Opin Mol Ther 9:432–438

17. David AL, Torondel B, Zachary I et al (2008) Local delivery of VEGF adenovirus to the uterine artery increases vasorelaxation and uterine blood flow in the pregnant sheep. Gene Ther 15:1344–1350

18. Katz AB, Keswani SG, Habli M et al (2009) Placental gene transfer: transgene screening in mice for trophic effects on the placenta. Am J Obstet Gynecol 201(499):e491–e498

19. Laurema A, Vanamo K, Heikkila A et al (2004) Fetal membranes act as a barrier for adenoviruses: gene transfer into exocoelomic cavity of rat fetuses does not affect cells in the fetus. Am J Obstet Gynecol 190:264–267

20. Xing A, Boileau P, Cauzac M et al (2000) Comparative in vivo approaches for selective adenovirus-mediated gene delivery to the placenta. Hum Gene Ther 11:167–177

21. Senut M, Suhr ST, Gage FH (1998) Gene transfer to the rodent placenta in situ. A new strategy for delivering gene products to the fetus. Clin Invest 101:1565–1571

22. Koyama S, Kimura T, Ogita K et al (2006) Transient local overexpression of human vascular endothelial growth factor (VEGF) in mouse feto-maternal interface during mid-term pregnancy lowers systemic maternal blood pressure. Horm Metab Res 38:619–624

23. Zenclussen ML, Anegon I, Bertoja AZ et al (2006) Over-expression of heme oxygenase-1 by adenoviral gene transfer improves pregnancy outcome in a murine model of abortion. J Reprod Immunol 69:35–52

24. Coutelle C, Douar A-M, College WH, Froster U (1995) The challenge of fetal gene therapy. Nat Med 1:864–866

25. Senut M, Gage FH (1999) Prenatal gene therapy: can the technical hurdles be overcome? Mol Med Today 5:152–156

26. Porada CD, Park P, Almeida-Porada G, Zanjani ED (2004) The sheep model of in utero gene therapy. Fetal Diagn Ther 19:23–30

27. Coutelle C, Themis M, Waddington SN et al (2005) Gene therapy progress and prospects: fetal gene therapy—first proofs of concept—some adverse effects. Gene Ther 12:1601–1607

28. O'Brien B, Bianchi DW (2005) Fetal therapy for single gene disorders. Clin Obstet Gynecol 48:885–896

29. Santore MT, Roybal JL, Flake AW (2009) Prenatal stem cell transplantation and gene therapy. Clin Perinatol 36:451–471, xi

30. David A, Themis M, Waddington S et al (2003) The current status and future direction of fetal gene therapy. Gene Ther Mol Biol 7:181–209

31. Zanjani ED, Anderson WF (1999) Prospects for in utero human gene therapy. Science 285:2084–2088

32. RAC (2000) Prenatal gene transfer: scientific medical and ethical issues. A report of the

Recombinant DNA Advisory Committee. Hum Gene Ther 11:1211–1229

33. Couzin J (1998) RAC confronts in utero gene therapy proposal. Science 282:27

34. Hacein-Bey-Abina S, von Kalle C, Schmidt M et al (2003) A serious adverse event after successful gene therapy for X-linked severe combined immunodeficiency. N Engl J Med 348:255–256

35. Raper SE, Chirmule N, Lee FS et al (2003) Fatal systemic inflammatory response syndrome in a ornithine transcarbamylase deficient patient following adenoviral gene transfer. Mol Genet Metab 80:148–158

36. Li Z, Dullmann J, Schiedlmeier B et al (2002) Murine leukemia induced by retroviral gene marking. Science 296:497

37. Themis M, Waddington SN, Schmidt M et al (2005) Oncogenesis following delivery of a non-primate lentiviral gene therapy vector to fetal mice. Mol Ther 12:763–771

Chapter 2

Candidate Diseases for Prenatal Gene Therapy

Anna L. David and Simon N. Waddington

Abstract

Prenatal gene therapy aims to deliver genes to cells and tissues early in prenatal life, allowing correction of a genetic defect, before irreparable tissue damage has occurred. In contrast to postnatal gene therapy, prenatal application may target genes to a large population of stem cells, and the smaller fetal size allows a higher vector to target cell ratio to be achieved. Early gestation delivery may allow the development of immune tolerance to the transgenic protein, which would facilitate postnatal repeat vector administration if needed. Moreover, early delivery would avoid anti-vector immune responses which are often acquired in postnatal life.

The NIH Recombinant DNA Advisory Committee considered that a candidate disease for prenatal gene therapy should pose serious morbidity and mortality risks to the fetus or neonate, and not have any effective postnatal treatment. Prenatal gene therapy would therefore be appropriate for life-threatening disorders, in which prenatal gene delivery maintains a clear advantage over cell transplantation or postnatal gene therapy. If deemed safer and more efficacious, prenatal gene therapy may be applicable for nonlethal conditions if adult gene transfer is unlikely to be of benefit. Many candidate diseases will be inherited congenital disorders such as thalassaemia or lysosomal storage disorders. However, obstetric conditions such as fetal growth restriction may also be treated using a targeted gene therapy approach. In each disease, the condition must be diagnosed prenatally, either via antenatal screening and prenatal diagnosis, for example, in the case of hemophilias, or by ultrasound assessment of the fetus, for example, congenital diaphragmatic hernia.

In this chapter, we describe some examples of the candidate diseases and discuss how a prenatal gene therapy approach might work.

Key words: Prenatal gene therapy, Congenital disease, Pre-eclampsia, Fetal growth restriction

1. Introduction

Gene therapy came onto the therapeutic scene in the 1980s and the first human gene therapy trials began over 15 years ago (1). But in spite of continuous technological progress most clinical results have been disappointing. The reasons for this are many and include difficulties in targeting the appropriate organ, a robust immune response to the therapy in adults and low level expression

Charles Coutelle and Simon N. Waddington (eds.), *Prenatal Gene Therapy: Concepts, Methods, and Protocols*,
Methods in Molecular Biology, vol. 891, DOI 10.1007/978-1-61779-873-3_2, © Springer Science+Business Media, LLC 2012

of the therapeutic gene product. Many of these difficulties may be avoidable by applying the therapy prenatally, and recent preclinical work has indeed shown proof of principle for phenotypic cure of congenital disease in animal models using this approach. Selecting the right diseases for therapeutic intervention will be critical for clinical translation. Prenatal gene therapy has been proposed to be appropriate for life-threatening disorders, in which prenatal gene delivery maintains a clear advantage over cell transplantation or postnatal gene therapy and for which there are currently no satisfactory treatments available (2). In this chapter we consider the various congenital diseases and obstetric disorders that might be suitable for this therapeutic approach.

2. Is There a Need for Prenatal Gene Therapy?

Congenital disease places a huge burden on the community and the health service. A study of pediatric inpatient admissions in 1996 in a US children's hospital found that wholly genetic conditions accounted for one-third of hospital admissions and for 50% of the total hospital charges for that year (3). Thus a preventative strategy such as prenatal gene therapy could have an important social and economic impact.

A criticism leveled at a prenatal approach is that gene transfer to an individual after birth may be as effective, and probably safer, than prenatal treatment. Indeed current conventional treatment of some genetic diseases is highly effective. For example, hereditary hemochromatosis is caused mainly by mutations in the human hemochromatosis protein but can also be caused by mutations in hemojuvelin, hepcidin antimicrobial peptide, and the transferrin receptor 2, amongst others. The disease, which is characterized by iron overload, results in a variety of pathological changes including liver cirrhosis, cardiomyopathy, diabetes, and arthritis. Highly effective treatment consists of regular bloodletting which reduces iron concentrations to normal. For those who cannot tolerate phlebotomy iron chelation agents such as Deferoxamine are available. For many genetic diseases however, treatment is palliative rather than curative, resulting in patients living longer but with a reduced quality of life. This has been particularly seen in cystic fibrosis, in which life expectancy has risen from school age in 1955 to the mid 30s today (Cystic Fibrosis Foundation). To achieve this however, patients require daily chest physiotherapy, antibiotic treatment, dietary supplementation, insulin for diabetes mellitus, and in many cases, lung transplants, which require immunosuppressive therapy. Effective treatment in utero could cure genetic disease, or at least provide partial correction that may have a huge impact on disease progression.

3. The Advantages of a Prenatal Approach for Some Diseases

1. Prenatal gene therapy may offer particular benefits in particular early onset genetic disorders in which irreversible pathological damage occurs to organs before or shortly after birth (4). For many such diseases, the organ may be difficult to target after birth, for example the lung in cystic fibrosis, the brain in urea cycle disorders, or the skin in epidermolysis bullosa, and prenatal treatment may take advantage of developmental changes to access organs that are inaccessible after birth.

2. Certain obstetric disorders may also be suitable for a prenatal gene therapy approach. Problems of deficient uteroplacental circulation underlie the common obstetric condition, fetal growth restriction (FGR) in which the fetus fails to achieve its genetic growth potential and which is currently untreatable. A targeted local gene transfer to improve the circulation is one approach that is being studied with some success.

3. The fetus has a size advantage in a number of ways for some congenital diseases, especially where transgenic protein expression levels may be important. Production of clinical grade vector is time-consuming and expensive and the small size of the fetus could lead to increased vector biodistribution at the same vector dose as an adult.

4. The fetus has a functionally immature immune system compared to an adult, which may be to its advantage for treatment of some diseases. Worldwide up to 50% of adults have preexisting humoral immunity to adenovirus and adeno-associated virus serotypes from which commonly used gene therapy vectors are derived (5). Even in the absence of a preexisting immune sensitivity, vector administration to adults often results in the development of an immune response that reduces the duration and the level of transgene expression. For example, after intramuscular injection of adenovirus vector containing the dystrophin gene into adult Duchenne muscular dystrophy transgenic mice, antibodies to the dystrophin protein were detected (6). This complication is particularly important when gene therapy is aiming to correct a genetic disease in which complete absence of a gene product is observed.

4. Candidate Diseases

As with any potential therapeutic modality, the risks of prenatal gene therapy are not well characterized and the efficacy is still undetermined. In this regard, the report of the NIH Recombinant

DNA Advisory Committee (7) proposed that initial application of prenatal gene therapy should be limited only to those diseases that:

- Are associated with serious morbidity and mortality risks for the fetus either in utero or postnatally.

- Do not have an effective postnatal therapy, or have a poor outcome using available postnatal therapies.

- Are not associated with serious abnormalities that are not corrected by the transferred gene.

- Can be definitively diagnosed in utero and have a well defined genotype/phenotype relationship.

- Have an animal model for in utero gene transfer that recapitulates the human disease or disorder.

Some of the diseases that may be suitable for fetal treatment are listed in Table 1.

Preclinical studies of prenatal gene transfer are encouraging. Fetal application of gene therapy in mouse models of congenital disease such as hemophilia A (8) and B (9), congenital blindness (10), Crigler–Najjar type 1 syndrome (11) and Pompe disease (glycogen storage disease type II) (12) have shown phenotypic correction of the condition. In the following sections we consider groups of candidate diseases and discuss the factors that will influence the success of a prenatal gene therapy approach.

4.1. The Hemophilias

Inherited blood disorders would be a relatively simple target for prenatal gene therapy as the fetal circulation can be reached during its circulation through the umbilical vein (UV) at the placental cord insertion or the intrahepatic UV, or even via the peritoneal cavity, a route used successfully to transfuse anemic fetuses. Congenital blood disorders are relatively common in some populations, and prenatal screening and diagnostic services are available. Translation of prenatal gene therapy into man is probably most advanced when considering congenital blood disorders such as the hemophilias. Research has progressed from demonstrating proof of principle in mouse models, into larger animal models such as the sheep and non-human primate, where delivery techniques, long-term transgene expression and safety can be better addressed (see Table 2).

Deficiency in factor VIII (FVIII) and FIX proteins of the blood coagulation cascade, result in hemophilias A and B, respectively, and have a combined incidence of around 1 in 8,000 people (13). Current treatment uses replacement therapy with human FVIII or FIX which is expensive but effective (14). Beneficial effects occur after achieving only 1% of the normal levels of clotting factor. Unfortunately, a proportion of patients develop antibodies to therapy leading to ineffective treatment and occasional anaphylaxis (15). The complications of hemophilia treatment which include the major risk of HIV and hepatitis B infections, although less of an

Table 1
Candidate diseases for prenatal gene therapy

Disease	Therapeutic gene product	Target cells/organ	Age at onset	Incidence	Life expectancy
Cystic fibrosis	CF transmembrane conductance regulator	Airway and intestinal epithelial cells	In utero	1:2,000–4,000	Mid-thirties
Duchenne muscular dystrophy	Dystrophin	Myocytes	2 years	1:4,500	25 years
Lysosomal storage disease	Glucocerebrosidase in Gaucher disease	Hepatocytes	9–11 years	1:9,000 overall	<2 years
Spinal muscular atrophy	Survival motor neuron protein	Motor neurons	In utero in type 0, 6 months in type 1	1:10,000	2 years
Urea cycle defects	Ornithine transcarbamylase in ornithine transcarbamalase deficiency	Hepatocytes	2 days	1:30,000 overall	2 days (severe neonatal onset)
Hemophilia	Human factor VIII or IX clotting factors	Hepatocytes	1 year	1:6,000	Adulthood with treatment
Homozygous α-thalassaemia	Globin	Erythrocyte precursors	In utero	1:2,700	Lethal
Severe combined immunodeficiency (SCID)	γc cytokine receptor (X-linked SCID); adenosine toxicity	Haematopoietic precursor cells	Birth	1:1,000,000	<6 months if no bone marrow transplant
Epidermolysis bullosa	Type VII collagen	Keratinocytes	Birth	1:40,000	Milder forms to adulthood
Severe fetal growth restriction	Vascular endothelial growth factor	Uterine arteries	In utero	1:100	Adulthood if individual survives the neonatal period
Congenital diaphragmatic hernia	Lung growth factors	Alveoli	In utero	1:2,200	Adulthood if individual survives neonatal surgery

Table 2
Translating fetal gene therapy into man, using hemophilia treatment as an example

Animal model	Vector and transgene	Vector dose	Delivery route	Outcome
Mouse	Adenovirus encoding luciferase reporter gene	1×10^7 pfu/fetus	Intrahepatic	High-level luciferase expression in neonatal liver (193)
Hemophilia A mouse*	Adenovirus encoding murine FVIII	3.3×10^5 pfu/fetus	IP	Short-term correction of factor VIII deficiency in neonatal period (8)
Mouse	Adenovirus encoding luciferase reporter gene	1×10^7 pfu/fetus	Intrahepatic to fetus and reinjection of 3 month-old neonate	Liver expression of luciferase up to 1 month of neonatal life. Antibodies to vector and protein developed after vector reinjection at 3 months postnatally (194)
Mouse	Adenovirus encoding hFIX	2×10^{11} pfu/fetus	Vitelline vessel	Induced immune tolerance to exogenous hFIX protein but only short-term transgenic hFIX protein expression in neonatal mouse plasma (195)
Hemophilia B mouse*	HIV Lentivirus encoding hFIX	2.0×10^{10} p/kg fetus	Vitelline vessel	Permanent cure of hemophilia with immune tolerance to exogenous hFIX (9)
Hemophilia B mouse*	AAV-1 encoding hFIX	5×10^9 vg/fetus	IM (hindlimb) to fetus and reinjection postnatally	Induction of tolerance and long-term therapeutic hFIX expression (23)

Early gestation sheep	Adenovirus	1.5×10^{12} pfu/kg fetus	USS-guided IP or hepatic injection	Short-term high-level transduction of the liver after IP compared with intrahepatic injection with therapeutic levels of hFIX expression in the plasma (26)
Late-gestation macaque	AAV2 encoding eGFP	8.6×10^{10} p/fetus	USS-guided hepatic injection	Vector genomes in peripheral blood at birth (136)
Late-gestation macaque	Lentivirus encoding eGFP	1×10^{7} p/fetus	USS-guided IP injection	High-level liver transduction for at least 9 months postnatally (136–138, 196)
Early and late-gestation sheep	scAAV8 encoding hFIX	1×10^{12} vg/kg fetus	IP injection by USS	hFIX expression in blood up to 6 months but no immune tolerance (28)
Late-gestation macaque	scAAV5 and 8 encoding hFIX	1.0–1.95×10^{13} vg/kg fetus	USS-guided UV injection	hFIX expression in blood and liver for at least 1 year, non-neutralizing immune response (197)

Studies have moved from using short-term expressing vectors such as adenovirus to long-term expressing vectors such as AAV and lentivirus, in more clinically relevant animals such as the sheep and non-human primate

hFIX human factor IX clotting factor, *AAV* adeno-associated virus vector, *eGFP* enhanced green fluorescent protein, *USS* ultrasound, *IM* intramuscular, *IP* intraperitoneal

* animal model of disease, *vg* vector genomes, *p* particles, *pfu* plaque forming units, *sc* self-complementary vector

issue now that blood donors are screened effectively, have in some cases been far worse than the diseases themselves, increasing their morbidity and mortality (16). The clotting proteins are required in the blood and can be secreted functionally from a variety of tissues, thus the actual site of production is not so important as long as therapeutic plasma levels are realized.

Adult gene therapy strategies have concentrated on application to the muscle or the liver, achieving sustained FIX expression in adult hemophiliac dogs or mice after intramuscular or intravascular injection of adeno-associated virus (AAV) vectors (17–19). Chao et al. observed that the AAV serotype is important. In mice, injection of AAV serotype 1 resulted in tenfold higher levels of canine FIX when compared with serotype 2 (19). In clinical trials using AAV2hFIX vectors in hemophilia B patients, only short-term and low level FIX expression was observed, however, which was probably caused by a cell-mediated immune response to transduced hepatocytes (20, 21). A clinical trial using a self-complementary AAV8 vector serotype, which may stimulate less of an immune response, is currently underway in adults with hemophilia B (22).

Waddington et al. demonstrated permanent phenotypic correction of immune-competent hemophiliac mice by intravascular injection of a lentivirus vector encoding the human Factor IX (hFIX) protein at 16 days of gestation (term = 22 days) (9). Plasma factor levels remained at 10–15% of normal in treated animals for their lifetime. Sabatino et al. subsequently demonstrated induction of tolerance after AAV-1-hFIX administration in Factor IX-deficient fetal mice (23).

Translation to large animals has been slower because of the need for longer-term gene transfer and a higher vector dose when compared to small animals, but recent studies demonstrate the potential for this route of delivery. Long-term transduction of hematopoietic stem cells in the bone marrow and blood could be demonstrated 5 years after delivery of retroviral vectors into the peritoneal cavity of early gestation fetal sheep at laparotomy (24). In early gestation, delivery of adenovirus vectors into the umbilical vein of fetal sheep via hysterotomy resulted in widespread transduction of fetal tissues (25). Using ultrasound-guided injection, systemic vector spread and widespread tissue transduction was demonstrated after first trimester intraperitoneal injection of adenovirus vectors into fetal sheep, although direct injection of the umbilical vein was limited by the procedure-related high mortality in late first trimester (26).

More recently, using a self-complementary AAV8 vector expressing the hFIX gene which has a high transduction capacity in mice and macaques (27), long-term hFIX expression has been demonstrated after ultrasound-guided intraperitoneal injection of fetal sheep in early and late gestation (28). No functional antibodies could be detected against the vector or transgene product and there was no liver toxicity observed. Antibodies to the therapeutic gene were detectable when the animals were challenged at 6 months of age postnatally with the hFIX recombinant protein, showing that induction of immune tolerance was not achieved. This is probably due to the fall in hFIX expression that was undetectable by 1 year after birth and a higher initial vector dose may be required to maintain hFIX levels. Umbilical vein delivery in fetal non-human primates of a tenfold higher dose of the same self-complementary AAV system in late gestation produced clinically relevant levels of hFIX sustained for over a year, with liver-specific expression and a non-neutralizing immune response (29).

Hemophilia A and B do not usually manifest until after birth. Deficiency of some clotting factors, however, can lead to life-threatening neonatal central nervous system hemorrhage. For example, congenital factor VII deficiency, the most common auto-somal bleeding disorder, is typified by severe or lethal bleeding in around 20% of patients with a homozygous or compounds heterozygous genotype. They present with <1% of normal levels and often present bleeding in the central nervous system (30, 31). A therapy delivered during the fetal period could avoid long-term pathology and provide therapeutic transgene expression for life. Moreover, increasing expression even above 1% would be considered enough to substantially improve the risk and incidence of spontaneous hemorrhage (31, 32).

A second bleeding diathesis, which would benefit from a fetal or early neonatal gene therapy approach is congenital factor X deficiency. As per factor VII, factor X is synthesized by hepatocytes and constitutes a central component of the coagulation cascade. In its severest form it can present at birth as severe and fatal intracra-nial hemorrhage (33). The Rosen laboratory generated a strain of mouse with a disruption of the factor X gene, which manifests as partial embryonic lethality or fatal neonatal bleeding (34). The same laboratory went on to perform fetal injection of wild-type hepatocytes into factor X-deficient recipients and were able to achieve phenotypic rescue (35).

4.2. The Thalassaemias and Sickle Cell Disorders

Inherited abnormalities of hemoglobin (Hb), a tetramer of two α-like and two β-like globin chains, are a common and global problem. Over 330,000 affected infants are born annually world-wide, 83% with sickle cell disorders and 17% with thalassaemias (36). Screening strategies can be premarital and/or antenatal depending on socio-cultural and religious customs in different populations and countries. Prenatal diagnosis is available in many countries from 11 weeks of gestation using chorionic villus sampling or amniocentesis from 15 weeks.

The β-thalassaemias, including the hemoglobin E disorders, are not only common in the Mediterranean region, South-East Asia, the Indian subcontinent, and the Middle East but have spread to much of Europe, the Americas, and Australia owing to migration of people from these regions. Approximately 1.5% of the global population are heterozygotes or carriers of the β-thalassemias. In the most severe form of β-thalassaemia, Cooley's anemia or β-thalassaemia major, there is a profound anemia that if untreated, leads to death in the first year of life. Even a mild correction of the globin chain imbalance in a fraction of maturing erythroblasts reduces the morbidity caused by ineffective erythropoiesis, and improves outcome (37). Affected children become dependent on regular blood transfusions, leading to iron deposition, organ failure, and death. Chelation therapy to remove the circulating iron is not effective. Postnatal allogeneic haematopoietic stem cell transplantation (HSCT) has been developed over the last 30 years and can cure the condition with recent results of 90% survival and 80% thalassaemia-free survival (38). However, HSCT is only available in approximately 30% of cases because of the lack of a suitable matched donor (37), and it is associated with complications such as Graft Versus Host Disease.

In alpha-thalassemia there is a deficit in the production of the Hb α globin chains. Underproduction of α globin chains gives rise to excess β-like globin chains which form tetramers, called Hb Bart's in fetal life and HbH in adult life. Compound heterozygotes and some homozygotes for α-thalassaemia have a moderately severe anemia with HbH in the peripheral blood. Finally, some individuals who make very little or no α globin chains, have a severe anemia, termed Hb Bart's hydrops fetalis syndrome which is commonly diagnosed prenatally and which, if untreated causes death in the neonatal period (39).

Sickle cell disorders are caused by Hb gene variants that similar to thalassaemia reduce mortality from falciparum malaria in carriers, and leads to high carrier levels in malaria endemic countries. The abnormal HbS cells in the circulation leads to recurrent painful sickle cell crises. Current treatment relies on a number of strategies to prevent crises from occurring using, for example, prophylactic antibiotics, pneumococcal vaccination and good hydration, and effective crisis management using oxygen and pain-relief (40).

Effective gene therapy in thalassaemia and sickle cell disease will depend on understanding the regulation of the globin genes. In β-thalassaemia, for example, the tissue and developmental-specific expression of the individual globin genes is governed by interactions between the upstream β-globin locus control region (β-LCR) and the globin promoters (41). Amelioration or even cure of mouse models of human sickle cell disease (42) and β-thalassemia major (43–45) has been achieved using lentivirus vectors that contain complex regulatory sequences from the LCR region. Injection of this optimized lentiviral vector into the yolk-sac vessels of fetal mice at mid-gestation resulted in human alpha-globin gene expression in the liver, spleen, and peripheral blood in newborn mice and expression peaked at 3–4 months reaching 20% in some recipients (46). Expression declined at 7 months of age (normal lifespan 2–3 years) possibly due to insufficient HSC transduction or the late stage of mouse gestation at which the vector was introduced. Work is continuing using a lentivirus vector system that has a natural tropism for the haematopoietic system by way of a ubiquitous chromatin opening element (UCOE) augmented spleen focus forming virus (SFFV) promoter/enhancer (47, 48). This provides reproducible and stable function in bone marrow stem and all differentiated, peripheral haematopoietic cell lineages (49) and may improve long-term expression.

The potential for an ex vivo gene transfer approach for the treatment of thalassaemia was recently demonstrated in an adult patient with severe β-thalassaemia who had been dependent on monthly transfusions since early childhood. After autologous transplantation of gene-modified haematopoietic stem cells using a lentivirus encoding β-globin, he became transfusion independent for at least 21 months. Blood hemoglobin was maintained between 9 and 10 g/dl, of which one-third contained vector-encoded β-globin. The therapeutic effect is due to dominance of a myeloid-based cell clone. Integration of the vector into the host chromosomal site AT-hook 2 (HMGA2) of this clone resulted in transcriptional activation of high mobility group protein (HMGA2) expression in erythroid cells and caused truncation of an inhibitory microRNA-binding site in this gene. The observed cell expansion could be a stochastic and fortuitous event or the result of dysregulation of the HMGA2 gene expression in myeloid stem/progenitor cells (50). Further research is needed to determine whether this clone remains homeostatic or whether its development may be a prelude to multistep leukemogenesis, a problem that has been observed in ex vivo autologous stem cell gene therapy for severe combined immunodeficiency (51).

4.3. Cystic Fibrosis

The fetal lung is an ideal target for prenatal gene therapy because transduction of the fluid-filled fetal lungs may be achieved more easily than in postnatal life, where there is an air-tissue interface. Postnatal lung damage also reduces gene transfer (52). Candidate diseases for lung-directed prenatal gene transfer include cystic fibrosis (CF), alpha-1 antitrypsin deficiency, surfactant protein B deficiency, and pulmonary hypoplasia.

Cystic fibrosis (CF) is a common lethal autosomal recessive disease in which tissue injury begins in the prenatal period (53). The potential targets for CF lung manifestations are the ciliated epithelial cells and ducts of the submucosal glands, where the wild-type CF transmembrane conductance regulator (CFTR) protein is expressed. Correction of as few as 6% of the defective cells may be sufficient to correct the chloride transport defect (54). Although CF is a multisystem disease, much of the morbidity and mortality derives from the diseased lung. Here the classical gene therapy target is the ciliated epithelial cells and ducts of the submucosal glands in the lungs where the wild-type CFTR protein is normally expressed (55). Gastrointestinal manifestations of CF are now increasingly recognized as an important contributor to morbidity in those patients who reach adulthood (56), as well as affecting 15% of neonates with the life-threatening condition of meconium ileus (57). With the advent of prenatal screening for CF, the possibility of offering treatment to couples whose fetus is affected becomes more real (58).

Around 400 CF patients have been given gene therapy postnatally using viral and nonviral gene transfer agents through mainly nebulized systems (59). Early trials established the safety of adenovirus and nonviral vectors but CFTR expression was hindered by the low transduction efficiency of both vector classes on the respiratory epithelium, partly due to the location of the adenovirus receptor in the basolateral membrane of the respiratory epithelium which is isolated from the lumen by tight junctions (60). In addition, a robust immune response caused a dose-dependent inflammation and pneumonia related to the immunogenicity of the viral proteins that prevented repeat administration (61). Based on these shortcomings other vectors were developed for clinical use, such as AAV2 which in clinical trials has reduced toxicity and immunogenicity (62, 63). Unfortunately, these phase I/II trials were in general unsuccessful due to neutralizing antibodies that prevented reliable repeat vector administration. Other vector systems that have been investigated for gene transfer to the lung include AAV1 and five in the rat (64) and helper-dependent adenovirus vector, incorporating a human epithelial cell-specific expression cassette in the rabbit lung (65).

The only present clinical trial for CF uses nonviral gene transfer agents such as polyethylenimine (PEI), cationic lipid 67 (GL67), and DNA nanoparticles which may generate less of an immune response. Proof-of-principle studies on the nasal epithelium show a 25% correction of the molecular defect (66) and expression of hCFTR is seen in sheep transfected with a human CFTR plasmid, complexed with GL67 (67).

One of the barriers to effective gene transfer to the airways in the adult or neonate with CF is that inflammation and damage of the lung precludes effective gene delivery. This could be overcome if gene therapy is applied at a stage of prenatal life where no or minimal lung damage has occurred. Importantly, the fetal lungs are fluid filled and transfection of the fluid-filled fetal airways may be more easily achieved. Fluorocarbon liquids such as perflubron have been used to push vector into the distal fetal airways from injection at the trachea (68) and have been shown to enhance adenovirus-mediated gene expression in normal and diseased rat lungs (52). The proliferating cell population in CF airways are mainly basal cells (69) and these would be the best target in any gene therapy approach.

Initial studies appeared promising, with a report that CFTR-knockout mice could be cured by prenatal adenovirus administration into the amniotic fluid (70). Since the fetus draws amniotic fluid into the lungs during fetal breathing movements, intra-amniotic delivery could provide an efficient route of gene transfer to the airways. Two further studies using the same vector, delivery method and mouse strain as well as a different CFTR-knockout mouse strain have, however, been unable to replicate these findings (71, 72). The high spontaneous survival rate of the CF-mouse strain used in Larson's experiments may explain the initial enthusiasm for the results observed (70). In addition, the inability to cure CF in this model might be due to the strain of mice used, the vector construct which only gives short-term gene transfer (73), or on account of insufficient fetal breathing movements. Nevertheless, gene transfer to human fetal lungs is achievable in a xenograft model in SCID mice with long-term expression in the surface epithelial and submucosal gland cells observed up to 4 weeks and 9 months after administration of adeno-associated and lentiviral vectors, respectively (74, 75).

Transgene expression in the fetal mouse lung can be improved by increasing fetal breathing movements using a combination of intra-amniotic theophylline administration and exposure of the dam to elevated CO_2 levels (76). Theophylline has a similar effect on breathing movements in fetal sheep (77). Much of the vector was diluted in the amniotic fluid volume and not concentrated in the organ(s) required for CF therapy as strong gene transfer to the skin occurred after intra-amniotic delivery (76).

In large animals such as the fetal sheep we were unable to produce significant airways gene transfer after intra-amniotic adenovirus vector injection in the first trimester although the nasal passages were transduced (26). Fetal breathing movements are not present in the first trimester human or sheep fetus, and the large amniotic fluid volume even at this gestation means that a more targeted approach to the lung may be required in clinical practice.

Several studies have applied adeno-associated virus vectors (AAV), many using the amniotic route. Injection of AAV2 into rabbit amniotic fluid transduced the trachea and pulmonary epithelium of the fetus (78) and prolonged gene expression was seen in further studies in mice, rats, and macaques (79). Gene delivery to the lung parenchyma can also be achieved by indirect means using AAV, by intraperitoneal injection, for example (80). Similarly, injection of AAV1 and AAV2 into mouse muscle, peritoneal cavity, and intravenously gave lung expression of the transgenic protein (81).

Local injection of the lung parenchyma is an alternative to the amniotic route but gives only local gene transfer in fetal rats (82, 83) and non-human primates (84). The stage of gestation is important, with transgene expression more local to the lung after vector injection in early second trimester (pseudoglandular stage) when compared to the late first trimester (embryonic stage) (85).

In larger animals, injection of the fetal trachea by transthoracic ultrasound-guided injection (86) targets gene transfer to the medium to small airways (68). Increased transgene expression in the fetal trachea and bronchial tree was seen after complexation of the virus with DEAE-dextran, which confers a positive charge to the virus, and pretreatment of the airways with sodium caprate, which opens tight junctions in the airway epithelia thereby improving vector access to the coxsackie-adenovirus receptors (68, 87). For gastrointestinal CF pathology widespread intestinal transduction was achieved using ultrasound-guided gastric injection in the early gestation fetal sheep (88) that had an associated low morbidity and mortality. Transgene expression was enhanced after pretreatment of the fetal gut with sodium caprate after adenovirus complexation with DEAE-dextran. In addition, instillation of the fluorocarbon perflubron after virus delivery resulted in tissue transduction from the fetal stomach to the colon.

Another potentially important application of prenatal gene therapy to the lung lies in fetuses affected with congenital diaphragmatic hernia (CDH), a condition where lung hypoplasia causes significant morbidity and mortality. Adenovirus-mediated prenatal CFTR expression enhanced saccular density and air space in the lungs of a rat model of CDH (89). Short-term expression of growth factors at a critical stage of lung growth may be useful for this serious condition. After surgical creation of CDH in fetal sheep, nonviral vector expression of keratinocyte growth factor in the trachea lead to increased surfactant protein B synthesis in the lungs suggesting better maturation of the regrowing lung (90).

4.4. Diseases of the Nervous System

Early lethal genetic diseases of the nervous system are individually rare, yet collectively lead to a large disease burden, and in some populations have a high prevalence (91). Conditions can directly affect the nerves themselves, such as spinal muscular atrophy (SMA), a disease primarily of the peripheral nervous system. Alternatively, enzyme deficiencies can lead to a damaging buildup of lysosomal substrates that damage neurons, as well as other organs in the body. Examples include the lysosomal storage diseases such as acute neuronopathic (Type II) Gaucher disease, neuronal ceroid lipofuscinoses, and Niemann–Pick disease type C. In some cases, these conditions are not recognized during fetal life on prenatal ultrasound examination. For example, there are a few case reports that some fetuses with SMA have increased nuchal translucency; however, a recent study in 12 women with affected fetuses did not find any association (92). In Niemann–Pick disease type C, however, in utero splenomegaly, hepatomegaly, ascites, fetal growth restriction, and oligohydramnios (reduced liquor volume) are common (93). Screening programs are available in populations with high prevalence such as the Ashkenazi Jews, where triple disease screening for Tay–Sachs disease, type 1 Gaucher disease, and cystic fibrosis (CF) are commonly performed together (94). Prenatal diagnosis is available for these conditions via chorionic villus sampling assuming the gene defect is known.

4.4.1. The Lysosomal Storage Diseases

The lysosomal storage diseases are inherited deficiencies of lysosomal enzymes that lead to intracellular substrate accumulation. In mucopolysaccharidosis type VII (MPS type VII), for example, a deficiency of β-glucuronidase activity leads to accumulation of glycosaminoglycans in lysosomes (95) leading to enlarged liver and spleen, growth and mental retardation, and death from cardiac failure. Lysosomal storage diseases can manifest during intrauterine life as nonimmune hydrops (96). Although rare, MPS VII has been a disease of choice to investigate gene therapy because of the availability of a mouse and dog model. Correction of the MPS phenotype theoretically requires only low levels of the therapeutic gene product (97). Neonatal injection of a retrovirus vector in MPS VII dogs and mice resulted in hepatocyte transduction, with uptake of the enzyme from the circulation by other organs. The treated animals did not develop cardiac disease or corneal clouding and skeletal, cartilage, and synovial disease was ameliorated (98). Nonviral mediated gene transfer to the liver of MPS I and VII mice also improved the phenotype (99). Still, the major challenge remains to target the brain which currently requires multiple brain injections with accompanying risks (100, 101) and immunosuppression to prevent pan-encephalitis that develops secondary to an immune response to the transgene (101). Widespread correction of the pathological lesions in an MPS VII mouse was recently observed with adeno-associated virus gene transfer (102), a vector which elicits less of an immune response. Prenatal gene delivery is an alternative strategy. Injection of adenovirus into the cerebral ventricles of fetal mice led to widespread and long-term gene expression throughout the brain and the spinal cord (103). In the same study, delivery of a therapeutic gene to the cerebral ventricles of fetal MPS type VII mice prevented damage in most of the brain cells before and until 4 months after birth. A similar study using an AAV vector had comparable results but with longer expression (104).

From a translational perspective, direct vector administration into the fetal brain or ventricles for prenatal gene transfer is unappealing. There are technical difficulties in injecting the fetal brain through the skull using minimally invasive injection techniques, although this has been achieved in non-human primate (105) and sheep (A.L. David, unpublished work) under ultrasound guidance. In contrast, ultrasound-guided access to the human fetal circulation is commonly used for fetal blood sampling and transfusions in clinical practice, with minimal fetal loss rate or complications (106). This triggered the hunt for vectors that could cross the blood/brain barrier.

Recently, AAV vectors of serotypes 2/9 have been shown to have an astonishing ability to transduce cells of the nervous system, achieved not by intracranial but via intravenous injection in neonatal mice (107, 108), cats (108), and non-human primates (109). The ability of the vector to cross the blood–brain barrier may depend on specific populations of receptors within the brain that facilitate transfer for particular AAV serotypes (110). A recent study describing fetal intravenous injection of AAV 2/9 in either single-stranded or self-complementary format showed comprehensive transduction of the central nervous system, including all areas of the brain and retina, and the peripheral nervous system including the myenteric plexus. Interestingly, the single stranded version, containing a woodchuck hepatitis virus posttranscriptional regulatory element (WPRE) achieved far higher and more comprehensive levels of expression than the self-complementary vector lacking WPRE (111).

Prenatal gene transfer has also been applied with some success in glycogen storage disease type II (GSDII), which is caused by a deficiency in acid α-glucosidase (GAA). This leads to lysosomal accumulation of glycogen in all cell types and abnormal myofibrillogenesis in striated muscle with death from respiratory failure. Delivery of the AAV-*GAA* vector by intraperitoneal injection to the mouse embryo in knockout models gave high-level transduction of the diaphragm and restoration of its normal contractile function (12).

Neuronal ceroid lipofuscinoses (NCLs), known collectively as Batten disease, are autosomal recessive lysosomal storage diseases which have lead to significant central nervous system pathology. Infantile NCL, caused by mutations in the CLN1 gene, results in deficiency in palmitoyl protein thioesterase 1 (PPT1). Patients with the disease are born with no pathological manifestations but by 12 months of age they show signs of mental retardation, motor dysfunction, and visual problems (112) and survive only to 6 years of age, on average. A mouse model of this disease shows many of the same pathological symptoms and premature death occurs by 8.5 months (113). Although there have been no attempts as yet at treating this model by fetal gene therapy, there is a strong case to be made for this in a preclinical and clinical setting.

4.4.2. Spinal Muscular Atrophy

Spinal muscular atrophy (SMA) is characterized by degeneration of the lower motor neurons in the anterior horn of the spinal cord and the brainstem. Although rare (incidence 1:10,000) (114), SMA is invariably fatal after a course of progressive muscle weakness and atrophy. It is caused by homozygous loss or mutation in the telomeric survival motor neuron gene 1 (SMN 1) with subsequent neuronal cell death through apoptosis. Affected individuals can be partially protected by the presence of an increased copy number of the SMN2 gene (114), a nearly identical copy gene of

SMN1, that produces only 10% of full-length SMN RNA/protein. This suggests that SMN2 may play a disease-modifying role and could be a target for gene therapy of the disease.

The childhood forms which are all autosomal recessive can be divided into three types depending on their severity (115). The fetal form of the disease, type 0, presents in utero with diminished fetal movements and arthrogryposis (116). Neuronal degeneration and loss in SMA type I begins during intrauterine life which makes prenatal gene therapy an attractive option (117). Vectors derived from adenovirus, herpes simplex virus (HSV), adeno-associated virus (AAV), and lentivirus are capable of transducing neurons in vitro and in vivo (115). Neuroprotective factors such as cardiotrophin 1 (118) or anti-apoptotic proteins such as Bcl-xL can be used, for example, adenovirus Bcl-xL has been shown to inhibit neuronal cell death in rat cell cultures (119). SMN gene replacement is also possible. Multiple intramuscular injections of a RabG–EIAV lentivirus vector containing the human SMN gene increased SMN protein levels in SMA type 1 fibroblasts and in SMA mice and reduced motor neuron death (120).

Lower motor neurons can be targeted by direct injection of the spinal cord which, although successfully achieved in the rat and mouse postnatally (121, 122), is technically risky in the adult human. Injection of an AAV8 containing the human SMN gene into the CNS of SMA mice improved mortality (123) although they still died prematurely despite continual, high-level expression from the viral vector, which may have been due to a failure to correct the autonomic system that regulates cardiac function.

An alternative is remote delivery, and motor neuron gene expression has been achieved by intramuscular or intra-axonal injection with subsequent retrograde axonal transport in small animals (124, 125). A therapeutic effect was documented after intramuscular injection of adenovirus-cardiotrophin 1 in a mouse model of SMA (118). It is not clear, however, if remote delivery will be effective in larger animals or an affected human where the peripheral nerves are much longer and retrograde transport is impaired secondary to the disease. Fetal application in this context may provide the advantage of a shorter and healthy axon and recent results in mice suggest that a fetal approach is feasible. Using lentivirus vectors, that are efficient at infecting nondividing cells Rahim et al. observed transduction of multiple dorsal root ganglia and efferent nerves following intrathecal injection of an EIAV (equine infectious anemia virus) lentivirus vector into fetal mice at 16 dpc (126).

Systemic delivery using AAV vectors is probably the way forward and might also correct the cardiac dysfunction that occurs. Foust et al. incorporated SMN cDNA into the 2/9 vector serotype and showed that neonatal intravenous injection of this vector into the corresponding mouse model of SMA resulted in an unprecedented improvement in survival and motor function (127). Using a self-complementary AAV9 vector containing a codon-optimized SMN1 sequence injected intravenously on day 1 postnatal, Dominguez et al. achieved 100% rescue of a mouse model of severe SMA, completely corrected motor function and reduced the weight loss associated with this model (128).

4.4.3. The Urea Cycle Defects

The inherited inborn errors of metabolism result from enzyme deficiencies in different metabolic pathways. One of the first metabolic disorders targeted by gene therapy is the defect in the urea cycle, ornithine transcarbamylase deficiency (OTC), an X-linked condition which results in accumulation of ammonia with resultant repeated episodes of hyperammonemia within 1 week of life, damaging the central nervous system and jeopardizing life (129, 130). A phase I trial targeting the liver through intra-arterial adenovirus injection ended with low level gene transfer and a fatal immune reaction in one of the 17 patients (131, 132). Subsequent investigation in small animals focused on less immunogenic vectors and showed that long-term correction of the metabolic defect in OTC deficiency could be achieved using a helper-dependent adenovirus vector (133) and AAV (134). Because of the difficulties with postnatal OTC deficiency gene therapy and the severity and very early onset of the complete form, fetal gene therapy may be a good approach.

A notable success in small animal models is in the long-term correction of bilirubin UDP-glucuronyltransferase deficiency in fetal rats using a lentivirus vector (11). Humans who suffer from this defect are classified as having Crigler–Najjar type 1 syndrome and suffer severe brain damage early in childhood due to the inability to conjugate and excrete bilirubin. A rat model of Criggler–Najjar was injected with a lentivirus vector carrying the gene for bilirubin UDP-glucuronyltransferase. The treated rats sustained a 45% decrease in serum bilirubin levels for more than a year, a level that would be considered therapeutic in the human (11). Despite the long-term expression, these rats developed antibodies against bilirubin UDP-glucuronyltransferase (135), which may be related to the fact that the fetal injection was done late in fetal life and to the unusual immunogenicity of the transgenic protein.

Intravascular vector delivery in small animals can give excellent liver transduction but in larger animals such as the fetal sheep, the intraperitoneal route seems to be the best route to target gene transfer to the fetal liver (26), and no immune response to the transgenic protein was detected after injection of adenovirus vectors in early gestation. Direct intrahepatic injection resulted in low level gene transfer with necrosis of the liver around the injection site, which was thought to be due to a direct toxic effect of adenovirus vector on hepatocytes (26). Studies of intrahepatic injection using other vectors such as lentivirus or AAV show better results.

In the non-human primate, intrahepatic or intraperitoneal (IP) vector injection resulted in widespread gene transfer and particularly to the liver, with no transplacental transfer to the mother (136–139). In one of these studies, however, IP injection of lentivirus vector at the end of the first trimester showed that a subset of female but not male germ cells were transduced (137). In the ovary, meiosis begins in the innermost areas of the cortex during the 12th and 13th weeks of gestation, while proliferating primordial germ cells forming the oocytes are found in the most superficial areas of the cortical region of the developing ovary, and these may be vulnerable to lentiviral gene transfer when delivered early in gestation via the IP route. Since IP injection is a relatively safe and well-studied ultrasound-guided fetal injection method (140), this is likely to be the route of choice when compared to liver injection, which is used rarely in fetal medicine for diagnosis of congenital liver disease. The risk of germline gene transfer in female fetuses will need to be evaluated carefully.

Metabolic diseases other than ornithine transcarbamylase deficiency and bilirubin UDP-glucuronyltransferase deficiency that could benefit from fetal gene therapy to the liver are phenylketonuria, galactosemia, and long-chain acyl-CoA dehydrogenase deficiency.

4.5. Muscular Dystrophy

Targeting the muscle for gene delivery could be a successful strategy for treatment of muscular dystrophies. Duchenne muscular dystrophy (DMD) is the commonest form and is X-linked. Abnormal or absent dystrophin leads to progressive muscle weakness in early childhood, culminating in death secondary to respiratory or cardiac failure during the third decade of life. For a one step prenatal gene therapy, the striated muscles in the limbs and chest, and the cardiac muscle would need to be transduced. Prenatal diagnosis is available, and carriers can be screened for the presence of a male fetus using noninvasive prenatal diagnosis, avoiding the need for invasive tests and the associated miscarriage risk in 50% (141).

In adult clinical trials, dystrophin gene transfer to striated muscle using viral (142) and nonviral vectors (143) has been hampered by low efficacy because of the development of cellular and humoral immunity to the transgenic dystrophin gene (144, 145). This could be avoided by prenatal application, which would also target a rapidly proliferating population of myocytes that are present in the fetus. Satellite cells that are capable of regenerating muscle fibers are transduced after intramuscular lentivirus vector delivery to fetal mice (146). Importantly, inducible dystrophin expression begun during fetal life corrected the phenotype in a DMD mouse model, where postnatal expression did not (147), supporting a fetal approach.

Intramuscular fetal injection of an adenovirus containing the full-length murine dystrophin gene in the mdx mouse model of DMD conferred effective protection from cycles of degeneration and regeneration normally seen in affected muscle fibers (148), but gene transfer level was low. More efficient gene transfer to all necessary muscle groups was seen after delivery of lentivirus vectors to fetal mice using multiple routes of injection. Systemic delivery targeted the heart, direct injection transduced the limb musculature and intraperitoneal injection reached the diaphragm and innermost costal musculature. Expression lasted for over 15 months and did not stimulate any immune response (149).

Large animal muscle gene transfer has been investigated. Gene delivery to the hindlimb musculature of the early gestation fetal sheep using ultrasound-guided injection of adenovirus vectors resulted in highly efficient gene transfer with a low procedure complication rate (26). A clinically relevant method for respiratory muscle gene transfer has also been evaluated in early gestation fetal sheep and showed that transduction of intercostal muscles occurred after ultrasound-guided creation of a hydrothorax into which adenovirus vectors were introduced (150).

There has been considerable recent success using AAV vectors to transduce fetal musculature. Early studies on AAV showed long-term local transgenic protein expression following direct injection into fetal mouse muscle (151) and transduction of the diaphragm after IP injection (152). Using this route to administer AAV1, Rucker and colleagues restored acid α-glucosidase activity to the diaphragm in a mouse model of Pompe disease (12). This was the first demonstration that fetal gene transfer could correct a model of congenital muscle pathology. More recently, studies on intraperitoneal delivery of AAV8 into normal fetal mice show high levels of marker gene expression in all the muscle groups affected by congenital muscular dystrophies (153), and in the mdx mouse model of DMD, delivery of an AAV containing dystrophin significantly improved the dystrophic phenotype (154). Postnatal application of AAV6 containing full length and micro-dystrophins in neonatal mice can almost entirely prevent and partially reverse the pathology associated with DMD, but only near the site of injection (155). In late-gestation macaques, umbilical vein injection of AAV9 results in very high levels of transgenic protein expression in many tissues including skeletal and cardiac muscle (29). Systemic delivery of AAV vectors in the fetus may finally provide a solution to target the necessary muscle groups for muscular dystrophy therapy. However, the packaging capacity of AAV is restricted to delivery of truncated dystrophin minigenes which may negatively counteract the efficiency of this vector system.

Exon skipping is another strategy that is proving quite successful. An antisense oligonucleotide is used to modify splicing, such as skipping of the mutated exon 51, which allows a partly functional dystrophin protein to be produced from the muscle. This therapy has been successful in the *mdx* mouse and a dog model of DMD, and there are currently three phase III trials internationally (156).

4.6. The Genodermatoses

The genodermatoses are a group of genetic skin diseases that may be associated with significant morbidity and mortality. Examples include the epidermolysis bullosa (EB) disorders, the ichthyotic disorders, and disorders of pigmentation such as oculocutaneous albinism. Methods of prenatal diagnosis are varied. Where the molecular defect is known, amniocentesis or chorionic villus sampling is commonly used. X-linked ichthyosis is associated with low levels of unconjugated estriol, one of the markers used for Down's syndrome screening, and this can prompt prenatal diagnosis in the mother. When the gene mutation is unknown, however, fetal skin biopsy is necessary that unfortunately carries a slightly higher fetal loss than other invasive tests (1–3%), may result in scarring, and may need to be performed at quite late gestations (157). Ultrasonography can be used in the diagnosis of a few of these disorders. In harlequin ichthyosis, for example, typical sonographic features include echogenic amniotic fluid, large joint and digital contractures and facial dysmorphism, including flat face and wide mouth with thickened lips.

The genodermatoses may be good candidates for prenatal gene therapy, where gene transfer to the skin via the amniotic fluid may provide obvious advantage to cumbersome postnatal therapy. Transgenic protein expression is seen in the skin after intra-amniotic delivery of adenoviral vectors to mice (12 days postconception (dpc)) (76), and sheep in the early first trimester (day 33 of 145 days of gestation) using ultrasound-guided injection (26). In a mouse model of Herlitz junctional epidermolysis bullosa, a lethal skin disease, a combination of adenovirus and AAV vectors injected into the amniotic cavity of fetal mice (14 dpc) led to expression of the laminin-5 transgenic protein although only a minor increase in the lifespan of treated mice was seen (158). In all these studies, only the most superficial layers of the skin, the periderm and epidermis were transduced. Several strategies have been used in small animals to target the deeper layers, such as intra-amniotic injection with subsequent electroporation (159) or application of microbubble-enhanced ultrasound (shot-gun method) (160, 161). Translation to clinical practice will be challenging. Earlier in gestation, epidermal stem cell populations are accessible for gene transfer using the intra-amniotic delivery route. Injection of lentivirus vectors between day 8 and 12 dpc in fetal mice resulted in long-term transgenic protein expression in basal epidermal stem cells into adulthood (162). Using a skin-specific keratin 5 promoter instead of the cytomegalovirus promoter also improved epidermal gene transfer.

4.7. Primary Immune Deficiencies

The primary immune deficiencies result from inherited mutations in genes required for the production, function or survival of specific leukocytes such as T, B or NK lymphocytes, neutrophils and antigen-presenting cells, or are caused by cytotoxic metabolites. The leukocytes are produced from the pluripotent hematopoietic stem cells (HSC) in the bone marrow, and therefore allogeneic bone marrow transplantation (BMT) from a healthy donor into an affected patient can restore the immune system. Successful BMT has been achieved in deficiencies such as Wiskott–Aldrich Syndrome (WAS), Chronic Granulomatous Disease (CGD) and Adenosine Deaminase Deficiency where there is toxicity rather than a defect of the proliferation gene and in X-linked Severe Combined Immunodeficiency (SCID). With the exception of X-SCID all of the other primary immune deficiencies require extra therapeutic steps such as pre-transplant conditioning, marrow cytoreduction to "make space" in the marrow for the transplanted HSC and immune ablation to prevent rejection of the donor HSC. These strategies carry risks for the patient and in some cases, a haploidentical donor is unavailable. Gene therapy has therefore been developed for treatment of some patients.

Postnatal gene therapy using in vitro transduced autologous HSCs with subsequent transplantation into the same patient has been used successful in adenosine deaminase deficient SCID (163, 164), X-linked SCID (165, 166), and CGD (167). Despite the encouraging results, 4 out of 26 subjects subsequently developed a T-cell leukemia-like condition which may have been related to integration of the retroviral vector near a suspected proto-onco-gene (168). Newer approaches to decrease this risk have used lentivirus vectors that have been studied in non-human primates (169). Also, semi-viral systems have been developed with the aim to offer stable gene transfer along with a favorable pattern of integration (170, 171). These semi-viral systems are still limited by their low transduction efficiency as in the case of the sleeping beauty transposon (170, 171).

Prenatal gene therapy in early gestation before the maturation of the immune system could, theoretically, eliminate the need for marrow conditioning or the restriction to an HLA-matched donor. Prenatal treatment with hematopoetic stem cell transplantation has been attempted for a variety of immunodeficiencies and hemoglobinopathies using IP transfer of paternal or maternal hematopoietic cells or fetal liver (172); however, the clinical successes were mainly in cases of X-linked SCID (173, 174) where no immune response to the transplanted cells could be mounted.

An autologous stem cell gene transfer approach (175) using fetal stem cells from a number of sources within the fetus including the blood, liver, amniotic fluid (AF), and placenta could be adopted. Fetal liver or blood sampling at an early gestational age carries a significant risk of miscarriage (176, 177). It is now apparent that pluripotent stem cells can be readily derived from fetal samples collected at amniocentesis (178) or chorionic villus sampling (179, 180), procedures that have a low fetal mortality. Human AFS cells have the potential to differentiate into a variety of cell types and can be transduced easily without altering their characteristics (178, 181, 182). Recent work in sheep described good fetal survival after autologous AF mesenchymal stem cell (MSC) transplantation using ultrasound-guided amniocentesis, and subsequent IP injection of selected, expanded, and transduced AFMSCs into the donor fetus. Widespread cell migration and engraftment, particularly in the liver, heart, muscle, placenta, umbilical cord, and adrenal gland was seen (183).

Prenatal diagnosis of the primary immune deficiencies is available where the gene mutation is known. For example, it has been applied in families that have been identified to be at risk of these conditions, such as those harboring mutations in both of the recombination-activating genes RAG1 and RAG2 that are involved SCID syndromes (184).

4.8. Diseases of the Sensory Organs

Prenatal gene delivery looks promising in eye and ear diseases but translation to man will be challenging. In animal models of Leber congenital amaurosis, a severe retinal dystrophy, fetal gene therapy using AAV or lentivirus vectors resulted in an efficient transduction of retinal pigment epithelium and restoration of visual function (10, 185). Similarly, AAV was able to efficiently transfect the developing cochlea in fetal mice (186). All these previous studies relied on injection into the developing sensory organ itself, which will be difficult to achieve in clinical practice. Vector delivery via the amniotic cavity early during embryonic development depends critically on the stage of gestation. For example, intra-amniotic delivery of lentivirus specifically at day 8 dpc resulted in gene transfer to the mouse retina (187) but later delivery time points were only able to target the lens and cornea. Greater tissue specificity and safety can probably be accomplished by the use of tissue-specific promoters, or regulated transgene expression, but there will still be the need for accurate prenatal diagnosis at a time of gestation equivalent to 3–5 weeks in human pregnancy, something that will be difficult to achieve with current diagnostic techniques.

4.9. Fetal Growth Restriction

Prenatal gene therapy is also being investigated for treatment of obstetric disorders. Severe fetal growth restriction (FGR) affects 1:500 pregnancies and is a major cause of neonatal morbidity and mortality. The underlying abnormality in many cases is uteroplacental insufficiency, whereby the normal physiology process of trophoblast invasion that converts the uterine spiral arteries into a high-flow large conduit for blood, fails to occur. Currently, there is no therapy available that can improve fetal growth or delay delivery to allow fetal maturity. FGR is commonly diagnosed on routine fetal ultrasound when the fetal growth velocity falls below the expected gestational age charts. Abnormally low uterine artery Doppler blood flow and increased vascular resistance is also classically seen in mid-gestation.

A targeted approach to the uteroplacental circulation is needed, since intravascular infusion of sildenafil citrate, a nitric oxide donor, causes a drop in systemic blood pressure and had detrimental effects on growth-restricted sheep fetuses (188). In the pregnant sheep local over-expression of VEGF mediated via adenovirus injection into the uterine arteries increased uterine artery blood flow and significantly reduced vascular contractility (189). VEGF expression was confined to the perivascular adventitia of the uterine arteries, together with new vessel formation, supporting the local effect of gene transfer. Further work suggests these effects are long term, lasting from mid-gestation (80 days) through to term (145 days) (190) with reduced intima to media ratio suggesting vessel remodeling, and adventitial angiogenesis demonstrated. Recent work in an FGR sheep model in which uterine blood flow is reduced by 35% in mid-gestation, demonstrates that uterine artery injection of the same dose of Ad. VEGF significantly improved fetal growth in late gestation (191).

4.10. Fetal Structural Malformations

Fetal structural malformations are another potentially important application of prenatal gene therapy. Although individually rare, collectively up to 1% of all fetuses are affected by a structural malformation that for some are lethal or are associated with significant morbidity. Congenital diaphragmatic hernia (CDH), for example, is a condition where there is a defect of the diaphragm resulting in herniation of some or all of the intra-abdominal organs into the fetal chest. This compresses the fetal lungs preventing adequate growth, which results in poor lung function at birth. With surgical correction of the diaphragmatic defect, many neonates do well. Current management of severe CDH involves fetoscopic placement of an inflatable balloon in the fetal trachea to block outflow of the tracheal fluid, which encourages lung growth (192). There is, however, an underlying lung defect which may contribute to the lung pathology, and gene transfer may play a part in correcting this problem.

Short-term expression of growth factors at a critical stage of lung growth may be useful for this serious condition. In a rat model of CDH, adenovirus-mediated prenatal CFTR expression enhanced saccular density and air space in the lungs (89). After surgical creation of CDH in fetal sheep, non-viral vector expression of keratinocyte growth factor in the trachea lead to increased surfactant protein B synthesis in the lungs suggesting better maturation of the regrowing lung (90).

5. Conclusions

Presently, candidate diseases for prenatal gene therapy would be monogenic diseases that are lethal, during the perinatal period or in early childhood, as well as obstetric conditions or structural malformations of the fetus, in which treatment before birth provides a therapeutic advantage when compared with neonatal or adult gene transfer. Clinically, severe phenotypes may be rescued, in particular increasing the likelihood of intact neurological and other key functions at birth. The broader application of therapy during fetal life will require the establishment of suitable screening programmes for particular conditions, and suitable accurate diagnostic procedures, to allow therapy to be applied in time to have a therapeutic effect.

References

1. Blaese RM, Culver KW, Miller AD, Carter CS, Fleisher T, Clerici M, Shearer G, Chang L, Chiang Y, Tolstoshev P, Greenblatt JJ, Rosenberg SA, Klein H, Berger M, Mullen CA, Ramsey WJ, Muul L, Morgan RA, French Anderson W (1995) Science 270:475–480

2. Wilson JM, Wivel NA (1999) Hum Gene Ther 10:689–692

3. McCandless SE, Brunger JW, Cassidy SB (2004) Am J Hum Genet 74:121–127

4. Coutelle C, Douar A-M, Colledge WH, Froster U (1995) Nat Med 1:864–866

5. Bessis N, GarciaCozar FJ, Boissier MC (2004) Gene Ther 11:S10–S17

6. Gilchrist SC, Ontell MP, Kochanek S, Clemens PR (2002) Mol Ther 6:359–368

7. Recombinant DNA Advisory Committee (2000) Hum Gene Ther 11:1211–1229

8. Lipshutz GS, Sarkar R, Flebbe-Rehwaldt L, Kazazian H, Gaensler KML (1999) Proc Natl Acad Sci U S A 96:13324–13329

9. Waddington SN, Nivsarkar MS, Mistry AR, Buckley SM, Kemball-Cook G, Mosley KL, Mitrophanous K, Radcliffe P, Holder MV, Brittan M, Georgiadis A, Al-Allaf F, Bigger BW, Gregory LG, Cook HT, Ali RR, Thrasher A, Tuddenham EG, Themis M, Coutelle C (2004) Blood 104:2714–2721

10. Dejneka NS, Surace EM, Aleman TS, Cideciyan AV, Lyubarsky A, Savchenko A, Redmond TM, Tang W, Wei Z, Rex TS, Glover E, Maguire A, Pugh EN, Jacobson SG, Bennett J (2004) Mol Ther 9:182–188

11. Seppen J, van der Rijt R, Looije N, van Til NP, Lamers WH, Elferink RPJO (2003) Mol Ther 8:593–599

12. Rucker M, Fraites TJ, Porvasnik SL, Lewis MA, Zolotukhin I, Cloutier DA, Byrne BJ (2004) Development 131:3007–3019

13. Furie B, Limentani SA, Rosenfield CG (1994) Blood 84:3–9

14. Di Minno MN, Di Minno G, Di Capua M, Cerbone AM, Coppola A (2010) Haemophilia 16:e190–e201

15. Lusher JM (2000) Baillieres best practice and research. Clin Haematol 13:457–468

16. Soucie JM, Nuss R, Evatt B, Abdelhak A, Cowan L, Jill H, Kolakoski M, Wilber N (2000) Blood 96:437–442

17. Herzog RW, Yang EY, Couto LB, Hagstrom JN, Elwell D, Fields PA, Burton M, Bellinger DA, Read MS, Brinkhous DM, Podsakoff GM, Nichols TC, Kurtzman GJ, High KA (1999) Nat Med 5:56–63

18. Snyder RO, Miao C, Meuse L, Tubb J, Donahue BA, Lin H-F, Stafford DW, Patel S, Thompson AR, Nichols T, Read MS, Brinkhous DM, Kay MA (1999) Nat Med 5:64–70

19. Chao H, Monahan PE, Liu Y, Samulski RJ, Walsh CE (2001) Mol Ther 4:217–222

20. Manno CS, Pierce GF, Arruda VR, Glader B, Ragni M, Rasko JJ, Ozelo MC, Hoots K, Blatt P, Konkle B, Dake M, Kaye R, Razavi M, Zajko A, Zehnder J, Rustagi PK, Nakai H, Chew A, Leonard D, Wright JF, Lessard RR, Sommer JM, Tigges M, Sabatino D, Luk A, Jiang H, Mingozzi F, Couto L, Ertl HC, High KA, Kay MA (2006) Nat Med 12:342–347

21. Manno CS, Chew AJ, Hutchison S, Larson PJ, Herzog RW, Arruda VR, Tai SJ, Ragni MV, Thompson A, Ozelo M, Couto LB,

Leonard DGB, Johnson FA, McClelland A, Scallan C, Skarsgard E, Flake AW, Kay MA, High KA, Glader B (2003) Blood 101: 2963–2972

22. Allay JA, Sleep S, Long S, Tillman DM, Clark R, Carney G, Fagone P, McIntosh JH, Nienhuis AW, Davidoff AM, Nathwani AC, Gray JT (2011) Hum Gene Ther 22: 595–604

23. Sabatino DE, MacKenzie TC, Peranteau WH, Edmondson S, Campagnoli C, Liu YL, Flake AW, High KA (2007) Mol Ther 15: 1677–1685

24. Porada CD, Tran N, Eglitis M, Moen RC, Troutman L, Flake AW, Zhao Y, Anderson WF, Zanjani ED (1998) Hum Gene Ther 9:1571–1585

25. Yang EY, Cass DL, Sylvester KG, Wilson JM, Adzick NS (1999) J Pediatr Surg 34:235–241

26. David AL, Cook T, Waddington S, Peebles D, Nivsarkar M, Knapton H, Miah M, Dahse T, Noakes D, Schneider H, Rodeck C, Coutelle C, Themis M (2003) Hum Gene Ther 14: 353–364

27. Nathwani AC, Gray JT, Ng CYC, Zhou J, Spence Y, Waddington SN, Tuddenham EGD, Kemball-Cook G, McIntosh J, Boon-Spijker M, Mertens K, Davidoff AM (2006) Blood 107:2653–2661

28. David AL, McIntosh J, Peebles DM, Cook T, Waddington SN, Weisz B, Wigley V, Abi-Nader K, Boyd M, Davidoff AM, Nathwani AC (2010) Hum Gene Ther 22:419–426

29. Mattar CN, Nathwani AC, Waddington SN, Dighe N, Kaeppel C, Nowrouzi A, McIntosh J, Johana NB, Ogden B, Fisk NM, Davidoff AM, David AL, Peebles DM, Valentine MB, Appelt JU, von Kalle C, Schmidt M, Biswas A, Choolani M, Chan JKY (2011) Mol Ther 19(11):1950–1960

30. Peyvandi F, Kaufman RJ, Seligsohn U, Salomon O, Bolton-Maggs PH, Spreafico M (2006) Haemophilia 12 (Suppl 3):137–142

31. McVey JH, Boswell E, Mumford AD, Kemball-Cook G, Tuddenham EG (2001) Hum Mutat 17:3–17

32. Rosen ED, Xu H, Liang Z, Martin JA, Suckow M, Castellino FJ (2005) Thromb Haemost 94:493–497

33. Ermis B, Ors R, Tastekin A, Orhan F (2004) Brain Dev 26:137–138

34. Dewerchin M, Liang Z, Moons L, Carmeliet P, Castellino FJ, Collen D, Rosen ED (2000) Thromb Haemost 83:185–190

35. Rosen ED, Cornelissen I, Liang Z, Zollman A, Casad M, Roahrig J, Suckow M, Castellino FJ (2007) J Thromb Haemost 1:19–27

36. Modell B, Darlison M (2008) Bull World Health Organ 86:480–487

37. Lucarelli G, Andreani M, Angelucci E (2002) Blood Rev 16:81–85

38. Angelucci E (2010) Hematol Am Soc Hematol Educ Program 2010:456–462

39. Harteveld CL, Higgs DR (2010) Orphanet J Rare Dis 5:13

40. Meremikwu MM, Okomo U (2011) Clin Evid (Online) pii:2402

41. Quek L, Thein SL (2007) Br J Haematol 136:353–365

42. Pawliuk R, Westerman KA, Fabry ME, Payen E, Tighe R, Bouhassira EE, Acharya SA, Ellis J, London IM, Eaves CJ, Humphries RK, Beuzard Y, Nagel RL, Leboulch P (2001) Science 294:2368–2371

43. Rivella S, May C, Chadburn A, Riviere I, Sadelain M (2003) Science 294:2368–2371

44. Persons DA, Allay ER, Sawai N, Hargrove PW, Brent TP, Hanawa H, Nienhuis AW, Sorrentino BP (2003) Blood 102:506–513

45. Puthenveetil G, Scholes J, Carbonell D, Qureshi N, Xia P, Zeng L, Li S, Yu Y, Hiti AL, Yee JK, Malik P (2004) Blood 104:3445–3453

46. Han XD, Lin C, Chang J, Sadelain M, Kan YW (2007) Proc Natl Acad Sci U S A 104: 9007–9011

47. Antoniou M, Harland L, Mustoe T, Williams S, Holdstock J, Yague E, Mulcahy T, Griffiths M, Edwards S, Ioannou PA, Mountain A, Crombie R (2003) Genomics 82:269–279

48. Williams S, Mustoe T, Mulcahy T, Griffiths M, Simpson D, Antoniou M, Irvine A, Mountain A, Crombie R (2005) BMC Biotechnol 5:17

49. Zhang F, Thornhill SI, Howe SJ, Ulaganathan M, Schambach A, Sinclair J, Kinnon C, Gaspar HB, Antoniou M, Thrasher AJ (2007) Blood 110:1448–1457

50. Cavazzana-Calvo M, Payen E, Negre O, Wang G, Hehir K, Fusil F, Down J, Denaro M, Brady T, Westerman K, Cavallesco R, Gillet-Legrand B, Caccavelli L, Sgarra R, Maouche-Chretien L, Bernaudin F, Girot R, Dorazio R, Mulder GJ, Polack A, Bank A, Soulier J, Larghero J, Kabbara N, Dalle B, Gourmel B, Socie G, Chretien S, Cartier N, Aubourg P, Fischer A, Cornetta K, Galacteros F, Beuzard Y, Gluckman E, Bushman F, Hacein-Bey-Abina S, Leboulch P (2010) Nature 467:318–322

51. Howe SJ, Mansour MR, Schwarzwaelder K, Bartholomae C, Hubank M, Kempski H, Brugman MH, Pike-Overzet K, Chatters SJ, de Ridder D, Gilmour KC, Adams S, Thornhill SI, Parsley KL, Staal FJT, Gale RE, Linch DC,

Bayford J, Brown L, Quaye M, Kinnon C, Ancliff P, Webb DK, Schmidt M, von Kalle C, Gaspar HB, Thrasher AJ (2008) J Clin Invest 118:3143–3150

52. Weiss DJ, Bonneau L, Liggitt D (2001) Mol Ther 3:734–745

53. Ornoy A, Arnon J, Katznelson D, Granat M, Caspi B, Chemke J (1987) Am J Med Genet 28:935–947

54. Johnson LG, Olsen JC, Sarkadi B, Moore KL, Swanstrom R, Boucher RC (1992) Nat Genet 2:21–25

55. Kreda SM (2005) Mol Cell Biol 16: 2154–2167

56. Dray X, Bienvenu T, Desmazes-Dufeu N, Dusser D, Marteau P, Hubert D (2004) Clin Gastroenterol Hepatol 2:498–503

57. Boué A, Muller F, Nezelof C, Oury JF, Duchatel F, Dumez Y, Aubry MC, Boue J (1986) Hum Genet 74:288–297

58. Scotet V, Dugueperoux I, Audrezet MP, Blayau M, Boisseau P, Journel H, Parent P, Ferec C (2008) Prenat Diagn 28:197–202

59. Boyd AC (2006) In: Bush A, Alton EWFW, Davies JC, Griesenbach U, Jaffe A (eds) Progress in Respiratory Research, Vol. 34 Gene and Stem Cell Therapy. Karger, Basel, pp 221–229

60. Pickles RJ, Fahrner JA, Petrella JM, Boucher RC, Bergelson JM (2000) J Virol 74:6050–6057

61. Harvey BG, Maroni J, O'Donoghue KA, Chu KW, Muscat JC, Pippo AL, Wright CE, Hollman C, Wisnivesky JP, Kessler PD, Rasmussen HS, Rosengart TK, Crystal RG (2002) Hum Gene Ther 13:15–63

62. Flotte TR, Zeitlin PL, Reynolds TC, Heald AE, Pedersen P, Beck S, Conrad CK, Brass-Ernst L, Humphries M, Sullivan K, Wetzel R, Taylor G, Carter BJ, Guggino WB (2003) Hum Gene Ther 14:1079–1088

63. Moss RB, Rodman D, Spencer LT, Aitken ML, Zeitlin PL, Waltz D, Milla C, Brody AS, Clancy JP, Ramsey B, Hamblett N, Heald AE (2004) Chest 125:509–521

64. Fleurence E, Riviere C, Lacaze-Masmonteil T, Franco-Motoya ML, Waszak P, Bourbon J, Danos O, Douar AM, Delacourt C (2005) Hum Gene Ther 16:1298–1306

65. Koehler DR, Frndova H, Leung K, Louca E, Palmer D, Ng P, McKerlie C, Cox P, Coates AL, Hu J (2005) J Gene Med 7:1409–1420

66. Griesenbach U, Alton EWFW (2007) Gene Ther 14:1439–1447

67. Griesenbach U, Geddes DM, Alton EWFW (2006) Gene Ther 13:1061–1067

68. Peebles D, Gregory LG, David A, Themis M, Waddington S, Knapton HJ, Miah M, Cook T, Lawrence L, Nivsarkar M, Rodeck C, Coutelle C (2004) Gene Ther 11:70–78

69. Voynow JA, Fischer BM, Roberts BC, Proia AD (2005) Am J Respir Crit Care Med 172:1013–1018

70. Larson JE, Morrow SL, Happel L, Sharp JF, Cohen JC (1997) Lancet 349:619–620

71. Davies SA, Varathalingam A, Painter H, Lawton AE, Sumner-Jones SG, Nunez-Alonso GA, Chan M, Munkonge F, Alton EWFW, Hyde SC, Gill DR (2008) Mol Ther 16(5):812–818

72. Buckley SMK, Waddington SN, Jezzard S, Bergau A, Themis M, MacVinish LJ, Cuthbert AW, Colledge WH, Coutelle C (2008) Mol Ther 16(5):819–824

73. Egan M, Flotte T, Afione S, Solow R, Zeitlin PL, Carter BJ, Guggino WB (1992) Nature 358:581–584

74. Lim F-Y, Martin BG, Sena-Esteves M, Radu A, Crombleholme TM (2002) J Pediatr Surg 37:1051–1057

75. Lim F-Y, Kobinger GP, Weiner DJ, Radu A, Wilson JM, Crombleholme TM (2003) J Pediatr Surg 38:834–839

76. Buckley SMK (2005) Mol Ther 12:484–492

77. Moss IR, Scarpelli EM (1981) Pediatr Res 15:870–873

78. Boyle MP, Enke RA, Adams RJ, Guggino WB, Zeitlin PL (2001) Mol Ther 4:115–121

79. Garrett DJ, Larson JE, Dunn D, Marrero L, Cohen JC (2003) BMC Biotechnol 3:16

80. Lipshutz GS, Gruber CA, Cao Y, Hardy J, Contag CH, Gaensler KML (2001) Mol Ther 3:284–292

81. Bilbao R, Reay DP, Li J, Xiao X, Clemens PR (2005) Hum Gene Ther 16:678–684

82. Henriques-Coelho T (2007) Mol Ther 15:340–347

83. Toelen J, Rik G, Sbragia L, Christophe D, Debyser Z, Deprest J (2006) Am J Obstet Gynecol 95:S22

84. Tarantal AF, McDonald RJ, Jimenez DF, Lee CI, O'Shea CE, Leapley AC, Won RH, Plopper CG, Lutzko C, Kohn DB (2005) Mol Ther 12:87–98

85. Tarantal AF, Lee CI, Ekert JE, McDonald R, Kohn DB, Plopper CG, Case SS, Bunnell BA (2001) Mol Ther 4:614–621

86. David AL, Peebles DM, Gregory L, Themis M, Cook T, Coutelle C, Rodeck C (2003) Fetal Diagn Ther 18:385–390

87. Gregory LG, Harbottle RP, Lawrence L, Knapton HJ, Themis M, Coutelle C (2002) Mol Ther 7:19–26

88. David AL, Peebles DM, Gregory L, Waddington SN, Themis M, Weisz B,

Ruthe A, Lawrence L, Cook T, Rodeck CH, Coutelle C (2006) Hum Gene Ther 17: 767–779

89. Larson JE, Cohen CJ (2006) Am J Physiol Lung Cell Mol Physiol 291:L4–L10

90. Saada J, Oudrhiri N, Bonnard A, De Lagausie P, Aissaoui A, Hauchecorne M, Oury JF, Aigrain Y, Peuchmaur M, Lehn JM, Lehn P, Luton D (2010) J Gene Med 12:413–422

91. Meikle PJ, Hopwood JJ, Clague AE, Carey WF (1999) JAMA 281:249–254

92. Zadeh N, Hudgins L, Norton ME (2011) Prenat Diagn 31:327–330

93. Spiegel R, Raas-Rothschild A, Reish O, Regev M, Meiner V, Bargal R, Sury V, Meir K, Nadjari M, Hermann G, Iancu TC, Shalev SA, Zeigler M (2009) Am J Med Genet A 149A:446–450

94. Eng CM, Schechter C, Robinowitz J, Fulop G, Burgert T, Levy B, Zinberg R, Desnick RJ (1997) JAMA 278:1268–1272

95. Sly WS, Quinton BA, McAlister WH, Rimoin DL (1973) J Pediatr 82:249–257

96. Wilkins I (2004) In: Creasy RK, Resnik R (eds) Maternal–fetal medicine principles and practice. Saunders, Philadelphia, p 572

97. Sands MS, Davidson BL (2006) Mol Ther 13:839–849

98. Mango RL, Xu L, Sands MS, Vogler C, Seiler G, Schwartz T, Haskins ME, Ponder KP (2004) Mol Genet Metab 82:4–19

99. Aronovich EL, Bell JB, Belur LR, Gunther R, Koniar B, Erickson DC, Schachern PA, Matise I, McIvor RS, Whitley CB, Hackett PB (2007) J Gene Med 9:403–415

100. Berges BK, Yellayi S, Karolewski BA, Miselis RR, Wolfe JH, Fraser NW (2006) Mol Ther 13:859–869

101. Ciron C, Desmaris N, Colle MA, Raoul S, Joussemet B, Verot L, Ausseil J, Froissart R, Roux F, Cherel Y, Ferry N, Lajat Y, Schwartz B, Vanier MT, Tardieu M, Moullier P, Heard JM (2006) Ann Neurol 60:204–213

102. Cearley CN, Wolfe JH (2007) J Neurosci 27:99409928

103. Shen JS, Meng XL, Yokoo T, Sakurai K, Watabe K, Ohashi T, Eto Y (2005) J Gene Med 7:540–551

104. Karolewski BA, Wolfe JH (2006) Mol Ther 14:14–24

105. Tarantal AF, Chu F, O'Brien WD, Hendrickx AG (1993) J Ultrasound Med 12:285–295

106. Daffos F, Capella-Pavlovsky M, Forestier F (1983) Am J Obstet Gynecol 146:985–987

107. Foust KD, Nurre E, Montgomery CL, Hernandez A, Chan CM, Kaspar BK (2009) Nat Biotechnol 27:59–65

108. Duque S, Joussemet B, Riviere C, Marais T, Dubreil L, Douar AM, Fyfe J, Moullier P, Colle MA, Barkats M (2009) Mol Ther 17:1187–1196

109. Bevan AK, Duque S, Foust KD, Morales PR, Braun L, Schmelzer L, Chan CM, McCrate M, Chicoine LG, Coley BD, Porensky PN, Kolb SJ, Mendell JR, Burghes AH, Kaspar BK (2011) Mol Ther 19(11):1971–1980. doi:10.1038/mt.2011.157

110. Manfredsson FP, Rising AC, Mandel RJ (2009) Mol Ther 17:403–405

111. Rahim AA, Wong AMS, Hoefer K (2011) FASEB J. Epub 25(10):3505–3518

112. Vanhanen SL, Sainio K, Lappi M, Santavuori P (1997) Dev Med Child Neurol 39: 456–463

113. Macauley SL, Wozniak DF, Kielar C, Tan Y, Cooper JD, Sands MS (2009) Exp Neurol 217:124–135

114. Wirth B, Brichta L, Schrank B, Lochmuller H, Blick S, Baasner A, Heller R (2006) Hum Genet 119:422–428

115. Federici T, Boulis NM (2006) Muscle Nerve 33:302–323

116. Dubowitz V (1999) Eur J Paediatr Neurol 3: 49–51

117. Fidzianska A, Rajalowska J (2002) Acta Neuropathol (Berl) 104:363–368

118. Lesbordes JC, Cifuentes-Diaz C, Miroglio A, Joshi V, Bordet T, Kahn A, Melki J (2003) Hum Mol Genet 12:1233–1239

119. Matsuoka N, Yukawa H, Ishii K, Hamada H, Akimoto M, ashimoto N, Miyatake S (1999) Neurosci Lett 270:177–180

120. Azzouz M, Le T, Ralph GS, Walmsley LE, Monani UR, Lee DC, Wilkes F, Mitrophanous KA, Kingsman SM, Burghes AH, Mazarakis ND (2008) J Clin Invest 114:1726–1731

121. Boulis NM, Bhatia V, Brindle TI, Holman HT, Krauss DJ, Blaivas M, Holf JT (1999) J Neurosurg 90:99–108

122. Tenenbaum L, Chtarto A, Lehtonen E, Velu T, Brotchi J, Levivier M (2004) J Gene Med 6:S212–S222

123. Passini MA, Bu J, Roskelley EM, Richards AM, Sardi SP, O'Riordan CR, Klinger KW, Shihabuddin LS, Cheng SH (2010) J Clin Invest 120:1253–1264

124. Kahn A, Hasse G, Akli S, Guidotti JE (1996) Comptes rendus des séances de la Société de biologie et de ses filiales 190:9–11

125. Yamamura J, Kageyama S, Uwano T, Kurokawa M, Imakita M, Shiraki K (2000) Gene Ther 7:934–941

126. Rahim AA, Wong AM, Buckley SM, Chan JK, David AL, Cooper JD, Coutelle C, Peebles

DM, Waddington SN (2010) Biochem Soc Trans 38:1489–1493

127. Foust KD, Wang X, McGovern VL, Braun L, Bevan AK, Haidet AM, Le TT, Morales PR, Rich MM, Burghes AH, Kaspar BK (2010) Nat Biotechnol 28:271–274

128. Dominguez E, Marais T, Chatauret N, Benkhelifa-Ziyyat S, Duque S, Ravassard P, Carcenac R, Astord S, Pereira DM, Voit T, Barkats M (2011) Hum Mol Genet 20: 681–693

129. Stratford-Perricaudet L, Levrero M, Chasse J-F, Perricaudet M, Briand P (1990) Hum Gene Ther 1:241–256

130. Maestri NE, Clissold D, Brusilow SW (1999) J Pediatr 134:268–272

131. Raper SE, Yudkoff M, Chirmule N, Gao GP, Nunes F, Haskal ZJ, Furth EE, Propert KJ, Robinson MB, Magosin S, Simoes H, Speicher L, Hughes J, Tazelaar J, Wivel NA, Wilson JM, Batshaw ML (2002) Hum Gene Ther 13:163–175

132. Raper SE, Chirmule N, Lee FS, Wivel NA, Bagg A, Gao GP, Wilson JM, Batshaw ML (2003) Mol Genet Metab 80:148–158

133. Mian A, McCormack WM Jr, Mane V, Kleppe S, Ng P, Finegold M, O'Brien WE, Rodgers JR, Beaudet AL, Lee B (2004) Mol Ther 10:492–499

134. Moscioni D, Morizono H, McCarter RJ, Stern A, Cabrera-Luque J, Hoang A, Sanmiguel J, Wu D, Bell P, Gao GP, Raper SE, Wilson JM, Batshaw ML (2006) Mol Ther 14:25–33

135. Seppen J (2006) Gene Ther 13:672–677

136. Lai L, Davison BB, Veazey RS, Fisher KJ, Baskin GB (2002) Hum Gene Ther 13:2027–2039

137. Lee CC, Jimenez DF, Kohn DB, Tarantal AF (2005) Hum Gene Ther 16:417–425

138. Tarantal AF (2006) Hum Gene Ther 17:1254–1261

139. Tarantal AF, O'Rourke JP, Case SS, Newbound GC, Li J, Lee CI, Baskin CR, Kohn DB, Bunnell BA (2001) Mol Ther 3:128–138

140. Gallot D, Boiret N, Vanlieferinghen P, Laurichesse H, Micorek JC, Berger M, Lemery D (2004) Fetal Diagn Ther 19:170–173

141. Raymond FL, Whittaker J, Jenkins L, Lench N, Chitty LS (2010) Prenat Diagn 30:674–681

142. van Deutekom JC, van Ommen GJ (2003) Nat Rev Genet 4:774–783

143. Romero NB, Benveniste O, Leturq F, Hogrel JY, Morris GE, Barois A, Eymard B, Payan C, Ortega V, Boch AL, Lajean L, Thioudellet C, Mourot B, Escot C, Choqel A, Recan D,

Kaplan JC, Dockson G, Klatzmann D, Molinier-Frenckel V, Guillet JG, Squiban P, Herson S, Fardeau M (2004) Hum Gene Ther 15:1065–1076

144. Wells DJ, Ferrrer A, Wells KE (2002) Expert Rev Mol Med 4:1–23

145. Tang Y, Cummins J, Huard J, Wang B (2010) Expert Opin Biol Ther 10:395–408

146. MacKenzie TC, Kobinger GP, Louboutin JP, Radu A, Javazon EH, Sena-Esteves M, Wilson JM, Flake AW (2005) J Gene Med 7:50–58

147. Ahmad A, Brinson M, Hodges BL, Chamberlain JS, Amalfitano A (2000) Hum Mol Genet 9:2507–2515

148. Reay DP, Bilbao R, Koppanati BM, Cai L, O'Day TL, Jiang Z, Zheng H, Watchko JF, Clemens PR (2008) Gene Ther 15:531–536

149. Gregory LG, Waddington SN, Holder MV, Mitrophanous KA, Buckley SMK, Mosley KL, Bigger BW, Ellard FM, Walmsley LE, Lawrence L, Al-Allaf F, Kingsman S, Coutelle C, Themis M (2004) Gene Ther 11: 1117–1125

150. Weisz B, David AL, Gregory LG, Perocheau D, Ruthe A, Waddington SN, Themis M, Cook T, Coutelle C, Rodeck CH, Peebles DM (2005) Am J Obstet Gynecol 193:1105–1109

151. Mitchell M, Jerebtsova M, Batshaw ML, Newman K, Ye X (2000) Gene Ther 7: 1986–1992

152. Bouchard S, MacKenzie TC, Radu AP, Hayashi S, Peranteau WH, Chirmule N, Flake AW (2003) J Gene Med 5:941–950

153. Koppanati BM, Li J, Xiao X, Clemens PR (2009) Gene Ther 16:1130–1137

154. Koppanati BM, Li J, Reay DP, Wang B, Daood M, Zheng H, Xiao X, Watchko JF, Clemens PR (2010) Gene Ther 17(11):1355–1362

155. Wang B, Li J, Xiao X (2000) Proc Natl Acad Sci U S A 97(25):13714–13719

156. Lu QL, Yokota T, Takeda S, Garcia L, Muntoni F, Partridge T (2011) Mol Ther 19:9–15

157. Luu M, Cantatore-Francis JL, Glick SA (2010) Int J Dermatol 49:353–361

158. Muhle C, Neuner A, Park J, Pacho F, Jiang Q, Waddington SN, Schneider H (2006) Gene Ther 13:1665–1676

159. Sato M, Tanigawa M, Kikuchi N (2004) Mol Reprod Dev 69:277

160. Endoh M, Koibuchi N, Sato M, Morishita R, Kanzaki T, Murata Y, Kaneda Y (2002) Mol Ther 5:501–508

161. Yoshizawa J, Li X-K, Fujino M, Kimura H, Mizuno R, Hara A, Ashizuka S, Kanai M, Kuwashima N, Kurobe M, Yamazaki Y (2004) J Pediatr Surg 39:81–84

162. Endo M, Zoltick PW, Peranteau WH, Radu A, Muvarak N, Ito M, Yang Z, Cotsarlis G, Flake AW (2008) Mol Ther 16:131–137

163. Aiuti A, Slavin S, Aker M, Ficara F, Deola S, Mortellaro A, Morecki S, Andolfi G, Tabucchi A, Carlucci F, Marinello E, Cattaneo F, Vai S, Servida P, Miniero R, Roncarolo MG, Bordignon C (2002) Science 296: 2410–2413

164. Gaspar HB, Bjorkegren E, Parsley K, Gilmour KC, King D, Sinclair J, Zhang F, Giannakopoulos A, Adams S, Fairbanks LD, Gaspar J, Henderson L, Xu-Bayford JH, Davies EG, Veys PA, Kinnon C, Thrasher AJ (2006) Mol Ther 14:505–513

165. Hacein-Bey-Abina S, Le Deist F, Carlier F, Bouneaud C, Hue C, De Villartay JP, Thrasher AJ, Wulffraat N, Sorensen R, Dupuis-Girod S, Fischer A, Davies EG, Kuis W, Leiva L, Cavazzana-Calvo M (2002) N Engl J Med 346:1185–1193

166. Gaspar HB, Sinclair J, Brouns G, Schmidt M, von Kalle C, Barington T, Jakobsen MA, Christensen HO, Al Ghonaium A, White HN, Smith JL, Levinsky RJ, Ali RR, Kinnon C, Thrasher AJ (2004) Lancet 364:2181–2187

167. Ott MG, Schmidt M, Schwarzwaelder K, Stein S, Siler U, Koehl U, Glimm H, Kuhlcke K, Schilz A, Kunkel H, Naundorf S, Brinkmann A, Deichmann A, Fischer M, Ball C, Pilz I, Dunbar C, Du Y, Jenkins NA, Copeland NG, Lutthi U, Hassan M, Thrasher AJ, Hoelser D, von Kalle C, Seger R, Grez M (2006) Nat Med 12:401–409

168. Hacein-Bey-Abina S, von Kalle C, Schmidt M, McCormack MP, Wulffraat N, Leboulch P, Lim A, Osborne CS, Pawliuk R, Morillon E, Sorensen R, Forster A, Fraser P, Cohen JI, de Saint Basile G, Alexander I, Wintergerst U, Frebourg T, Aurias A, Stoppa-Lyonnet D, Romana S, Radford-Weiss I, Gross F, Delabesse E, Macintyre E, Sigaux F, Soulier J, Leiva LE, Wissler M, Prinz C, Rabbitts TH, Le Deist F, Fischer A, Cavazzana-Calvo M (2003) Science 302:415–419

169. Hanawa H, Hematti P, Keyvanfar K, Metzger ME, Krouse A, Donahue RE, Kepes S, Gray J, Dunbar CE, Persons DA, Nienhuis AW (2004) Blood 103:4062–4069

170. Hackett PB, Ekker SC, Largaespada DA, McIvor RS (2008) Adv Genet 54:189–232

171. Hollis RP, Nightingale SJ, Pepper KA, Yu XJ, Barsky LW, Crooks GM, Kohn DB (2006) Exp Hematol 34:1333–1343

172. Muench MO (2005) Bone Marrow Transplant 35:537–547

173. Flake AW, Roncarolo MG, Puck JM, Almeida-Porada G, Evans MI, Johnson MP, Abella EM, Harrison DD, Zanjani ED (1996) N Engl J Med 335:1806–1810

174. Westgren M, Ringden O, Bartmann P, Bui T, Lindton B, Mattson J, Uzunel M, Zetterquist H, Hansmann M (2002) Am J Obstet Gynecol 187:475–482

175. Shaw SWS, David AL, De Coppi P (2011) Curr Opin Obstet Gynecol 23:109–116

176. Schoeberlein A, Holzgreve W, Dudler L, Hahn S, Surbek DV (2004) Am J Obstet Gynecol 191:1030–1036

177. Orlandi F, Damiani G, Jakil C, Lauricella S, Bertolino O, Maggio A (1990) Prenat Diagn 10:425–428

178. De Coppi P, Bartsch G Jr, Siddiqui MM, Xu T, Santos CC, Perin L, Mostoslavsky G, Serre AC, Snyder EY, Yoo JJ, Furth ME, Soker S, Atala A (2007) Nat Biotechnol 25:100

179. Portmann-Lanz CB, Schoeberlein A, Huber A, Sager R, Malek A, Holzgreve W, Surbek DV (2006) Am J Obstet Gynecol 194:664–673

180. Lee LK, Ueno M, Van Handel B, Mikkola HK (2010) Curr Opin Hematol 17:313–318

181. Bollini S, Pozzobon M, Nobles M, Riegler J, Dong X, Piccoli M, Chiavegato A, Price AN, Ghionzoli M, Cheung KK, Cabrelle A, O'Mahoney PR, Cozzi E, Sartore S, Tinker A, Lythgoe MF, De Coppi P (2011) Stem Cell Rev 7(2):364–380

182. Ditadi A, De Coppi P, Picone O, Gautreau L, Smati R, Six E, Bonhomme D, Ezine S, Frydman R, Cavazzana-Calvo M, André-Schmutz I (2009) Blood 113:3953–3960

183. Shaw SWS, Bollini S, Abi-Nader K, Gastadello A, Mehta V, Filippi E, Cananzi M, Gaspar HB, Qasim W, De Coppi P, David AL (2010) Cell Transplant 20(7):1015–1031

184. Tabori U, Mark Z, Amariglio N, Etzioni A, Golan H, Biloray B, Toren A, Rechavi G, Dalal I (2004) Clin Genet 65(4):322–326

185. Williams ML, Coleman JE, Haire SE, Aleman TS, Cideciyan AV, Sokal I, Palczewski K, Jacobson SG, Semple-Rowland SL (2006) PLoS Med 3:e201

186. Bedrosian JC, Gratton MA, Brigande JV, Tang W, Landau J, Bennett J (2006) Mol Ther 14:328–335

187. Endo M (2007) Mol Ther 15:579–587

188. Acosta R, Lee JJ, Oyachi N, Buchmiller-Crair TL, Atkinson JB, Ross MG (2002) J Matern Fetal Neonatal Med 11:153–157

189. David AL, Torondel B, Zachary I, Wigley V, Abi-Nader K, Mehta V, Buckley SMK, Cook

T, Boyd M, Rodeck CH, Martin J, Peebles DM (2008) Gene Ther 15:1344–1350

190. Mehta V, Abi-Nader KN, Peebles DM, Benjamin E, Wigley V, Torondel B, Filippi E, Shaw SWS, Boyd M, Martin J, Zachary I, David AL (2011) Gene Ther doi: 10.1038/gt.2011.158

191. Carr DJ, Aitken RP, Milne JS, Peebles DM, Martin JM, Zachary IC, Wallace JM, David AL (2011) Reprod Sci 18:269A

192. Deprest J, Gratacos E, Nicolaides K (2004) Ultrasound Obstet Gynecol 24:121–126

193. Lipshutz GS, Flebbe-Rehwaldt L, Gaensler KML (1999) J Surg Res 84:150–156

194. Lipshutz GS, Flebbe-Rehwaldt L, Gaensler KML (2000) Mol Ther 2:374–380

195. Waddington SN, Buckley SMK, Nivsarkar M, Jezzard S, Schneider H, Dahse T, Dallman MJ, Kemball-Cook G, Miah M, Tucker N, Dallman MJ, Themis M, Coutelle C (2003) Blood 101:1359–1366

196. Tarantal AF, Lee CI, Ekert JE, O'Rourke JP, McDonald RJ, Plopper CG, Permenter M, Kohn DB, Bunnell BA (2001) Mol Ther 3:S129

197. Mattar CN, Nathwani AC, Waddington SN, Dighe N, Kaeppel C, Nowrouzi A, McIntosh J, Johana NB, Ogden B, Fisk NM, Davidoff AM, David A, Peebles D, Valentine MB, Appelt JU, von Kalle C, Schmidt M, Biswas A, Choolani M, Chan JK (2011) Mol Ther 19(11):1950–1960

Chapter 3

Vector Systems for Prenatal Gene Therapy: Choosing Vectors for Different Applications

Charles Coutelle and Simon N. Waddington

Abstract

This chapter gives a comparative review of the different vector systems applied to date in prenatal gene therapy experiments highlighting the need for versatility and choice for application in accordance with the actual aim of the study. It reviews the key characteristics of the four main gene therapy vector systems and gives examples for their successful application in prenatal gene therapy experiments.

Key words: Gene therapy vectors, Genetic diseases, Prenatal gene therapy, Non-viral vectors, Adenoviral vectors, Lentiviral vectors, AAV vectors, Safety and efficiency profiles

1. Why Are Vectors Needed?

Gene therapy relies on the introduction of therapeutic nucleic acid sequences into the cells of a patient who requires treatment. However, nucleic acids do not easily enter cells as evolution has equipped higher organisms with numerous protective barriers against foreign invasion of viruses and other pathogens. The first barriers are the skin and mucosa tissues preventing entry into the body. In the body, the distances to target cells, dilution in body fluids, serum nucleases, as well as the protective layers of connective tissue, mucous, cilia, and the plasma membrane covering and shielding these cells restrict access. But even once inside the cell, the foreign DNA has to escape degradation-pathways in order to reach and cross the nuclear membrane into nucleus where it can be expressed. Besides these more passive physical barriers, the formidable and flexible humoral and cellular immune defense system has evolved to actively attack and destroy any intruder. Viruses and other pathogens in turn have also evolved to circumvent these barriers using a high mutation rate to evade immune surveillance and by adapting to host cell mechanisms such as receptor recognition

Charles Coutelle and Simon N. Waddington (eds.), *Prenatal Gene Therapy: Concepts, Methods, and Protocols*, Methods in Molecular Biology, vol. 891, DOI 10.1007/978-1-61779-873-3_3, © Springer Science+Business Media, LLC 2012

for cell entry and by reprogramming the cell's synthetic machinery in order to produce their proteins and propagate their DNA.

For gene therapy, these cellular defense systems mean that we have to develop specific carriers—vectors—when aiming to overcome all these hindrances and deliver our therapeutic nucleic acid constructs to the target cells. These vectors have turned out to be among the most critical prerequisites for successful gene transfer. This is particularly true when delivery to a specific cell type and/or when particular levels and sustained gene expression are required for a therapeutic effect. Furthermore, adverse effects from the vectors need to be avoided or at least minimized in order to benefit from their therapeutic application. These demands have led to a wealth of research and accumulated knowledge on the development of constantly improved diverse vector systems. Over time, these efforts have proven futile the various "fashion" debates favoring one vector system and pronouncing others "dead"; usually on the basis of the observation of some adverse effects. It has now become clear that no vector will serve all purposes; that each of them has advantages and disadvantages; and that the strength of this field lies in its diversity and versatility, whereby each therapeutic goal will determine, which vector system would be best suited for a particular purpose. Having said this, there are of course criteria, which will define a particular vector as "best" for a defined purpose.

2. Different Vectors for Different Diseases

Gene therapy for nongenetic conditions may often require only temporary expression of the therapeutic gene. This goal puts much less constraints on the vector than when it is used for treatment of genetic diseases, which by their very nature will in most cases require a life-long therapy. As genetic diseases are the result of gene mutations, the ideal gene therapy for them would be genuine gene correction, i.e., the exchange of the mutated nucleotide or sequence with the "normal" one, although even then, "normal" may come in different allelic variants. This field is presently evolving very rapidly particularly since the development of Zinc finger targeted nucleases (1) and meganucleases (2) and more recently Tal nuclease (3) to recognize virtually any gene sequence of interest (1, 4). These targeting methods could also be applied for inserting of whole therapeutic expression cassettes into "safe" genomic sites (1). Proof of principle for a therapeutic application of targeted gene correction in an animal disease model of hemophilia has just been established (5) and two phase 1 clinical trials using Zinc finger nucleases have been initiated (1). However, despite these promising developments gene correction is still some years away from reaching a broader human therapeutic application, which will require improved efficiency and safety with respect to off-target and nonhomologous end-joining events (1, 5). Presently and within the near future, the majority gene therapy is based on supplementation, i.e., the addition of a therapeutic nucleic acid sequence to the genetic complement of the treated cells, either as episomal constructs or by random integration into the host genome.

For autosomal recessive genetic diseases, gene supplementation should be sufficient as long as therapeutic levels of expression, and where necessary appropriate regulation, can be achieved. After all, the heterozygote carriers of many diseases, who only have one normal allelic gene, are usually unaffected. The same would be true for dominant diseases caused by haploinsufficiency, i.e., diseases in which the gene product of only one normal allele is not sufficient for normal function. However, in most dominant diseases, the mutations result in, what is often referred to as a "gain of function," brought about by the production of a toxic gene product from the mutated allele. In this case, gene supplementation is not sufficient to correct the disease and lacking the ability for real gene correction, inactivation of the mutated gene, for instance, by a vector expressing an inhibitory shRNA or by targeted knockout, may be a successful strategy.

The demand on gene therapy for most genetic diseases is to provide permanent expression of the required gene product at therapeutic levels. This would ideally be achieved by a single gene application into the stem cells of the tissue where the therapeutic gene is normally expressed and where it should stay and exert its therapeutic effect over the patient's lifetime at the physiological level of cell differentiation. This would be best achieved by integration into the genome of these stem cells. In tissues with little or no cell turnover, a vector may remain in a stable episomal form and provide sustained transgene expression in the differentiated cells. Alternatively, repeated application to this tissue or even a different tissue from where the gene product is secreted may be more feasible and effective. In addition, in some instances, particular regulation of the therapeutic gene may be necessary, and in any case the vector should be constructed in a way to minimize the risk of potential adverse effects by ectopic and/or over expression (6). In prenatal gene therapy, this may also require to exclude adverse effects due to expression at nonphysiological concentrations at a certain time or location during development as observed, for example, with transgenically expressed growth factors (7).

The different vector systems provide different options to address the particular requirement of the target diseases. Prenatal application of gene therapy targeted to extraembryonic tissues, such as the placenta, is designed to exert its effect only temporary and therefore vector systems, without the ability of long-term cellular maintenance, are applicable or even preferable. This will also apply to the treatment of genetic or nongenetic conditions, such as developmental aberrations, where short-term intervention may be sufficient for correction or prevention of disease manifestation. By contrast, for life-long manifesting genetic diseases, vectors with the ability to establish themselves in the target cell and provide sustained expression of the therapeutic protein may be preferable.

3. Choosing Vectors for Short- and Long-Term Expression

In the following sections, we review the presently available main gene therapy vector systems and their properties with a view to their choice for different applications in prenatal gene therapy. More systematic and detailed overviews of their biological features and molecular properties determining their construction and production are given in Chapters 4–7.

For practical purposes, vectors are often classified as integrating and non-integrating. However, from the viewpoint of this book, it may be more appropriate to look at them according to their ability to either provide only short-term gene expression over a few days or, at least in principle, to provide sustained gene expression over several months, years, or even a lifetime. "In principle," because as we will see, this not only depends on the nature of the vector but also on the context of the cells into which the vector is introduced. As a general rule, non-integrating vectors such as adenovirus and most non-viral vectors are usually used as short-term expressing vectors.

3.1. Non-viral Vectors: Low Toxicity but Need Improvement of Efficiency

Conventional non-viral vectors do not normally integrate into the mammalian host genome and, excluding artificial chromosome constructs, which are interesting research tools but have so far not achieved therapeutic practicability, they have no inherent mechanism of independent replication in these cells. Therefore, their genome is diluted out and becomes continuously lost during host cell division. These vectors are either relatively simple, chemically synthesized, small, linear DNA or RNA molecules (oligonucleotides, siRNA, decoy RNA, and ribozymes), which interfere directly with a particular process of host cell gene expression, or they are circular DNA plasmid vectors, which act by expressing a therapeutic gene product, usually a protein but sometimes a gene-regulatory RNA.

Plasmid vectors are based on bacterial plasmids, which can be easily manipulated with standard genetic engineering techniques and can be constructed and applied without an inherent size limit. They are generated by bio-fermentation in bacteria for which they require a bacterial backbone comprising an origin of replication and usually also one or more antibiotic resistance genes to allow easy selection. The therapeutic gene product is encoded in the expression cassette, which can be inserted easily into a multicloning site in the basic vector construct. Lack of genome integration and replication in mammalian cells, which results in vector loss during host cell division, as well as fast innate immune reactions induced by the high unmethylated CpG containing bacterial DNA backbone, contribute to the rapid decline of transgene expression from these vectors (8, 9).

Non-viral vectors are usually extremely inefficient in transfecting mammalian target cells and thus require specific physical or chemical means for effective cell entry. Besides the expression-repressing effects of the bacterial vector sequences, the physical (topical injection, electroporation, ultrasound, and hydrodynamic application) as well as the chemical (lipofection and polyfection) means of vector delivery can cause some degree of cell damage or toxicity, which will also limit the efficiency of transfection. However, the methods for physical transfection and the composition of the chemical transfectants are very flexible and can be varied and optimized to minimize damage and allow rapid recovery after vector delivery.

Conventional non-viral vectors do obviously not appear useful for longer-term gene therapy approaches. However, proof of principle for a short-term interference in fetal development has been demonstrated by blocking the normal closure of the ductus arteriosus vessel at birth by in utero liposome-delivery of a plasmid expressing decoy RNA, which sequesters an essential mRNA-binding protein involved in this process (10). In another short-term application of liposome–plasmid complexes, KGF was expressed in fetal lungs of sheep with an artificial diaphragmatic hernia in order to support the correction of this developmental defect using tracheal occlusion (11).

Furthermore, reductions of CpG sequences (9) in plasmid vectors or elimination of the entire bacterial DNA in the last stages of vector production to generate minicircle vectors (12–14) and introduction of scaffold matrix attachment sequences (S/MAR) (15) are new approaches to increase the duration of expression from non-viral vectors. Together they can turn these constructs into long-term and even permanently expressing vector systems (16, 17) which may also be applied in prenatal gene therapy, both in vivo and for ex vivo applications to modify autologous stem cells.

3.2. Adenoviral Vectors (18): Highly Efficient but Strongly Immunogenic

The other major transient vector group is derived from non-integrating, pathogenic but not oncogenic adenovirus strains, which causes various, mostly respiratory and gastrointestinal illness in humans (19, 20). Adenoviruses are able to target and enter a broad range of mammalian cells very efficiently. Their DNA is then transferred into the host cell nucleus where it is replicated and takes over gene expression to direct intracellular virus protein production. The new virus genomes are packaged into these proteins and the newly generated viruses destroy the host cell and spread to neighboring cells and throughout the infected organism. Usually, this spread is rapidly contained by immunological defense mechanisms of the infected host organism. The ability to enter mammalian cells and direct virus encoded gene expression in them very efficiently has made these viruses an attractive basis for gene therapy vector construction.

Wild-type adenovirus is highly immunogenic. The induction of immunological and toxic reactions manifesting as rhinitis, conjunctivitis, or gastroenteritis among others in its host is part of the viruses reproductive strategy to spread maximal infection within a population. In the individual host the infection is usually relatively quickly overcome by elimination of the virus and development of immunity against re-infection with this particular strain. Both the structural capsid and fiber proteins on the virus surface, and the functional virus proteins that direct viral protein synthesis are immunogenic. The surface proteins induce a B cell reaction and the formation of neutralizing antibodies within a few days. These antibodies eliminate any free virus particles in the host and prevent renewed infection. Virus that does enter the host cells expresses its functional proteins. They become processed and presented by the MHC class I system of the host's antigen presenting cells. This triggers a cellular immune response by which T cells recognize and destroy the infected host cells, thereby putting an end to further virus expression and production (21).

For the purpose of using the adenovirus as a vector, it has been disabled by deletion of most essential genes, which direct DNA replication and virus protein production. This vector DNA can express transgenes inserted into its genome to replace the deleted viral genes; and it also contains the packaging signal enabling the formation of recombinant virus particles with separately provided viral proteins. The construction and production of such disabled viral vectors follows a pattern that is paradigmatic for other viral gene therapy vectors: since the modified vectors do no longer produce the proteins required for viral DNA replication and virus formation, these functions have to be provided in trans by genes residing in the genome of either an in vitro production cell line or of a viral construct (helper-virus) that is transferred into the production cell line together with the gene therapy vector construct. In this way, the producer cells provide all the proteins for vector DNA replication and packaging. By virtue of these provided proteins, recombinant virus particles are formed that are able to infect the target cells of the organism to be treated. These vectors are harvested and purified away from any helper-virus before application in gene therapy. They can express the therapeutic transgenic proteins in their target cells but remain confined to these cells, as they are unable to replicate and generate new virus particles or to spread to other cells (22). Among the many adenovirus serotypes, serotype 5 has been used most extensively for such constructs.

Adenovirus-derived vector systems are subject to the same immune responses, which will eliminate successfully transduced host cells within a week or two by T-cell reaction and prevent therapeutic success of repeated vector application by neutralizing antibodies. To avoid cellular immune reactions, vectors have been constructed with increasing deletion of viral gene sequences coding for the functional virus proteins (23). By eliminating virtually all of these sequences in the helper dependent (HD) or high capacity adenovirus (24) constructs, long-term expression over several months has indeed been achieved (25). However, these vectors are difficult to produce in high quality and quantity (26).

Recognition by neutralizing antibodies at repeat administration can be avoided by exchanging the immunogenic external proteins of the therapeutic vector with those from a different serotype strain each time the virus is applied. Particular care should be taken for extensive purification of vector preparations in order to eliminate empty virus capsids, which may otherwise lead to incorrect virus doses and increased toxicity and immunogenicity of the preparation (26).

The standard first- and second-generation adenovirus vectors will provide in vivo gene expression for about 1–3 weeks. As with all vector systems, the duration and level of gene expression will depend on the choice of promoter and other regulatory elements. In general, virus promoters will usually give the highest levels of gene expression but they are also more prone to inactivation by cellular mechanism of the host compared to species or tissue-specific promoters. The lack of vector integration into the host genome will also shorten sustenance of gene expression particularly in rapidly dividing cells. Non-integration has, however, the advantage, that these vectors are not genotoxic and carry little risk of germ line transmission.

First- and second-generation adenovirus vectors are not suited for applications which aim for long-lasting expression. However, since they transduce a remarkably broad range of cell types and provide high levels of transgene expression, we and others have used them very successfully in marker gene experiments to assess various surgical and ultrasound-guided gene delivery techniques in the mouse and sheep fetus, respectively (27). These vectors may also prove very useful if only short-term, high-level transgene expression is required, for instance, in extraembryonic tissue such as the placenta (28) and fetal membranes (29). By contrast, an early claim to postnatal reversal of the CF-phenotype in CFTR-knockout mice after in utero CFTR adenovirus administration (30) was not reproducible in independent investigations (31, 32).

However, after application of HD-adenovirus vectors up to 5 months expression of β-galactosidase with a slow decrease due to rapid muscle growth has been observed after in utero delivery to fetal mice and HD-adenovirus vector was also used to express dystrophin in mdx mice (33). Combining such vectors with serotype or capsid changes (34) or short-term immunosuppression (35) during application in order to avoid preexisting or developing antibody reactions may make them applicable for repeated application and longer-term gene expression.

3.3. Retroviral Vectors (18): Permanent Persistence, Low Immunogenicity but Potentially Oncogenic

Integrating vectors based on modified murine gamma-retro viruses (MLV) or HIV-lentivirus vectors are the conventional choices when long-term vector persistence and expression is sought. The envelope of these vectors attaches to specific cell receptors on the target cells and enables entry of the virus RNA/protein complex into the cytoplasm of the host cell where the single-stranded RNA genome serves as template for the virus reverse transcriptase. The resulting double-stranded cDNA forms a preintegration complex, which enters the nucleus and by action of the viral integrase it is inserted into the host genome. The integrated virus genome is frequently shut down and may be maintained silently as a provirus for the rest of the host's life, but can be reactivated to form infectious virus particles by endogenous or exogenous events. On activation, the integrated virus DNA is transcribed and translated to produce new virus, which spreads to integrate into other cells of the host and/or to infect other host organisms. One of the main practical differences between the gamma-retroviruses and lentiviruses is that entry of the former into the nuclear compartment is passive, requiring the nuclear membrane to break down during cell division, while the lentiviral cDNA preintegration complex (PIC) can interact with the cellular nuclear transport mechanism enabling its active transport through the nuclear pore complex (36). The other important difference concerns preferences of integration sites in the host genome. Both viruses integrate without site specificity but with a certain preference for actively transcribed genes. Integrated gamma-retroviruses were found frequently near transcription start regions, and HIV appears to favor chromosome regions rich in actively expressed genes (37). In contrast to adenovirus infection, no fulminant host reaction that ensures virus spread to other hosts but also leads to immunological elimination of the infecting virus in its immediate host is provoked by retroviruses. Here the virus can remain silent or be transmitted over the lifetime of the infected individual according to the different infections patterns of the various retroviruses.

Vectors constructed from these viruses are deleted in their essential genes required for natural virus production and propagation, thereby creating space for the insertion of an expression cassette for a therapeutic or marker gene sequence. As described for adenovirus vectors, these constructs require, therefore, either a producer cell line or a helper-virus for bioproduction. By modifying or exchanging the envelope gene with sequences from different viruses (pseudotyping), it is possible to target different cell receptors and thereby generate vectors with various cell tropisms (38). Inside the target cell, the gene therapy vector can express its transgene sequence, but it cannot multiply and spread to other cells. Integration into the host genome offers the advantage of lasting transgene expression but also poses the threat of oncogenesis (39) by activating transcription of oncogenes or inactivation of antioncogenes. Vectors and their producer cell lines are constructed in ways that aim to prevent the generation of replication competent virus, which would spread and by sheer multitude of insertions dramatically increase the risk of oncogenesis. In order to ensure the desired transgene expression, the design of such constructs has also to safeguard against shutdown of transgene expression from the vector. In addition, safer vectors have been developed and are constantly being improved to avoid unwanted activation of endogenous gene sequences.

The long-term persistence of retroviral vectors makes them theoretical highly suitable for in utero application. However, initial attempts of in utero application using the Mouse Moloney Leukemia virus (MLV), which requires cell division and appeared therefore ideal for transfection of the rapidly growing fetal tissues, gave disappointingly low levels of transduction (40, 41). This was most likely due to inadequate titre of these early virus preparations. Lentivirus vectors, on the other hand, particularly after pseudotyping with Vesicular stomatitis virus G (VSVG) protein, have given remarkably high transfection efficiencies following different routes of application to the mouse fetus and providing a prolonged therapeutic correction of Crigler Najjar disease (42) and thrombotic thrombocytopenic purpura (43) as well as the life-long cure of Hemophilia IX (44) in mouse models of these diseases. However, we and others (45) have found transduction of fetal sheep with VSVG lentivirus quite ineffective and it has been suggested that this may be overcome by pseudotyping with the Jaagsiekte sheep retrovirus envelope glycoprotein, although transduction in initial in vitro experiments was not very impressive (46).

It should also be noted that EIAV-based lentiviral constructs have shown a high propensity for tumor induction in the mouse fetal system, which was, however, not observed when HIV-based vectors were applied (47). This is discussed in more detail in Chapter 16.

3.4. Adeno-Associated Virus Vectors (18): Good Safety Profile but Low Insert Capacity

Vectors derived from the adeno-associated virus (AAV) (48) appeared initially very promising since, although frequently found in humans, AAV2 does not cause any known human pathology and the wild type was shown to integrate preferentially into an apparently safe region of the human genome on chromosome 19. However, the initial AAV2-derived vectors could only accommodate about 4.7 kb of foreign DNA and had a vey limited target cell range. They had lost the ability of their parent virus for site-specific integration but remained mainly episomal with some tendency to integrate randomly into the host genome and they required several weeks after application to reach their maximal level of transgene expression. They required Adeno- or Herpes-helper-viruses for production and were very difficult to produce free from helper-virus at high titre and purity. Given the rapid cell division of fetal tissues, it seemed obvious that these early generation vectors would be lost before reasonable expression levels could be reached and therefore they were not considered particularly suitable for in utero gene delivery (49). Even in postnatal gene therapy, it took an unusually long time for AAV vectors to come of age. Production of pure vector at high titre was particularly difficult and expensive and initial clinical trials for hemophilia B using AAV2 were discontinued because of transient detection of vector sequences in the semen of treated patients (50) and of mild hepatotoxic immune reactions (50, 51). The AAV-vector efficacy and production situation has dramatically changed in the last few years by the development of methods to enhance formation of self-complementary (sc) vector (52), to broaden the cell tropism of AAV vectors (53), and, very importantly, to improve AAV vector production (54).

Several promising adult clinical trials are now underway with these vectors (55). However, the limited size capacity of AAV is still a significant drawback to therapeutic AAV-mediated transfer of larger gene sequences and methods to increase the transgene capacity (56–58) appear all to work through intracellular reconstruction of transgene fragments (59).

First therapeutic in utero applications of AAV 1 vectors to rodent models of human diseases have included Pompe disease (60), congenital blindness (61), mucopolysaccharidosis type VII (62), Duchenne muscular dystrophy (DMD) (63), and Hemophilia B (64). AAV2/8 has shown remarkable efficiency in utero, which appears to be an excellent platform for therapeutic application in early onset muscular disease (65). Although as mentioned above, the loss of AAV during cell division in dividing cells limits its application in utero, the high titre preparations that can now be delivered without toxicity can compensate this progressive loss. Early second trimester intrathoracic or intramyocardial injection of AAV2/5, 2/9, and 2/10 have been shown to mediate prolonged CMV-promoter controlled intramuscular expression of luciferase in nonhuman primates (66) and fetal injection of AAV2/8 encoding human factor IX cDNA has resulted in long-term factor IX expression in sheep (67) and macaques (68) (see also Chapter 12). This allows taking advantage of the fetal induction of tolerance against the transgenic therapeutic protein, which if required can be renewed by postnatal reapplication of vector. Very recently, we have also observed that intravenous injection of AAV2/9 results in efficient gene transfer to neurons in the central and peripheral nervous system in both mice and nonhuman primates (69, 70).

So far no adverse effects have been reported after in utero application of AAV vectors; however, it should be noted that hepatocellular tumors were observed after application of β-glucuronidase expressing AAV to neonatal mice (71, 72). The mechanism for these events remains speculative (73, 74) but obviously, in common with the other potential vector systems, they indicate the need for particular care and long-term follow-up animal studies in preparation for human fetal application.

References

1. Rahman SH, Maeder ML, Joung JK, Cathomen T (2011) Zinc-finger nucleases for somatic gene therapy: the next frontier. Hum Gene Ther 22:925–933

2. Galetto R, Duchateau P, Paques F (2009) Targeted approaches for gene therapy and the emergence of engineered meganucleases. Expert Opin Biol Ther 9:1289–1303

3. Li T, Huang S, Jiang WZ, et al (2010) TAL nucleases (TALNs): hybrid proteins composed of TAL effectors and FokI DNA-cleavage domain. Nucleic Acids Res 39:359–372

4. Urnov FD, Miller JC, Lee YL et al (2005) Highly efficient endogenous human gene correction using designed zinc-finger nucleases. Nature 435:646–651

5. Li H, Haurigot V, Doyon Y, et al (2011) In vivo genome editing restores haemostasis in a mouse model of haemophilia. Nature 475:217–221

6. Toscano MG, Romero Z, Munoz P et al (2011) Physiological and tissue-specific vectors for treatment of inherited diseases. Gene Ther 18:117–127

7. Gonzaga S, Henriques-Coelho T, Davey M et al (2008) Cystic adenomatoid malformations are induced by localized FGF10 overexpression in fetal rat lung. Am J Respir Cell Mol Biol 39:346–355

8. Krieg AM (1999) Direct immunologic activities of CpG DNA and implications for gene therapy. J Gene Med 1:56–63

9. Yew NS, Zhao H, Przybylska M et al (2002) CpG-depleted plasmid DNA vectors with enhanced safety and long-term gene expression in vivo. Mol Ther 5:731–738

10. Mason CA, Bigras JL, O'Blenes SB et al (1999) Gene transfer in utero biologically engineers a patent ductus arteriosus in lambs by arresting fibronectin-dependent neointimal formation. Nat Med 5:176–182

11. Saada J, Oudrhiri N, Bonnard A et al (2010) Combining keratinocyte growth factor transfection into the airways and tracheal occlusion in a fetal sheep model of congenital diaphragmatic hernia. J Gene Med 12(5):413–422

12. Darquet A, Cameron B, Wils P et al (1997) A new DNA vehicle for nonviral gene delivery: supercoiled minicircle. Gene Ther 4:1341–1349

13. Bigger BW, Tolmachov O, Collombet JM et al (2001) An araC-controlled bacterial cre expression system to produce DNA minicircle vectors for nuclear and mitochondrial gene therapy. J Biol Chem 276:23018–23027

14. Chen ZY, He CY, Ehrhardt A, Kay MA (2003) Minicircle DNA vectors devoid of bacterial DNA result in persistent and high-level transgene expression in vivo. Mol Ther 8:495–500

15. Argyros O, Wong SP, Niceta M et al (2008) Persistent episomal transgene expression in liver following delivery of a scaffold/matrix attachment region containing non-viral vector. Gene Ther 15:1593–1605

16. Argyros O, Wong SP, Fedonidis C et al (2011) Development of S/MAR minicircles for enhanced and persistent transgene expression in the mouse liver. J Mol Med 89:515–529

17. Hagedorn C, Wong SP, Harbottle R, Lipps HJ (2011) Scaffold/Matrix attached region-based nonviral episomal vectors. Hum Gene Ther 22:915–923

18. Kootstra NA, Verma IM (2003) Gene therapy with viral vectors. Annu Rev Pharmacol Toxicol 43:413–439

19. Strauss SE (1984) Adenovirus infections in humans. In: Ginsberg HS (ed) Adenoviruses. Plenum Press, New York, London, pp 451–496

20. Horowitz MS (1990) Adenoviruses NO. In: Fields BN, Knipe DM (eds) Virology. Raven Press, Ltd, New York, pp 1723–1740

21. Yang Y, Li Q, Ertl HCJ, Wilson JM (1995) Cellular and humoral immune responses to viral antigens create barriers to lung-directed gene therapy with recombinant adenoviruses. J Virol 69:2004–2015

22. Danthinne X, Imperiale MJ (2000) Production of first generation adenovirus vectors: a review. Gene Ther 7:1707–1714

23. Kochanek S (1999) High-capacity adenoviral vectors for gene transfer and somatic gene therapy. Hum Gene Ther 10:2451–2459

24. Kochanek S, Clemens PR, Mitani K et al (1996) A new adenoviral vector: Replacement of all viral coding sequences with 28 kb of DNA independently expressing both full-length dystrophin and ß-galactosidase. Proc Natl Acad Sci U S A 93:5731–5752

25. Schiedner G, Morral N, Parks RJ et al (1998) Genomic DNA transfer with a high-capacity adenovirus vector results in improved in vivo gene expression and decreased toxicity. Nat Genet 18:180–183

26. Kreppel F, Biermann V, Kochanek S, Schiedner G (2002) A DNA-based method to assay total and infectious particle contents and helper virus contamination in high-capacity adenoviral vector preparations. Hum Gene Ther 13:1151–1156

27. Waddington S, Buckley SMK, David AL et al (2007) Fetal gene transfer. Curr Opin Mol Ther 9:432–438

28. Katz AB, Keswani SG, Habli M et al (2009) Placental gene transfer: transgene screening in mice for trophic effects on the placenta. Am J Obstet Gynecol 201(499):e491–e498

29. Laurema A, Vanamo K, Heikkila A et al (2004) Fetal membranes act as a barrier for adenoviruses: gene transfer into exocoelomic cavity of rat fetuses does not affect cells in the fetus. Am J Obstet Gynecol 190:264–267

30. Larson J, Morrow SL, Happel L et al (1997) Reversal of cystic fibrosis phenotype in mice by gene therapy in utero. Lancet 349:619–620

31. Buckley SM, Waddington SN, Jezzard S et al (2008) Intra-amniotic delivery of CFTR-expressing adenovirus does not reverse cystic fibrosis phenotype in inbred CFTR-knockout mice. Mol Ther 16:819–824

32. Davies LA, Varathalingam A, Painter H et al (2008) Adenovirus-mediated in utero expression of CFTR does not improve survival of CFTR knockout mice. Mol Ther 16:812–818

33. Bilbao R, Reay DP, Wu E et al (2005) (2005) Comparison of high-capacity and first-generation adenoviral vector gene delivery to murine muscle in utero. Gene Ther 12:39–47

34. Roberts DM, Nanda A, Havenga MJ et al (2006) Hexon-chimaeric adenovirus serotype 5 vectors circumvent pre-existing anti-vector immunity. Nature 441:239–243

35. Fang B, Eisensmith RC, Wang H et al (1995) Gene therapy for Hemophilia B: host immunosuppression prolongs the therapeutic effect of adenovirus-mediated Factor IX expression. Hum Gene Ther 6:1039–1044

36. Zennou V, Petit C, Guetard D et al (2000) HIV-1 genome nuclear import is mediated by a central DNA flap. Cell 101:173–185

37. Mitchell RS, Beitzel BF, Schroder AR et al (2004) Retroviral DNA integration: ASLV, HIV, and MLV show distinct target site preferences. PLoS Biol 2:E234

38. Bartosch B, Cosset FL (2004) Strategies for retargeted gene delivery using vectors derived from lentiviruses NO. Curr Gene Ther 4:427–443

39. Baum C, Kustikova O, Modlich U et al (2006) Mutagenesis and oncogenesis by chromosomal insertion of gene transfer vectors. Hum Gene Ther 17:253–263

40. Pitt BR, Schwarz MA, Pilewski JM et al (1995) Retrovirus-mediated gene transfer in lungs of living fetal sheep. Gene Ther 2:344–350

41. Douar AM, Adebakin S, Themis M et al (1997) Foetal gene delivery in mice by intra-amniotic administration of retroviral producer cells and adenovirus. Gene Ther 4:883–890

42. Seppen J, van der Rijt R, Looije N et al (2003) Long-term correction of bilirubin UDP glucuronyltransferase deficiency in rats by in utero lentiviral gene transfer. Mol Ther 8:593–599

43. Niiya M, Endo M, Shang D et al (2008) Correction of ADAMTS13 Deficiency by In Utero Gene Transfer of Lentiviral Vector encoding ADAMTS13 Genes, Mol Ther 17:34–41

44. Waddington S, Nivsarkar M, Mistry A et al (2004) Permanent phenotypic correction of Haemophilia B in immunocompetent mice by prenatal gene therapy. Blood 104:2714–2721

45. Yu ZY, McKay K, van Asperen P et al (2007) Lentivirus vector-mediated gene transfer to the developing bronchiolar airway epithelium in the fetal lamb. J Gene Med 9:429–439

46. Sinn PL, Penisten AK, Burnight ER et al (2005) Gene transfer to respiratory epithelia with lentivirus pseudotyped with Jaagsiekte sheep retrovirus envelope glycoprotein. Hum Gene Ther 16:479–488

47. Themis M, Waddington SN, Schmidt M et al (2005) Oncogenesis following delivery of a non-primate lentiviral gene therapy vector to fetal mice. Mol Ther Mol Ther 12:763–771

48. Buning H, Perabo L, Coutelle O et al (2008) Recent developments in adeno-associated virus vector technology. J Gene Med 10:717–733

49. Douar A-M, Themis M, Coutelle C (1996) Fetal somatic gene therapy. Hum Mol Reprod 2:633–641

50. Manno CS, Pierce GF, Arruda VR et al (2006) Successful transduction of liver in hemophilia by

AAV-Factor IX and limitations imposed by the host immune response. Nature 12:342–347

51. Mingozzi F, Maus MV, Hui DJ et al (2007) CD8(+) T-cell responses to adeno-associated virus capsid in humans. Nat Med 13:419–422

52. Nathwani AC, Gray JT, Ng CY et al (2006) Self-complementary adeno-associated virus vectors containing a novel liver-specific human factor IX expression cassette enable highly efficient transduction of murine and nonhuman primate liver. Blood 107:2653–2661

53. Davidoff AM, Gray JT, Ng CY et al (2005) Comparison of the ability of adeno-associated viral vectors pseudotyped with serotype 2, 5, and 8 capsid proteins to mediate efficient transduction of the liver in murine and nonhuman primate models. Mol Ther 11:875–888

54. Cecchini S, Virag T, Kotin RM (2011) Reproducible high yields of recombinant adeno-associated virus produced using invertebrate cells in 0.02- to 200-liter cultures. Hum Gene Ther 22:1021–1030

55. Mueller C, Flotte TR (2008) Clinical gene therapy using recombinant adeno-associated virus vectors. Gene Ther 15:858–863

56. Mah C, Sarkar R, Zolotukhin I et al (2003) Dual vectors expressing murine factor VIII result in sustained correction of hemophilia A mice. Hum Gene Ther 14:143–152

57. Lu H, Chen L, Wang J, Huack B et al (2008) Complete correction of hemophilia A with adeno-associated viral vectors containing a full-size expression cassette. Hum Gene Ther 19:648–654

58. Allocca M, Doria M, Petrillo M et al (2008) Serotype-dependent packaging of large genes in adeno-associated viral vectors results in effective gene delivery in mice. J Clin Invest 118:1955–1964

59. Hirsch ML, Agbandje-McKenna M, Samulski RJ (2010) Little vector, big gene transduction: fragmented genome reassembly of adeno-associated virus. Mol Ther 18(1):6–8

60. Rucker M, Fraites TJ Jr, Porvasnik SL et al (2004) Rescue of enzyme deficiency in embryonic diaphragm in a mouse model of metabolic myopathy: Pompe disease. Development 131:3007–3019

61. Dejneka NS, Surace EM, Aleman TS et al (2004) In utero gene therapy rescues vision in a murine model of congenital blindness. Mol Ther 9:182–188

62. Karolewski BA, Wolfe JH (2006) Genetic correction of the fetal brain increases the lifespan of mice with the severe multisystemic disease mucopolysaccharidosis type VII. Mol Ther 14:14–24

63. Koppanati BM, Li J, Reay DP et al (2010) Improvement of the mdx mouse dystrophic phenotype by systemic in utero AAV8 delivery of a minidystrophin gene. Gene Ther 17:1355–1362

64. Sabatino DE, Mackenzie TC, Peranteau W et al (2007) Persistent expression of hF.IX after tolerance induction by in utero or neonatal administration of AAV-1-F.IX in hemophilia B mice. Mol Ther 15:1677–1685

65. Koppanati BM, Li J, Xiao X, Clemens PR (2009) Systemic delivery of AAV8 in utero results in gene expression in diaphragm and limb muscle: treatment implications for muscle disorders. Gene Ther 16:1130–1137

66. Tarantal AF, Lee CC (2010) Long-term luciferase expression monitored by bioluminescence imaging after adeno-associated virus-mediated fetal gene delivery in rhesus monkeys (*Macaca mulatta*). Hum Gene Ther 21:143–148

67. David AL, Peebles DM, Gregory L et al (2006) Clinically applicable procedure for gene delivery to fetal gut by ultrasound-guided gastric injection: toward prenatal prevention of early-onset intestinal diseases. Hum Gene Ther 17:767–779

68. Mattar CN, Nathwani AC, Waddington SN et al (2011) Stable human FIX expression after 0.9G intrauterine gene transfer of self-complementary adeno-associated viral vector 5 and 8 in macaques, Mol Ther 19:1950–1960

69. Rahim AA, Wong AM, Hoefer K, et al (2011) Intravenous administration of AAV2/9 to the fetal and neonatal mouse leads to differential targeting of CNS cell types and extensive transduction of the nervous system. FASEB J 25:3505–3518

70. Mattar CN, Waddington SN, Biswas A et al (2012) Systemic delivery of scAAV9 in fetal macaques facilitates neuronal transduction of the scAAV9 in fetal macaques facilitates neuronal transduction of the central and peripheral nervous systems, Gene Ther. In Press

71. Donsante A, Miller DG, Li Y et al (2007) AAV vector integration sites in mouse hepatocellular carcinoma. Science 317:477

72. Donsante A, Vogler C, Muzyczka N et al (2001) Observed incidence of tumorigenesis in long-term rodent studies of rAAV vectors. Gene Ther 8:1343–1346

73. Russell DW (2007) AAV vectors, insertional mutagenesis, and cancer. Mol Ther 15:1740–1743

74. Kay MA (2007) AAV vectors and tumorigenicity. Nat Biotechnol 25:1111–1113

Chapter 4

Vector Systems for Prenatal Gene Therapy: Principles of Adenovirus Design and Production

Raul Alba, Andrew H. Baker, and Stuart A. Nicklin

Abstract

Adenoviruses have many attributes, which have made them one of the most widely investigated vectors for gene therapy applications. These include ease of genetic manipulation to produce replication-deficient vectors, ability to readily generate high titer stocks, efficiency of gene delivery into many cell types, and ability to encode large genetic inserts. Recent advances in adenoviral vector engineering have included the ability to genetically manipulate the tropism of the vector by engineering of the major capsid proteins, particularly fiber and hexon. Furthermore, simple replication-deficient adenoviral vectors deleted for expression of a single gene have been complemented by the development of systems in which the majority of adenoviral genes are deleted, generating sophisticated Ad vectors which can mediate sustained transgene expression following a single delivery. This chapter outlines methods for developing simple transgene over expressing Ad vectors and detailed strategies to engineer mutations into the major capsid proteins.

Key words: Adenovirus, 293 cells, Titer, Plaque forming unit, Viral particle, Targeting, Replication deficient

1. Introduction

The human adenovirus (Ad) family consists of seven species designated A–G, and a total of 57 different serotypes have been identified to date. Ads are classified via hemagglutination profiles in different species and differences in capsid structure and receptor preference. Gene transfer vectors based on Ads have historically focused on those from species C, particularly serotype 5 (Ad5) and are the most widely investigated viral vectors for basic viral biology, human gene therapy, and therapeutic vaccination strategies. Consequently, Ads are very commonly utilized as viral vectors in gene therapy clinical trials. A major barrier to gene therapy with Ad5 is the proinflammatory and immune response to the vector. Furthermore, since receptor use and infection efficiency for different cell types is

Charles Coutelle and Simon N. Waddington (eds.), *Prenatal Gene Therapy: Concepts, Methods, and Protocols*,
Methods in Molecular Biology, vol. 891, DOI 10.1007/978-1-61779-873-3_4, © Springer Science+Business Media, LLC 2012

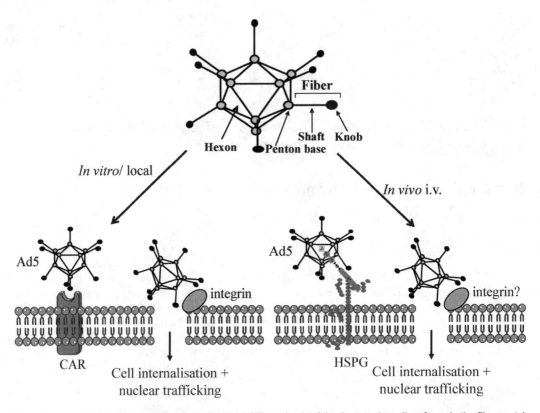

Fig. 1. Ad cell entry. In cells in vitro and following local delivery in vivo Ad tethers to the cell surface via the fiber protein interacting with CAR. However, following *i.v.* delivery Ad sequesters FX and the Ad:FX complex tethers to the hepatocyte surface via HSPGs. Following cell binding, activation of cell-surface integrins via the penton base induces endocytosis of the virus. Although this has not been explicitly demonstrated via the FX-mediated pathway.

divergent between individual Ad serotypes, recent research has focused on alternative serotypes, particularly the species B Ad serotype 35 (Ad35), expanding the repertoire of available Ads. Ads have an icosahedral, unenveloped capsid and are approximately 70–90 nm in diameter with a double stranded linear 36 kb DNA genome. The viral capsid consists of three main proteins: the hexon, penton base, and fiber (see Fig. 1). The majority of the capsid is the hexon and comprises 240 trimeric hexon capsomeres. Additionally, there are 12 pentameric penton bases and the trimeric fiber protein protrudes from the penton base at each of the 12 vertices of the capsid, consisting of a shaft and globular knob domain (see Fig. 1).

The adenoviral infection pathway (at least in vitro) is well defined, particularly for serotype 5 (species C). Ads infect host cells classically via receptor-mediated endocytosis mediated by both the fiber protein and the penton base. For species C Ads, initial cell tethering is mediated by an interaction between the fiber protein and the coxsackievirus and adenovirus receptor (CAR) (1, 2) (see Fig. 1). This initial binding of Ad5 is mediated by the fiber knob domain.

Recombinant knob protein or anti-knob antibodies efficiently inhibit adenoviral infection (3) and mutations in key amino acids in the fiber knob domain are able to ablate CAR interactions (for review, see ref. 4). The fiber knob domains from all Ad species, except those from species B, interact with CAR. Additionally, some species D Ads also use sialic acid (5–8) and species B Ads predominantly utilize CD46 (9) although an alternative receptor X has also been described (10).

Following intravascular (i.v.) delivery, the major site of Ad uptake is the liver and results in transduction of hepatocytes (11, 12). In the bloodstream, Ad5 interacts with host proteins including coagulation factor IX (FIX), FX, and complement binding protein (C4BP) to deliver virus to hepatocytes and this is thought to be mediated via interaction with heparan sulfate proteoglycans (HSPGs) for FIX and FX, or low density lipoprotein receptor-related protein for C4BP (13, 14). Sequestration of FX and delivery of the Ad5:FX complex to the liver is the predominant liver transduction mechanism for Ad5 via the i.v. route (15, 16). One of the main reasons Ads have been particularly well developed for gene therapy is the ease by which replication-deficient vectors can be made in the 293 cell line. The human embryonic kidney cell line, 293, is transformed with the entire left-hand end of the adenovirus serotype 5 genome (17) enabling development of recombinant Ads in a helper virus-free environment. Other cell lines have subsequently been developed, including PER.C6 and 911 cells with the aim of minimizing common DNA sequences between cell line and vector and to improve vector yields (18, 19).

By far the most widely utilized vectors for gene therapy are the so-called first generation, replication-deficient Ad which contain deletions of the E1A and E1B regions, enabling insertion of expression cassettes of up to approximately 5 kb. To address this size limitation, additional deletions have been developed in the E3 region (dispensable for Ad production as it is involved in suppressing immune responses to viral infection by inhibiting cell-surface expression of MHC class I molecules (20)). The main issues with "first generation" Ads is the transient transgene expression they produce, due to leaky production of viral proteins in transduced cells resulting in a profound proinflammatory response (21, 22). "Second generation" Ads were also produced via additional mutations in the E2 and/or E4 regions; however, although certain advantages in terms of reduced inflammatory response and enhanced transgene expression compared to standard E1-deleted vectors were demonstrated (23–25), challenges with expressing the Ad genes in trans in the 293 cell line were hampered by toxicity of the gene products and "second generation" Ads have not been pursued for routine use. Third generation [also called gutless or helper-dependent adenoviral vectors (hdAd)] are promising next-generation Ad vectors [reviewed in refs. 26, 27). HdAd are deleted for all viral open reading frames, contain only essential *cis* elements (inverted terminal repeats and contiguous packaging sequences) and are promising vectors for long-term transgene expression in vivo. There are two main advantages for HdAd. First, ablation of "leaky" production of adenoviral genes in these vectors abrogates host immune responses against vector transduced cells (28). Second, the size of DNA that can be encoded is increased to 36–38 kb. The limit in DNA packaging is dictated by the size of the capsid, meaning DNA molecules greater than approximately 38 kb (105% of native Ad genome size) are unable to be packaged into mature virions (29). Importantly, even when inserting expression cassettes of smaller size into HdAd they must be flanked by "stuffer" DNA to ensure the viral genome size is maintained, as particles with low molecular weight DNA are unstable and fail to mature (30). The production of HdAd is a complex and specialist procedure, available in only a few labs worldwide and is therefore outside the scope of this chapter.

It is recognized that Ad tropism modification for targeting would be beneficial by improving the therapeutic index of gene delivery. Research has focused on altering the receptor interactions of Ad5 vectors mainly with the aim of enabling i.v. delivery to be targeted selectively to diseased tissues and away from the liver and spleen to improve safety and efficiency. Moreover, such strategies may potentially also improve therapeutic efficacy for local delivery using lower vector loads. Although antibody-based targeting strategies have been pursued, questions over the in vivo stability of Ad:antibody complexes and the quality control issues with a two component strategy has meant that genetic modification of Ad capsid proteins is the main method utilized. Strategies to modify the fiber, penton base, and hexon have been pursued in attempts to alter Ad tropism. Initial studies focused on the fiber protein with the aim of removing CAR interaction and/or providing a targeting ligand and the main site utilized is the HI loop of the fiber (see Fig. 2).

The HI loop is exposed outside the knob and consists mainly of hydrophilic amino acids connecting the β-strands H and I (31). The HI loop of individual Ad serotypes differs in length highlighting that incorporation of additional protein sequence does not have an effect on fiber trimerization or knob domain folding. Many studies have utilized the HI loop to insert targeting peptides for a range of cell types, both in the presence and absence of additional mutations, which ablate the fiber knob from binding CAR [for review, see ref. 4). While it is clear that modification of Ad tropism in vitro is efficient via this method, this has not translated into efficient targeted in vivo gene delivery via the i.v. route. Other modifications to the RGD motif in the penton base to ablate integrin interaction, both alone and in combination with CAR-binding ablation, have also generally failed to modify Ad tropism via the i.v. route (for review, see ref. 4). With the knowledge that the

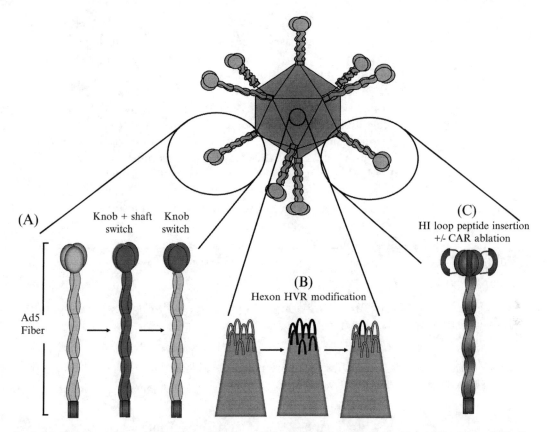

Fig. 2. Adenoviral targeting strategies. Schematic of the main methods for Ad5 genetic targeting. (a) Serotype switching by exchanging either the Ad5 fiber knob or entire fiber for that of an alternative serotype to provide new tropism. (b) Modification of the hexon HVRs by either pseudotyping strategies to replace either all or individual Ad5 HVRs with those of another serotype, or by point mutation at individual amino acids. (c) Insertion of peptide targeting ligands into the surface exposed HI loop of the Ad fiber knob.

hexon protein interaction with FX is the route of Ad5 gene transfer to the liver (15, 16) and strategies developed to genetically modify the hexon protein to ablate FX interaction and abrogate liver transduction (32), establishing selective targeted gene delivery via the i.v. route is potentially achievable.

Alternative strategies for Ad5 targeting have been via the identification of novel adenoviral serotypes with alternative tropism. Pseudotyping in which the fiber of Ad5 is exchanged for that of another serotype to potentially provide Ad5-based vectors with alternative tropism has also been pursued (see Fig. 2). The main focus of these alternative serotypes has been species B, in particular, Ad35 whose primary receptor is CD46 (9), a protein expressed in all human nucleated cells. The first studies utilized Ad35 fiber interaction with CD46 by pseudotyping onto Ad5 to bypass the challenges with transducing Ad5-refractory cells (hematopoietic cells) due to low CAR expression (33). Furthermore, pseudotyping

Ad5 with the rare species D 19p fiber produced vectors with significantly decreased liver transduction (34) which enabled retargeting to the rat kidney following insertion of renal targeting peptides into the HI loop (35).

Here, we provide detailed methods to generate "first generation" Ad5 vectors expressing transgenes, quality control assays, scale up for propagation and purification as well as titration. Additionally, methods are introduced to enable introduction of mutations into the Ad5 capsid proteins, fiber, and hexon. These methods should enable researchers in the field to establish Ad5 production in their laboratory and also to generate genetic mutations that will tailor the tropism of Ad5 for desired applications.

2. Materials

2.1. Preparation of Recombinant Adenoviruses

2.1.1. Homologous Recombination in Bacteria Between pAdEASY-1 and pSHUTTLE-CMV

1. *Pac*I restriction endonuclease [New England Biolabs (NEB), Hitchin, Herts, UK].

2. pSHUTTLE-CMV (Stratagene, CA, USA).

3. pAdEasy-1 (Stratagene, CA, USA).

4. 1% (v/v) agarose gel (UltraPure Agarose) (Invitrogen, CA, USA).

5. Ethidium bromide (Sigma, Steinheim, Germany).

6. Tris borate ethylenediaminetetraaceticacid (EDTA) buffer (TBE): 10× stock (per liter): 108 g Tris–base, 55.0 g boric acid, 40 mL 0.5 M EDTA (pH 8.0), pH 8.3. Working stock = 1×.

7. Wizard SV Gel PCR Clean Up System (Promega, WI, USA).

8. Restriction endonucleases (NEB, Hitchin, Herts, UK).

9. *BJ*5183 cells (Stratagene, #200154, CA, USA).

10. Electroporator (BIO-RAD #165-2100, Hercules, CA, USA).

11. Electroporation cuvettes 0.2 cm gap (Cell Project, #EP-102, Staffordshire, UK).

12. DH5α or TOP10 competent bacteria (Invitrogen, CA, USA).

13. SOC (Super Optimal Broth with Catabolite repression) media (Invitrogen, #15544-034, CA, USA).

14. LB broth [per liter sterile deionized water (dH$_2$O)]: 10 g of NaCl, 10 g of bactotryptone, 5 g of yeast extract, pH 7.0 with 5 N NaOH. Autoclave and add antibiotic once cooled to ≤45°C if required.

15. LB agar plates. Use LB broth recipe adding 20 g of agar. Pour into Petri dishes (25 mL/100 mm dish), when the solution is ≤45°C add the correct concentration of the appropriate antibiotic.

16. Kanamycin (Sigma, #K0254, Steinheim, Germany). Stock concentration, 50 mg/mL. Working concentration, 50 μg/mL.

17. Ampicillin (Sigma, #A9518, Steinheim, Germany). Stock solution, 100 mg/mL. Working stocks, 100 μg/mL.

18. QIAgen miniprep kit (QIAgen, #27106. Hilden, Germany).

2.1.2. Transfection into Low Passage 293 Cells for Adenovirus Generation

1. 10× Citric saline: 100 g of potassium chloride and 44 g of sodium citrate in 100 mL dH_2O. Sterilize by autoclaving. Stable for up to 1 year at room temperature. For a working solution, dilute the stock 10× in sterile phosphate-buffered saline (PBS). Make fresh as required.

2. Lipofectamine™2000.

3. AdEasy-1 recombinant viral DNA from Subheading 2.1.1.

4. *Pac*I restriction endonuclease (NEB, Hitchin, Herts, UK).

5. 100% Ethanol.

6. 6-Well plates.

7. Sterile PBS; Ca^{2+}- and Mg^{2+}-free.

8. Low passage (below passage 30 for generation of Ads or 50 for propagation) human kidney embryonic 293 cells (Microbix, Toronto, Canada).

9. 293 Culture media: minimal essential medium (MEM) containing 10% (v/v) fetal calf serum (FCS), 2 mM L-glutamine, 100 international units (IU)/mL penicillin, and 100 μg/mL streptomycin. Stable at 4°C for up to 1 month.

10. Opti-MEM media (without serum) (Invitrogen, CA, USA).

11. Potassium acetate (3 M, pH 5.2).

2.1.3. Generation of Crude Adenovirus Stocks

1. ArkloneP (trichlorotrifluoroethane) (Sigma, Steinheim, Germany).

2.1.4. Plaque Purification by End-Point Dilution

1. 96-Well plates.

2.1.5. Generation of High Titer Recombinant Adenoviral Stocks

1. T-150 cm² tissue culture flasks.

2. UltraClear ultracentrifuge tubes, Catalog #344061 (16×102 mm) (Beckman, CA, USA). We use a Beckman Optima L-80 ultracentrifuge and Beckman SW40Ti swing out bucket rotor.

3. Tris–EDTA (TE) 10× stock: 12.1 g Tris–HCl (10 mM), 3.72 g EDTA, pH 8 (1 mM) in 1 L dH_2O. Autoclave. Stable at RT for 1 year.

4. TD 10× buffer (1 L dH_2O): 80 g NaCl (750 mM), 3.8 g KCl (50 mM), 2.5 g $Na_2HPO_4\cdot12H_2O$ (10 mM), 30 g Tris–base

(250 mM). Adjust pH to 7.4 with 10 N HCl (approximately 16 mL). Autoclave and store at 4°C.

5. Cesium chloride solutions (see Note 1).

Density 1.25 g/mL: 36.16 g CsCl + 100 mL 1× TD buffer.

Density 1.40 g/mL: 62.20 g CsCl + 100 mL TD 1× buffer.

Density 1.34 g/mL: 51.20 g CsCl + 100 mL TD 1× buffer.

Sterilize by filtration (0.22 μm filter). Can be stored at room temperature for 1 year. If stored at 4°C, remember to bring it to RT before use.

6. Slide-A-Lyzer Dialysis Cassettes (extra strength), 10,000 molecular weight cut off (MWCO), 0.5–3.0 mL capacity, Catalog #66380 (Pierce Protein Research Products, Thermo Scientific, Rockford, IL, USA).

7. Viral preparation dialysis buffer: 27 mL of 5 M NaCl (135 mM), 10 mL of 1 M Tris, pH 7.5 (10 mM), and 0.5 mL of 2 M $MgCl_2$ (1 mM) in 1 L dH_2O. Sterilize individual reagents separately by autoclaving and store at room temperature for up to 6 months. Combine individual reagents to make final working stock of dialysis buffer, fresh as required.

8. Dialysis buffer with 10% glycerol: 900 mL of working dialysis buffer made as above plus 100 mL of sterile autoclaved glycerol. Make fresh as required.

9. MicroBCA (Bicinchoninic acid) Protein Assay Kit (ThermoScientific, IL, USA).

2.1.6. Assay for Replication Competent Adenovirus (RCA)

1. HeLa cells (European Collection of Animal Cell Cultures, Salisbury, UK).

2. Growth media for HeLa cells: MEM containing 10% FCS, 2 mM L-glutamine, 100 IU/mL penicillin, and 100 μg/mL streptomycin. Stable at 4°C for up to 1 month.

2.2. Generation of Adenoviral Vectors with Altered Capsid Proteins

2.2.1. Molecular Biology Reagents

1. Proof reading *Taq* polymerase (Herculase II Fusion DNA Polymerase, #600677-41, Stratagene, CA, USA).

2. Rapid DNA Ligation kit (#11635379001, Roche, Mannheim, Germany).

3. StrataClone Blunt PCR cloning kit (Stratagene, #240207, CA, USA).

4. Annealing buffer: 10 mM Tris, pH 7.5–8.0, 60 mM NaCl, 1 mM EDTA.

5. DH5α TOP10 competent bacteria (Invitrogen, CA, USA).

6. 1 kb and 100 bp molecular weight ladders (Promega, WI, USA).

2.2.2. Generation of an Ad5 Fiber Shuttle Plasmid

Oligonucleotides flanking the fiber gene and homologous regions.

1. Oligonucleotide 1 (Forward): 5′-GCACAAGAACGCCATAG TTGC-3′.

2. Oligonucleotide 2 (Reverse) 5′-TGCAGCACCACCGCCCT ATC-3′ (see Note 2).

3. Oligonucleotides for sequencing (1 every 400 nucleotides).

2.2.3. Incorporation of CAR-Binding Mutations

Oligonucleotides flanking the region of interest (Ad5 region between *Nde*I and *Mfe*I).

1. Oligonucleotide 1 (Forward): 5′-CCCGTGTATCCATATGA CAC-3′.

2. Oligonucleotide 2 (Reverse): 5′-ATAAGCTATGTGGTGGTG GGG-3′.

Oligonucleotides incorporating the desired point mutations.

3. Oligonucleotide 3 (Forward): 5′-CCAGCTCCA**GAGGCT**AA CTGTAGAC-3′.

4. Oligonucleotide 4 (Reverse): 5′-GTCTACAGTT**AGCCTC**TG GAGCTGG-3′.

5. Oligonucleotides for sequencing (1 every 400 nucleotides).

Underlined bold sequences correspond to the S408E and P409A mutations.

2.2.4. Incorporation of Peptides into the HI Loop of the Ad5 Fiber

Oligonucleotides flanking the region of interest (Ad5 region between *Nde*I and *Mfe*I).

1. Oligonucleotide 1 (Forward): 5′-CCCGTGTATCCATATGA CAC-3′.

2. Oligonucleotide 2 (Reverse): 5′-ATAAGCTATGTGGTGGTG GGG-3′.

Oligonucleotides incorporating the desired restriction site (*Not*I).

3. Oligonucleotide 5 (Forward): 5′-ACAGGA**GCGGCCGCCG** ACACAACTCCAAGTGCATAC-3′.

4. Oligonucleotide 6 (Reverse): 5′-TGTGTC**GGCGGCCGC**TCC TGTTTCCTGTGTACCGTT-3′.

Underlined bold sequences correspond to the *Not*I restriction site.

5. Oligonucleotides containing the peptide of interest.

6. Oligonucleotides for sequencing (1 every 400 nucleotides).

2.2.5. Generation of Ad5 Hexon Shuttle Plasmid

Oligonucleotides flanking the hexon gene and homologous regions.

1. Oligonucleotide 7 (Forward): 5′-CTCCTTATTCCACTGATC GCC-3′.

2. Oligonucleotide 8 (Reverse): 5′-ATCTGATCTCCGACAAGA GCG-3′.

3. Oligonucleotides for sequencing (1 every 400 nucleotides).

Oligonucleotides flanking the region of interest (Ad5 region between *Nde*I and *Bam*HI).

1. Oligonucleotide 9 (Forward): 5′-CCAATGAAACCATGTTA CGG-3′.

2. Oligonucleotide 10 (Reverse): 5′-CTCGTCCATGGGATCC ACC-3′.

Oligonucleotides incorporating the E451Q mutation in the hexon gene.

3. Oligonucleotide 11 (Forward): 5′-CAGATAAAAAT**C**AAAT AAGAGTTGG-3′.

4. Oligonucleotide 12 (Reverse): 5′-CCAACTCTTATTT**G**ATTT TTATCTG-3′.

5. Oligonucleotides for sequencing (1 every 400 nucleotides).

Underlined bold sequences correspond to the E451Q point mutation.

3. Methods

3.1. Preparation of Recombinant Adenoviruses

3.1.1. Generation of Ad5 Vectors Containing Transgenes in the E1 Region

For the insertion of transgenes in early region 1 (E1), we recommend the use of the commercially available pSHUTTLE-CMV (Stratagene). The cloning can be easily performed in the supplied multicloning site. For the insertion of different promoters, we recommend the use of pSHUTTLE (Stratagene).

3.1.2. Homologous Recombination in Bacteria Between pAdEasy-1 and pSHUTTLE-CMV

The most common Ad vector utilized is based on human Ad5 and our protocols are based on this serotype. Homologous recombination mediates combination of two fragments of DNA, one containing the whole Ad genome linearized at the region of interest (E1 region) and a second linearized shuttle plasmid (with the expression cassette incorporated in place of E1) with the modified region of interest flanked by approximately 1 kb of homologous DNA, in a bacterial strain expressing active recombination genes (*BJ*5183). In this chapter, we recommend the use of the commercially available AdEasy system. The available commercial shuttle plasmid is only designed for inserting expression cassettes into the E1 region to generate replication-deficient Ad vectors. However, later in the chapter, we describe the generation of alternative shuttle plasmids enabling modification of the fiber or hexon capsid proteins in the Ad genome, also based on the AdEasy system.

1. Digest pSHUTTLE-CMV with a suitable restriction endonuclease (e.g., *Pme*I) (see Note 3, Fig. 3).

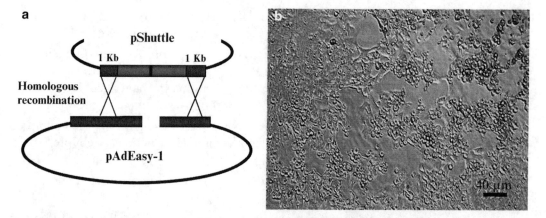

Fig. 3. Homologous recombination strategy. (a) The homologous recombination technique combines two fragments of DNA containing homologous regions. For the generation of Ad5 vectors, it is required to use two consecutive recombination events between a shuttle plasmid (which modifies the gene of interest) flanked by 1 kb of homologous Ad5 DNA and pAdEasy-1. This recombination event needs to be performed in the bacterial strain *BJ*5183. (b) An example of a plaque generated by Ad5 5–7 days after transfection of linearized pAdEasy-1 into HEK293 cells. Magnification 10×.

2. Digest pAdEasy-1 with *Pac*I to linearize the DNA ready for modification in the E1 region.

3. Electrophorese each digested DNA fragment to confirm the digestion is complete and purify the bands (7.5 kb shuttle plasmid + expression cassette molecular weight and 33.5 kb pAdEasy-1 band) using gel extraction kits following the manufacturer's protocol (see Note 4).

4. Thaw *BJ*5183 cells (50–100 µL) on ice and combine 100 ng of purified, linearized pAdEasy-1 genome plasmid with 100 ng of purified, linearized shuttle plasmid in chemically competent or electrocompetent *BJ*5183 cells. Maintain the cells on ice for 20 min to maximize the efficiency of recombination (see Note 5). For chemically competent *BJ*5183 cells, perform heat shock at 42°C for 45 s and maintain the cells on ice for 5 min. If you are using electrocompetent *BJ*5183 cells, set the following parameters on the electroporator: 200 Ω, 2.5 kV, and 25 µF and use prechilled electroporation cuvettes maintained on ice as the bacteria and DNA samples are added. Slide the cuvette into the chilled electroporation chamber until the cuvette connects with the electrical contacts. Pulse the sample once then quickly remove the cuvette. Immediately after the pulse, add 500–1,000 µL of SOC media and continue to the next step.

5. For chemically competent cells, add 500 µL of SOC media. For either chemically or electrocompetent bacteria, incubate with shaking at 225–250 rpm for 1 h at 37°C (see Note 6).

6. Subject the cells to centrifugation at $800 \times g$ for 2 min and discard the supernatant. Resuspend the cells in 100 µL SOC media and plate them onto agar plates with the appropriate

antibiotic using a sterile spreader to spread the mixture. Incubate the plates at 37°C overnight (see Note 7).

7. Select several tiny colonies (see Note 8) and add each separately to 5 mL LB media (supplemented with appropriate working concentration of antibiotic) and incubate at 37°C with shaking at 225–250 rpm overnight (see Note 9).

8. Extract the DNA from the bacteria using the QIAgen miniprep kit following the manufacturer's instructions (see Note 10).

9. To re-transform the plasmids, use 2–5 μL of miniprep DNA and 50 μL DH5α or TOP10 competent bacteria and repeat steps 4–10.

10. Check the minipreps by restriction endonuclease digestion and sequence the region of interest to confirm the integrity and identity of the DNA.

3.1.3. Transfection of Adenoviral Genomes into Low Passage 293 Cells for Derivation of Adenovirus

Adenoviral genomes need to be linearized and purified to exclude expression of the antibiotic resistance marker gene and the origin of replication of the plasmid. We use the AdEasy system (Stratagene (36)) and the suggested restriction endonuclease *Pac*I to achieve this step. The next stage in the production of a recombinant adenovirus is the transfection of the intact recombinant Ad plasmid into 70% confluent 293 cells. Purified and linearized Ad genomes can be transfected using commercially available Lipofectamine™2000 (see Note 10). This method produces a transfection efficiency of between 25 and 40% which is adequate for the initial amplification of Ad vectors.

1. Digest 100 μg of pAdEasy-1 with the expression cassette encoded in the E1 region generated in protocol 3.1.2 with 30 U of *Pac*I overnight in a final volume of 200 μL of sterilized water (see Note 11).

2. Purify and precipitate the DNA by adding two volumes of 100% ethanol and 20 μL of 3 M potassium acetate (pH 5.2) to the digested DNA. Freeze the sample at −80°C for 30 min and subject to centrifugation at 4°C for 15 min at $16,100 \times g$. Wash the pellet with 70% ethanol.

3. Discard supernatant, air dry the sample for 10 min at 37°C, and resuspend the DNA pellet in 100 μL of sterilized water (see Note 12).

4. Subculture low passage 293 cells (<P30) into a 6-well plate (see Note 13). Seed each well with 6×10^5 cells so that they reach approximately 70% confluence the following day. It is important that the cells are in mid-log phase for optimum transfection efficiency.

5. Pre-mix 3–4 μg of DNA containing the Ad genome with 450 μL of Opti-MEM media and incubate the sample for 5 min at room temperature (RT).

6. Pre-mix 2.5 µL of Lipofectamine™2000 with 450 µL of Opti-MEM media and incubate the sample for 5 min at RT.

7. Combine the DNA and Lipofectamine™2000 solutions (900 µL) and mix by inverting several times. Incubate at room temperature for 20–25 min to allow DNA–Lipofectamine™2000 complexes to form.

8. Remove culture media from the 6-well plates and add the DNA–Lipofectamine™2000 complexes from step 7 (900 µL) directly to each well (from step 4).

9. Incubate for 4 h at 37°C/5% CO_2.

10. On each well gently exchange the Opti-MEM transfection media for pre-warmed 293 cell culture media.

11. Replace media every 2–3 days or when acidity is evident.

12. Adenoviral plaques should appear after 6–10 days of incubation (see Fig. 3). Plaques can be visualized by phase contrast microscopy or by direct visualization with the naked eye in the 6-well plate as the plaque grows (see Note 14).

13. Once one or several Ad plaques are apparent, mark the well and allow the virus to propagate through the well until the cytopathic effect (CPE) is complete. Do not change the media during this time (approximately 2 additional days). Next proceed to protocol 3.1.4.

3.1.4. Generation of Crude Adenovirus Stocks

Once a viral plaque has arisen in the 293 cell monolayer, indicating that viral progeny are being produced (see Fig. 3), monitor the CPE as it spreads throughout the cell monolayer. Approximately, 90% of the virus will remain trapped within the cell. Therefore, a simple extraction with a commercial solvent such as ArkloneP or 3 cycles of freeze/thawing in dry ice/37°C water bath will release the adenovirus.

1. Once CPE is complete, collect the media containing the cells into a sterile tube (see Note 15). Wash the flask with 5 mL of complete media and pool. Pellet the cells in a sealed container at $250 \times g$ for 10 min at room temperature (see Note 16).

2. Decant media and resuspend pellet in 1 mL of sterile PBS. Add an equal volume of ArkloneP, then invert the tube for 10 s followed by shaking (not too vigorously) for 5 s. Repeat the mixing step (see Note 17).

3. Centrifuge the tube at $750 \times g$ for 10 min at room temperature. The tube will now contain three layers, the lower layer is ArkloneP, the middle layer contains cellular debris, and the top layer is the adenovirus in solution.

4. Remove the adenovirus with a pipette, taking care not to disturb the layer of cellular debris and store in sterile Eppendorfs in 50 µL aliquots at –80°C (see Note 18). This crude preparation

of adenovirus can be used to test for transgene expression and for the extraction of DNA for sequencing.

Using the AdEasy-1 system to generate adenoviral vectors ensures all viral progeny are identical since one intact DNA molecule is used for the initial packaging of the vector. However, viral plaques can arise as a result of multiple recombination events using other methods and obtaining adenoviral vectors via collaboration does not always enable knowledge of their provenance. Therefore, a simple plaque purification technique is a useful adjunct to viral quality control. We utilize plaque purification by serial dilution of crude viral stock into 293 cells.

1. Subculture low passage 293 cells into a 96-well plate (80 wells in 8 rows of 10) (see Note 19, Fig. 5) for approximately 50–60% confluence the following day (see Note 20).

2. Prepare serial dilutions of adenovirus from the crude stock (see Fig. 5, Note 21).

3. Beginning at row 8, replace the media with 100 μL of fresh media. Next replace the media in row 7 with the correct adenoviral dilution, followed by row 6 and so on. Under aseptic conditions and starting with the control wells and working through to the higher concentrations of adenovirus, the same tip can be used for all wells.

4. Incubate the plate in a humidified chamber for 16–18 h at 37°C in 5% CO_2.

5. Replace the media with 200 μL of fresh complete media (see Note 22).

6. Replace the media every 2–3 days for 8 days. Once the CPE is apparent in a well, mark it and stop replacing the media in that well. On the seventh day, prepare another 96-well plate as previously described, for a second round of plaque purification. Once the assay is complete, collect the cells and media from three positive wells at the highest dilution for which the CPE is apparent. Collect them by pipetting up and down three times and then store them in sterile Eppendorfs. These plaques will have arisen from a single adenovirus; therefore, freeze two of the samples and extract the other by either 2 cycles of freeze/thawing (dry ice to 37°C water bath) or with an equal volume of ArkloneP.

7. Take 50 μL of this adenovirus stock and repeat the plaque purification assay with the plate prepared on day 7. At this point, plaques taken from the highest dilutions on the next plate can be considered plaque pure.

3.1.6. Generation of Pure High-Titre Stocks of Recombinant Adenoviruses

High-titre stocks of recombinant adenoviruses are generated by large-scale multiplication of the initial pure recombinant in 293 cells. The cytopathological effect (CPE) of the adenovirus in these permissive cells causes them to detach from the tissue culture flask allowing them to be easily collected.

1. Subculture low passage 293 cells into $21 \times$ T-150 tissue culture flasks (20 for the Ad preparation and 1 control flask). One confluent T-150 flask can be passaged into $5 \times$ T-150 flasks and then into 21 T-150 flasks once confluence is achieved for infection (approximately 1 week later).

2. Once cells have reached 80–90% confluence, they can be infected with a multiplicity of infection (MOI) of 0.1–10 per flask (see Notes 23 and 24).

3. Change the media every 3 days until the CPE begins and the cells start to detach from the flask. After this if the cells need to be fed before the CPE is complete, then add 10–15 mL complete media to each flask.

4. Once the CPE is complete, collect the cells immediately. Centrifuge in sterile Falcon tubes at $250 \times g$ for 10 min at room temperature.

5. Decant the supernatant to waste and resuspend cells in a total volume of 12–15 mL of PBS. Add an equal volume of ArkloneP, and then invert the tube for 30 s followed by shaking (not too vigorously) for 10 s. Repeat the mixing step (see Note 17). Centrifuge the tube at $750 \times g$ for 15 min at room temperature.

6. Remove the top layer containing the adenovirus to a fresh tube, taking care not to disturb the lower layers.

7. Add a further 5 mL PBS to the ArkloneP and cell debris and repeat the extraction. Add the upper layer to the adenovirus already extracted. The adenoviral stocks can be immediately purified on a CsCl gradient or stored at −80°C.

   Freeze/thawing and ArkloneP extraction are crude methods for adenovirus extraction and preparations are contaminated with cellular proteins which may be cytotoxic in vitro and in vivo. Centrifugation on CsCl density gradients provides an efficient and simple method for both concentrating and purifying stocks of recombinant adenoviruses.

8. Rinse the ultracentrifuge tubes with 70% ethanol followed by sterile dH_2O to sterilize.

9. Add 2.5 mL of 1.25 g/mL density CsCl to each tube.

10. Add 2.5 mL of 1.40 g/mL density CsCl under the first gradient (see Note 25).

11. Drip the adenovirus solution to be purified on top very gently, do not disturb the gradient. Add PBS to fill the tube.

12. Subject the gradients to centrifugation in the SW40Ti rotor at room temperature for 1.5 h at $217{,}874 \times g$.

13. At the end of the centrifugation step, the recombinant adenovirus presents as a discrete white layer between the two CsCl layers (see Fig. 4, Note 26).

14. Using a 21-gauge needle and a 5-mL syringe, gently pierce the side of the tube just underneath the adenovirus band, with a gentle side-to-side sweeping motion collect the adenovirus into the syringe without collecting too much of the solution (see Note 27).

15. Prepare the second CsCl gradient by the addition of 5 mL of 1.34 g/mL density CsCl solution to a fresh ultracentrifuge tube. Repeat step 11 and centrifuge the second gradient in the SW40Ti rotor at room temperature for 18 h (or overnight) at $217{,}874 \times g$.

16. Remove the adenovirus band (step 14 above) (see Fig. 4).

17. Dialyse the adenovirus in Slide-A-Lyzer dialysis cassettes for 2 h against 5 L dialysis buffer in a sterile beaker on a magnetic stirrer.

18. Dialyse the adenovirus overnight against 5 L of fresh dialysis buffer containing 10% glycerol. Remove the virus from the Slide-A-Lyzer as per manufacturer's instructions. Aliquot the adenoviral stocks and store them at −80°C.

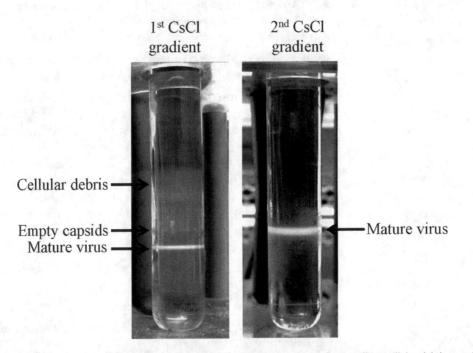

Fig. 4. Sequential CsCl gradients for purifying recombinant adenoviruses. In the first gradient cellular debris, empty Ad capsids and mature virus can be visualized. Following a second CsCl gradient ultracentrifugation step, only a mature virus band can be observed.

3.1.7. Titration of Adenovirus by End-Point Dilution

Once recombinant adenovirus has been purified on a CsCl gradient, it can be titered by serial dilution on 293 cells as per protocol 3.1.5.

1. Subculture low passage 293 cells into 80 wells of a 96-well plate (see Fig. 5) for approximately 50–60% confluence after 24 h (see Note 20).

2. Make serial dilutions of the adenovirus in complete media (see Fig. 5).

3. Add 100 μL of each adenovirus dilution to the appropriate wells and incubate for 16–18 h at 37°C in 5% CO_2.

4. After 18 h, replace the media containing the adenovirus with 200 μL fresh complete media and incubate at 37°C in 5% CO_2 (see Note 22). Change the media every 2–3 days or when acidity shows. Once the cytopathic effect is apparent in a well mark it and stop replacing the media in that well.

5. After 8 days incubation, count the number of wells containing plaques and fit the results into the equation below in order to obtain the titer of the adenoviral stocks in pfu/mL (37).

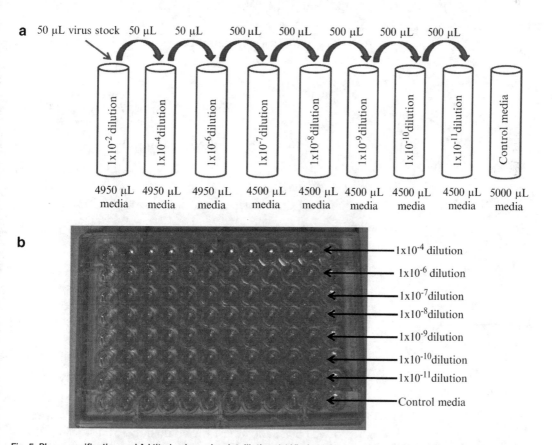

Fig. 5. Plaque purification and Ad titering by end-point dilution. (**a**) Viral stocks are subjected to the illustrated serial dilution steps in cell culture media ready for infection. (**b**) 293 cells are seeded into a 96-well plate as the image shows and the appropriate viral dilutions (or control media) added to each row.

6. The proportionate distance

$$= \frac{\% \text{ Positive above } 50\% - 50\%}{\% \text{ Positive above } 50\% - \% \text{ Positive below } 50\%}$$

and

$$\log \text{ID}_{50} \text{ (infectivity dose)} = \log \text{ dilution above } 50\%$$
$$+ (\text{Proportionate distance} \times -1)$$
$$\times \text{Dilution factor.}$$

For example: titration gives:

At 1×10^{-4} dilution, 10/10 wells positive.
At 1×10^{-6} dilution, 10/10 wells positive.
At 1×10^{-7} dilution, 10/10 wells positive.
At 1×10^{-8} dilution, 9/10 wells positive.
At 1×10^{-9} dilution, 3/10 wells positive.
At 1×10^{-10} dilution, 0/10 wells positive.
At 1×10^{-11} dilution, 0/10 wells positive.

$$\text{The proportionate distance} = \frac{90 - 50}{90 - 30} = 0.67.$$

$$\log \text{ID}_{50} = -8 + (0.67 \times -1) = -8.67.$$

$$\text{ID}_{50} = 10^{-8.67}.$$

$$\text{TCID}_{50} \text{ (tissue culture infectivity dose 50)} = \frac{1}{10^{-8.67}}.$$

$\text{TCID}_{50}/100 \, \mu\text{L} = 10^{8.67}.$
X dilution factor (10).
$\text{TCID}_{50}/\text{mL} = 10^{9.67}.$
$= 4.68 \times 10^9 \, \text{TCID}_{50}/\text{mL}.$
$1 \, \text{TCID}_{50} = 0.7 \, \text{pfu}.$
Final titer $= 3.28 \times 10^9 \, \text{pfu}/\text{mL}.$

3.1.8. Calculation of Total Viral Particle Titer

1. We measure viral particle titers using microBCA assay. First, generate a set of bovine serum albumin (BSA) standards ranging from 0.5 to 200 μg/mL as per the manufacturer's instructions.

2. Load 150 μL of each standard in duplicate to empty wells of a clear 96-well plate.

3. Next add 1, 3, 5 μL of virus stock in duplicate to empty wells of the 96-well plate and adjust the volume to 150 μL using PBS.

4. Add 150 μL of BCA working reagent to all samples and standards and incubate the plate at 37°C for 2 h.

5. Measure the absorbance of each well at 570 nm using a spectrophotometer.

6. Calculate the protein content by first subtracting the blank absorbance values from the gross readings for each sample and standard. Plot the standard curve and calculate the equation of the curve. Calculate the protein value of the virus samples in $\mu g/mL$ from the equation of the curve.

7. Convert the protein concentrations to viral particle titer/mL (vp/mL) according to the following formula: 1 μg protein $= 1 \times 10^9$ vp (38).

3.1.9. Infection in Nonpermissive Cells to Assess RCA Contamination

The 293 cell line contains adenovirus type 5 sequences which are homologous with sequences in Ad vectors; therefore, it is possible that RCA may contaminate stocks of recombinant Ad. Stocks of recombinant Ad can therefore be tested in nonpermissive cell lines such as HeLa.

1. Subculture HeLa cells into 80 wells of a 96-well plate (see Fig. 5). Aim for 50–60% confluence the following day.

2. Prepare serial dilutions of the adenoviral stocks (see Fig. 5). Replace the media in row 8 with fresh media. Replace the media in rows 1–7 with 100 μL of the appropriate adenoviral dilution (starting at the highest dilution and working through to the lowest using the same tip) and incubate overnight at 37°C in 5% CO_2.

3. The following day replace the media in all the wells with 200 μL of fresh media. Change the media every 2–3 days or when acidity is apparent.

4. Incubate the plate for 8 days at 37°C in 5% CO_2 and observe any CPE that occurs. If after 8 days no cytopathology is apparent, recombinant adenoviral stocks can be assumed to be free of RCA.

3.2. Generation of Adenoviral Vectors with Altered Capsid Proteins

3.2.1. Generation of an Ad5 Fiber Shuttle Plasmid

Shuttle plasmids are around 6–9 kb and only encode small regions of the Ad genome enabling ready manipulation of individual genes within the adenovirus genome. Shuttle plasmids should be constructed to contain the region for manipulation and approximately 1–3 kb of DNA sequence flanking this region to facilitate homologous recombination. Bioinformatic programs for DNA analysis are essential to select the correct strategy for cloning (we use Serial Cloner http://serialbasics.free.fr/Serial_Cloner.html). To generate the shuttle plasmid, it is crucial to localize the specific region for subcloning in order to identify a unique restriction endonuclease site in pAdEasy-1 which is present in the target viral gene. For the fiber gene, *Spe*I is unique in the Ad5 genome and therefore linearizes pAdEasy-1 in the fiber gene allowing homologous recombination in this specific region. Of note, it is essential to modify the target capsid gene and re-clone it into the pAdEasy-1 backbone prior to the insertion of expression cassettes in the E1 region as they may contain this specific restriction site. In addition, we recommend finding unique restriction sites in the target region which facilitates targeted direct cloning of the target gene into the shuttle plasmid. In the fiber gene region, the use of *Nde*I and *Mfe*I sites will excise almost the entire fiber gene sequence enabling replacement with a modified fiber using the same restriction sites. The fiber protein is anchored by the tail domain into the penton base protein of the adenoviral capsid which means the initial 48 nucleotides of the fiber must be maintained as wild-type sequence. Therefore, do not modify the gene upstream of the *Nde*I restriction site at the beginning of the Ad5 fiber gene in order to maintain correct anchoring of the fiber protein into the penton base protein. Here, we provide a detailed protocol to generate an Ad5 fiber shuttle plasmid.

1. First use oligonucleotides 1 and 2 to amplify 4 kb of Ad genome including the fiber gene with 1 kb of flanking DNA sequence from pAdEasy-1 by PCR using a proof reading polymerase (see Note 28).

2. Confirm the correct size DNA fragment has been generated via agarose gel electrophoresis.

3. Purify the DNA band using a commercially available gel extraction kit.

4. Clone the purified PCR product into a plasmid which does not contain the restriction sites that will be used in future cloning steps (i.e., *Nde*I and *Mfe*I for pSC-A). For cloning, we recommend the use of Strataclone Blunt PCR cloning kit (see Note 29). We use pSC-A for all our fiber cloning.

5. Sequence the entire fragment to confirm that it does not contain any mutations.

6. Now the fiber shuttle plasmid is ready for incorporation of specific modifications.

3.2.2. Incorporation of CAR-Binding Mutations

Site-directed mutagenesis can be readily performed using the gene assembling technique, in which the desired mutations are included in different oligonucleotides (see Fig. 6). In this chapter, we provide protocols to incorporate point mutations that ablate interaction with CAR (S408E and P409A) (39). Gene assembling consists of amplification of two or more DNA fragments sharing small homologous regions at the end of each fragment enabling their combination via a second PCR step with flanking oligonucleotides to amplify the entire assembled region containing the desired mutations.

1. For this protocol you require the flanking oligonucleotides 1 and 2 and the forward and reverse oligonucleotides 3 and 4 encoding the mutations S408E and P409A.

2. Perform two separate PCRs using pAdEasy-1; one reaction using oligonucleotides 1 and 4 and the second using oligonucleotides 2 and 3.

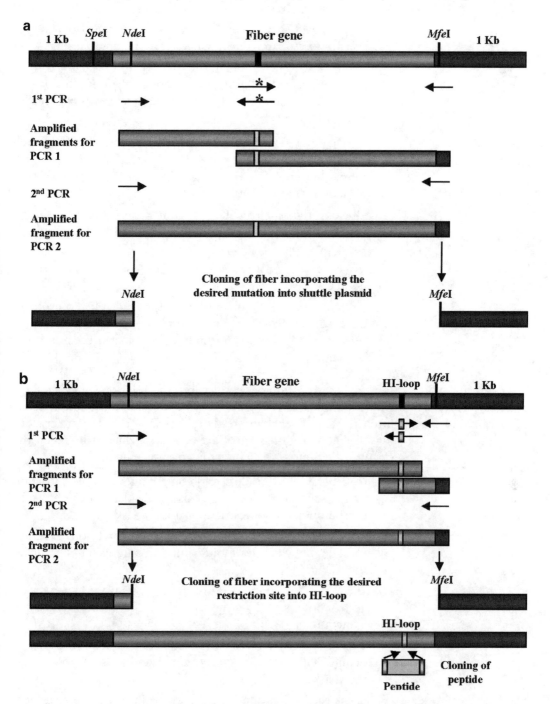

Fig. 6. Manipulation of the Ad5 fiber protein. (a) Cloning strategy for the generation of a shuttle plasmid (*top* schematic) illustrating the use of *Nde*I and *Mfe*I sites to replace the Ad5 fiber gene with a mutated version and the *Spe*I site in pAdEasy-1 allowing for homologous recombination in this region. For the incorporation of different point mutations (*asterisk*), the gene assembling strategy is utilized which generates a fragment containing the desired modification. This fragment is then cloned into the shuttle plasmid via the *Nde*I and *Mfe*I sites. (b) Strategy of cloning to introduce a unique restriction site into the HI loop for the incorporation of different peptides by direct ligation. The generation of the unique restriction site can be performed via the gene assembling technique. The assembled DNA fragment is cloned into the fiber shuttle plasmid via the *Nde*I and *Mfe*I sites. Peptides can be designed to incorporate cohesive ends for the direct ligation into the HI loop.

3. Check the PCR fragments are the correct size via agarose gel electrophoresis and purify each DNA fragment using a gel extraction kit.

4. Once the desired DNA fragments are purified, perform a second PCR using both bands mixed as template with oligonucleotides 1 and 2.

5. Check the PCR product is the correct size via agarose gel electrophoresis.

6. Purify the amplified band using commercially available gel extraction kits following the manufacturer's instructions.

7. Subclone the amplified DNA fragment into the fiber shuttle plasmid via the *Nde*I and *Mfe*I sites following excision of the wild-type fiber.

8. Sequence the entire fragment to confirm the correct modifications are present.

9. Linearize the shuttle plasmid with *Spe*I and refer to protocol 3.1.2 to perform a homologous recombination step with pAdEasy-1 to incorporate the modified fiber protein into the Ad genome.

3.2.3. Incorporation of Peptides into the HI Loop of the Ad5 Fiber

For the incorporation of peptides into the HI loop we generated a shuttle plasmid which incorporates the unique restriction site *Not*I into the HI loop enabling ready manipulation for the direct cloning of oligonucleotides encoding peptides (while maintaining the open reading frame) via a direct ligation technique.

1. To generate a shuttle plasmid containing a unique restriction site in the HI loop (see Fig. 6), use oligonucleotides 5 and 6 to insert a unique *Not*I restriction site.

2. Perform two separate PCRs using pAdEasy-1; one reaction using oligonucleotides 1 and 6 and a second using oligonucleotides 2 and 5.

3. Check the PCR fragments are the correct size via agarose gel electrophoresis and purify each DNA fragment using a gel extraction kit.

4. Once the desired DNA fragments are purified, perform a second PCR with oligonucleotides 1 and 2 using both bands mixed as template.

5. Check the PCR product is the correct size via agarose gel electrophoresis and purify the DNA fragment using a gel extraction kit.

6. Subclone the amplified DNA fragment into the fiber shuttle plasmid via the *Nde*I and *Mfe*I sites following excision of the wild-type fiber.

7. Sequence the entire fragment to confirm that the *Not*I site is present and in-frame.

8. Next, design two further oligonucleotides encoding your peptide flanked by *Not*I cohesive ends (see Note 30).

9. Anneal the oligonucleotides by adding equimolar concentrations of each oligonucleotide (100 μM) in 50 μL of annealing buffer. Heat the samples to 100°C for 10 min and then allow the oligonucleotides to slowly cool to RT for 15 min to facilitate annealing.

10. Digest 2–5 μg of the shuttle plasmid DNA with 10–20 U *Not*I, confirm the digestion by agarose electrophoresis, and purify the linearized plasmid using a gel extraction kit.

11. Perform a direct ligation into the linearized plasmid with 50 ng of annealed oligonucleotides.

12. Transform the plasmid into competent DH5α bacteria following standard protocols and screen for the insert in miniprep DNA clones (see Note 31).

13. Sequence the modified fiber gene to confirm that the modifications have been incorporated.

14. Linearize the shuttle plasmid with *Spe*I and refer to protocol 3.1.2 to perform a homologous recombination step with pAdEasy-1 to incorporate the peptide-modified fiber protein into the Ad genome.

3.2.4. Generation of an Ad5 Hexon Shuttle Plasmid

The hexon gene contains seven hypervariable regions (HVR1–7) which are exposed on the outer face of the Ad5 capsid. Since the recent interest in the hexon due to its role in the humoral immune response (40) and the biodistribution of the Ad5 vector (15), we provide a protocol for the modification of this protein. This follows a similar procedure to that described for the fiber gene (protocol 3.2.1) to generate a shuttle plasmid for the hexon gene. The hexon gene is amplified incorporating an additional flanking 1 kb DNA from the unique *Bam*HI and *Asi*SI restriction sites present in the Ad5 genome (see Fig. 7). We recommend these restriction sites to facilitate the downstream homologous recombination steps for the generation of a hexon-modified Ad vector. Depending on the part of the hexon gene you are aiming to modify, design restriction sites (e.g., here we use the example of modification of HVR-7 and using *Nde*I and *Bam*HI as restriction sites excises a portion of the hexon gene encoding HVR5–7).

1. Use oligonucleotides 7 and 8 to amplify the hexon gene and the flanking 1 kb DNA downstream of the *Asi*SI site and upstream of the *Nde*I restriction sites from AdEasy-1 using standard PCR techniques and proof reading polymerase.

2. Check the PCR product is the correct size using agarose gel electrophoresis.

3. Purify the band using a commercially available gel extraction kit.

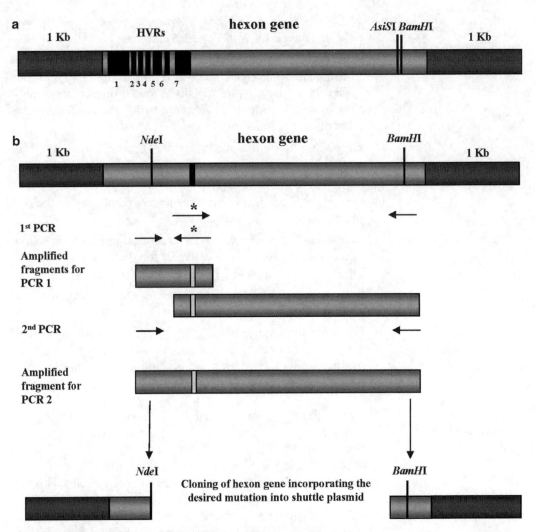

Fig. 7. Manipulation of the Ad5 hexon gene. (a) Schematic of the Ad5 hexon gene and flanking homologous regions (1 kb). The seven HVRs are indicated. *AsiS*I and *Bam*HI are suitable restriction sites in pAdEasy-1 to allow homologous recombination. (b) Cloning strategy for the incorporation of a point mutation (*asterisk*) in the Ad5 hexon gene. *Nde*I and *Bam*HI are suitable restriction sites to facilitate subcloning of the portion of the hexon gene containing HVRs 5–7 for the incorporation of the E451Q modification (*asterisk*) in HVR7. Point mutations in the Ad5 hexon gene are incorporated via the gene assembling technique. The assembled fragment incorporating the point mutation E451Q is then cloned into the hexon shuttle plasmid via the *Nde*I and *Bam*HI sites.

4. Clone the PCR product as described in protocol 3.2.1 using Stratagene Blunt PCR Cloning Kit into pSC-A.

5. Sequence the entire fragment to confirm no mutations have been introduced.

3.2.5. Incorporation of Point Mutations into the Hexon Gene

The hexon protein is of increased interest due to its role in mediating hepatocyte transduction from Ad5 via FX interaction following intravenous delivery (15). Here, we provide a protocol to introduce one of the mutations (E451Q) that have been proven to block hexon:FX interaction and reduce hepatocyte transduction for Ad5 after intravascular delivery (32). The protocol can be readily modified with alternative oligonucleotides to introduce other mutations. The protocol follows a similar strategy to that shown in 3.2.2 to introduce mutations into the fiber gene.

1. Perform two PCRs using pAdEasy-1 following a similar strategy to that described in protocol 3.2.2. The first reaction uses oligonucleotides 9 and 12 and the second reaction uses oligonucleotides 10 and 11.

2. Check the size of the PCR fragments using agarose gel electrophoresis and purify each DNA fragment using a gel extraction kit.

3. Once the desired DNA fragments are purified, perform a PCR using both bands from step 2 mixed as a template in combination with oligonucleotides 9 and 10 to assemble the modified hexon gene.

4. Check the PCR is the correct size using agarose gel electrophoresis.

5. Purify the amplified band using a commercially available gel extraction kit following the manufacturer's instructions.

6. Clone the modified hexon gene into the shuttle plasmid via the *Bam*HI/*Asi*SI sites following excision of the wild-type hexon gene.

7. Sequence the entire fragment to ensure that the desired mutations have been introduced.

8. Linearize the shuttle plasmid with *Asi*SI and refer to protocol 3.1.2 to perform a homologous recombination step with pAdEasy-1 to incorporate the modified hexon protein into the Ad genome.

4. Notes

1. You can check the density of the solutions by weighing or by the use of a refractometer and correct by adding 1× TD solution or CsCl.

2. The design of these primers will generate a band of approximately 4 kb.

3. A range of restriction endonucleases can be used as long as they digest outside the E1 region and the homologous flanking region.

4. It is essential to obtain clean DNA bands with yields of approximately 30–100 ng/μL to guarantee successful recombination.

If the obtained yield is low, elute the purified DNA fragments in a lower volume.

5. It is recommended to check the efficiency of digestion by transforming 100 ng of each individual plasmid. We also recommend ratios of 100:100, 100:200, and 100:300 ng DNA (pAdEasy-1: shuttle plasmid). Do not add DNA to more than 10% of the total volume of the cells.

6. LB media can also be used although it is recommended to use a rich media source such as SOC.

7. The agar plates can be pre-warmed to facilitate bacterial growth. Use the appropriate resistant antibiotic in agar plates (e.g., kanamycin for pAdEasy-1).

8. It is recommended to choose the smallest colonies on the agar plate since small colonies are more likely to contain large intact plasmids which require more time to replicate and hence slow down bacterial growth.

9. Selection of 6–10 colonies per construct is usually enough to identify a positive clone via this method.

10. The yields of adenoviral DNA are usually quite low. Thus, re-transformation of extracted miniprep DNA into an alternative competent bacterial strain with inactive recombination genes (e.g., DH5α or TOP10) is recommended. First, subject 50% (v/v) of the miniprep DNA to agarose gel electrophoresis to check DNA integrity. Only use CsCl banded DNA or DNA prepared using high-grade commercial kits for downstream transfection.

11. You can electrophorese 10 μL in an agarose gel to visualize the digestion.

12. DNA can be incubated at 37°C for 1 h if it is incompletely dissolved.

13. Passaging 293 cells via trypsin–EDTA for subculturing low passage 293 cells can be too harsh. An alternative is to use 1× citric saline. Briefly, for a T-150 flask make 50 mL of 1× citric saline (in PBS), wash the cells with 25 mL and decant to waste, wash the cells with a further 25 mL and leave approximately 2 mL on the cells. Leave the flask at room temperature until the cells detach, add 10 mL of media, and pipette the cells up and down to avoid clumping. Cells can be counted and subcultured at this point, they do not require pelleting and re-suspending in fresh media.

14. Viral plaques usually arise after approximately 7 days, if after 15 days there is no evidence of cytopathic effect discard the flasks and repeat the transfection. Prolonged incubation of transfected cells can increase the likelihood of RCA production due to homology between the AdEasy-1 sequences and the 293 cells.

15. If the cells are left too long following completion of CPE, this will result in a loss of titer.

16. Crude virus can be stored at −80°C at this point if necessary.

17. If the ArkloneP/cell suspension is mixed too vigorously, adenoviral fiber proteins can be sheared from the capsid, resulting in a loss of infectivity.

18. Avoid repeated freeze/thawing steps of adenoviral stocks as each step will result in an approximate 10% loss of titer. Aliquot all adenoviral stocks into suitable volumes.

19. You can also add 300 μL of sterile PBS to the outer wells in each row which are not used in the plaque purification assay in order to stop the plate drying out.

20. If 293 cells are plated at higher densities, then after 5–7 days the cell monolayer will begin to peel from the edges of the well, hampering accurate identification of CPE. Subculture a confluent T-150 flask into 15 mL of media, take 3 mL of cell containing media, and add 10 mL of fresh media to this. Add 100 μL of cell suspension per well.

21. Figure 5 shows the range of dilutions used for protocol 3.1.7 (titration) and this range of dilutions are recommended for accurately titering large-scale preparations of Ad, although it should be noted that very high titer Ad preparations may have to be diluted through another two logs for an accurate titer. However, for plaque purification for crude stocks which are likely to be of lower titer, it is recommended to generate and plate a range of Ad dilutions for 1×10^{-2} to 1×10^{-8} in tenfold dilutions per row.

22. To change media quickly and efficiently on plaque assays, without cross-contamination rest the plate edge on its upturned lid. Remove media with one pipette tip per well and eject to waste. Add fresh media using a single tip per row, ensuring that you do not touch the wells with the tip.

23. Viral infection can be performed alongside a media change. Mix 50–100 μL of crude adenovirus stock with 750 mL of complete media and then exchange media (25 mL/flask).

24. Avoid multiply passaging recombinant adenoviruses in 293 cells as this increases the likelihood of replication competent adenovirus generation. Try to always use an original plaque pure stock for infecting the cells when generating new stocks.

25. This is a delicate step which needs to be performed gently to guarantee the correct formation of the CsCl gradient. A simple way to test that discrete layers have formed when practicing making CsCl gradients is to mix a dye (e.g., trypan blue) with the density 1.40 g/mL CsCl. Discrete separation between the two layers can then be easily observed.

26. To avoid confusion between the bands, place 2.5 mL water into a fresh centrifuge tube and line it up with the tube containing the banded adenovirus. If there are no clear, separate bands, repeat the CsCl purification.

27. Take adenovirus in a minimum of 500 μL of solution per 10× T-150 flasks to avoid aggregation during dialysis.

28. Oligonucleotides should be designed to amplify more than 1 kb DNA flanking the fiber (or hexon) gene and this will increase the efficiency of the homologous recombination step. However, this will also increase the possible random mutagenesis rates by the DNA polymerase and therefore we recommend that 1 kb is an adequate overlap.

29. It is essential that the chosen plasmid does not contain *Nde*I and *Mfe*I restriction sites and also any other specific restriction endonuclease recognition sites that have been designed to be unique in any downstream steps, e.g., oligonucleotides for mutagenesis or for linearization prior to the homologous recombination step.

30. It is essential not to include the entire restriction endonuclease recognition sequence, only the nucleotides encoding the cohesive ends to enable the downstream direct ligation step.

31. Also transform the linearized shuttle plasmid without any insert alongside your ligation reactions to check the digestion efficiency of your plasmid.

References

1. Bergelson JM, Cunningham JA, Droguett G, Kurt-Jones EA, Krithivas A, Hong JS, Horwitz MS, Crowell RL, Finberg RW (1997) Isolation of a common receptor for coxsackie B viruses and adenoviruses 2 and 5. Science 275:1320–1323

2. Tomko RP, Xu R, Philipson L (1997) HCAR and MCAR: the human and mouse cellular receptors for subgroup C adenoviruses and group B coxsackieviruses. Proc Natl Acad Sci U S A 94:3352–3356

3. Henry LJ, Xia D, Wilke ME, Deisenhofer J, Gerard RD (1994) Characterisation of the knob domain of the adenovirus type 5 fiber protein expressed in *Escherichia coli*. J Virol 68:5239–5246

4. Nicklin SA, Wu E, Nemerow GR, Baker AH (2005) The influence of adenovirus fiber structure and function on vector development for gene therapy. Mol Ther 12:384–393

5. Wu E, Fernandez J, Kaye Fleck S, Von Seggern DJ, Huang S, Nemerow GR (2001) A 50-kDa membrane protein mediates sialic acid-independent binding and infection of conjunctival cells by adenovirus type 37. Virology 279:78–89

6. Arnberg N, Kidd AH, Edlund K, Nilsson J, Pring-Akerblom P, Wadell G (2002) Adenovirus type 37 binds to cell surface sialic acid through a charge-dependent interaction. Virology 302:33–43

7. Arnberg N, Edlund K, Kidd AH, Wadell G (2000) Adenovirus type 37 uses sialic acid as a cellular receptor. J Virol 74:42–48

8. Seiradake E, Henaff D, Wodrich H, Billet O, Perreau M, Hippert C, Mennechet F, Schoehn G, Lortat-Jacob H, Dreja H, Ibanes S, Kalatzis V, Wang JP, Finberg RW, Cusack S, Kremer EJ (2009) The cell adhesion molecule "CAR" and sialic acid on human erythrocytes influence adenovirus *in vivo* biodistribution. PLoS Pathog 5:e1000277

9. Gaggar A, Shayakhmetov DM, Lieber A (2003) CD46 is a cellular receptor for group B adenoviruses. Nat Med 9:1408–1412

10. Tuve S, Wang H, Ware C, Liu Y, Gaggar A, Bernt K, Shayakhmetov D, Li Z, Strauss R, Stone D, Lieber A (2006) A new group B

adenovirus receptor is expressed at high levels on human stem and tumor cells. J Virol 80:12109–12120

11. Huard J, Lochmuller H, Acsadi G, Jani A, Massie B, Karpati G (1995) The route of administration is a major determinant of the transduction efficiency of rat tissues by adenoviral recombinants. Gene Ther 2:107–115

12. Sullivan DE, Dash S, Du H, Hiramatsu N, Aydin F, Kolls J, Blanchard J, Baskin G, Gerber MA (1997) Liver-directed gene transfer into non-human primates. Hum Gene Ther 8: 1195–1206

13. Shayakhmetov DM, Gaggar A, Ni S, Li ZY, Lieber A (2005) Adenovirus binding to blood factors results in liver cell infection and hepatotoxicity. J Virol 79:7478–7491

14. Parker AL, Waddington SN, Nicol CG, Shayakhmetov DM, Buckley SM, Denby L, Kemball-Cook G, Ni S, Lieber A, McVey JH, Nicklin SA, Baker AH (2006) Multiple vitamin K-dependent coagulation zymogens promote adenovirus-mediated gene delivery to hepatocytes. Blood 108:2554–2561

15. Waddington SN, McVey JH, Bhella D, Parker AL, Barker K, Atoda H, Pink R, Buckley SM, Greig JA, Denby L, Custers J, Morita T, Francischetti IM, Monteiro RQ, Barouch DH, van Rooijen N, Napoli C, Havenga MJ, Nicklin SA, Baker AH (2008) Adenovirus serotype 5 hexon mediates liver gene transfer. Cell 132: 397–409

16. Kalyuzhniy O, Di Paolo NC, Silvestry M, Hofherr SE, Barry MA, Stewart PL, Shayakhmetov DM (2008) Adenovirus serotype 5 hexon is critical for virus infection of hepatocytes in vivo. Proc Natl Acad Sci U S A 105:5483–5488

17. Graham FL, Smiley J, Russel WC, Nairu R (1977) Characteristics of a human cell line transformed by DNA from human adenovirus type 5. J Gen Virol 36:59–72

18. Fallaux FJ, Kranenburg O, Cramer SJ, Houweling A, Van Ormondt H, Hoeben RC, Van Der Eb AJ (1996) Characterization of 911: a new helper cell line for the titration and propagation of early region 1-deleted adenoviral vectors. Hum Gene Ther 7:215–222

19. Fallaux FJ, Bout A, van der Velde I, van den Wollenberg DJ, Hehir KM, Keegan J, Auger C, Cramer SJ, van Ormondt H, van der Eb AJ, Valerio D, Hoeben RC (1998) New helper cells and matched early region 1-deleted adenovirus vectors prevent generation of replication-competent adenoviruses. Hum Gene Ther 9: 1909–1917

20. Feuerbach D, Burgert HG (1993) Novel proteins associated with MHC class-1 antigens in cells expressing the adenovirus protein E3/19K. EMBO J 12:3153–3161

21. Yang Y, Ertl HCJ, Wilson JM (1994) MHC class I restricted cytotoxic T lymphocytes to viral antigens destroy hepatocytes in mice transfected with E1 deleted recombinant adenoviruses. Immunity 1:433–442

22. Yang Y, Nunes FA, Berencsi K, Furth EE, Gonczol E, Wilson JM (1994) Cellular immunity to viral antigens limits E1-deleted adenoviruses for gene therapy. Proc Natl Acad Sci U S A 91:4407–4411

23. Dedieu JF, Vigne E, Torrent C, Jullien C, Mahfouz I, Caillaud JM, Aubailly N, Orsini C, Guillaume JM, Opolon P, Delaere P, Perricaudet M, Yeh P (1997) Long-term gene delivery into the livers of immunocompetent mice with E1/E4-defective adenoviruses. J Virol 71:4626–4637

24. Brough DE, Hsu C, Kulesa VA, Lee GM, Cantolupo LJ, Lizonova A, Kovesdi I (1997) Activation of transgene expression by early region 4 is responsible for a high level of persistent transgene expression fro adenovirus vectors in vivo. J Virol 71:9206–9213

25. Wang Q, Greenburg G, Bunch D, Farson D, Finer MH (1997) Persistent transgene expression in mouse liver following in vivo gene transfer with a delta E1/delta E4 adenovirus vector. Gene Ther 4:393–400

26. Alba R, Bosch A, Chillon M (2005) Gutless adenovirus: last-generation adenovirus for gene therapy. Gene Ther 12:S18–S27

27. Segura MM, Alba R, Bosch A, Chillon M (2008) Advances in helper-dependent adenoviral vector research. Curr Gene Ther 8:222–235

28. Muruve DA, Cotter MJ, Zaiss AK, White LR, Liu Q, Chan T, Clark SA, Ross PJ, Meulenbroek RA, Maelandsmo GM, Parks RJ (2004) Helper-dependent adenovirus vectors elicit intact innate but attenuated adaptive host immune responses in vivo. J Virol 78:5966–5972

29. Bett AJ, Prevec L, Graham FL (1993) Packaging capacity and stability of human adenovirus type 5 vectors. J Virol 67:5911–5921

30. Mitani K, Graham FL, Caskey CT, Kochanek S (1995) Rescue, propagation, and partial purification of a helper virus-dependent adenovirus vector. Proc Natl Acad Sci USA 92: 3854–3858

31. Xia D, Henry LJ, Gerard RD, Deisenhofer J (1994) Crystal structure of the receptor-binding domain of adenovirus type 5 fiber protein at 1.7 Å resolution. Structure 2:1259–1270

32. Alba R, Bradshaw AC, Parker AL, Bhella D, Waddington SN, Nicklin SA, van Rooijen N, Custers J, Goudsmit J, Barouch DH, McVey JH, Baker AH (2009) Identification of coagulation factor (F)X binding sites on the adenovirus

serotype 5 hexon: effect of mutagenesis on FX interactions and gene transfer. Blood 114: 965–971

33. Shayakhmetov DM, Papayannopoulou T, Stamatoyannopoulos G, Lieber A (2000) Efficient gene transfer into human CD34+ cells by a retargeted adenovirus vector. J Virol 74: 2567–2583

34. Denby L, Work LM, Graham D, Hsu C, von Seggern DJ, Nicklin SA, Baker AH (2004) Adenoviral serotype 5 vectors pseudotyped with fibers from subgroup D show modified tropism *in vitro* and *in vivo*. Hum Gene Ther 15:1054–1064

35. Denby L, Work LM, Seggern DJ, Wu E, McVey JH, Nicklin SA, Baker AH (2007) Development of renal-targeted vectors through combined *in vivo* phage display and capsid engineering of adenoviral fibers from serotype 19p. Mol Ther 15:1647–1654

36. He T-C, Zhou S, da Costa LT, Yu J, Kinzler KW, Vogelstein B (1998) A simplified system for generating recombinant adenoviruses. Proc Natl Acad Sci U S A 95:2509–2514

37. Lowenstein PR, Shering IAF, Bain D, Castro MG, Wilkinson GWG (1996) How to examine the interactions between adenoviral mediated gene transfer and different indentified target brain cell types *in vitro*. In: Lowenstein PR, Enquist LW (eds) Towards gene therapy for neurological disorders. Wiley, Chichester, UK, pp 93–114

38. Von Seggern DJ, Kehler J, Endo RL, Nemerow GR (1998) Complementation of a fibre mutant adenovirus by packaging cell lines expressing the adenovirus type 5 fibre protein. J Gen Virol 79:1461–1468

39. Kirby I, Davidson E, Beavil AJ, Soh CPC, Wickham TJ, Roelvink PW, Kovesdi I, Sutton BJ, Santis G (2000) Identification of contact residues and definition of the CAR-binding site of adenovirus type 5 binding protein. J Virol 74:2804–2813

40. Roberts DM, Nanda A, Havenga MJ, Abbink P, Lynch DM, Ewald BA, Liu J, Thorner AR, Swanson PE, Gorgone DA, Lifton MA, Lemckert AA, Holterman L, Chen B, Dilraj A, Carville A, Mansfield KG, Goudsmit J, Barouch DH (2006) Hexon-chimaeric adenovirus serotype 5 vectors circumvent preexisting anti-vector immunity. Nature 441: 239–243

Chapter 5

Vector Systems for Prenatal Gene Therapy: Principles of Retrovirus Vector Design and Production

Steven J. Howe and Anil Chandrashekran

Abstract

Vectors derived from the Retroviridae family have several attributes required for successful gene delivery. Retroviral vectors have an adequate payload size for the coding regions of most genes; they are safe to handle and simple to produce. These vectors can be manipulated to target different cell types with low immunogenicity and can permanently insert genetic information into the host cells' genome. Retroviral vectors have been used in gene therapy clinical trials and successfully applied experimentally in vitro, in vivo, and in utero.

Key words: Retrovirus, Gammaretrovirus, MLV, Lentivirus, HIV, SIN, WPRE, Preparation, Concentration, Titration, Packaging cell line, 293T, Envelope, Pseudotype, Targeting, Phoenix packaging cell lines

1. Introduction

1.1. Retroviridae

Members of the Retrovirus family have an RNA genome that is reverse transcribed into double-stranded DNA as part of the viral life cycle. Several different members of this family have been adapted to make them safe for use as gene delivery vectors, including alpharetroviruses (1), gammaretroviruses (2, 3), lentiviruses (4), and spumaviruses (5). The individual attributes for each vector make them excellent tools in different situations, but currently gammaretroviruses and lentiviruses are the only two that have been sufficiently refined and tested for use in clinical trials (6–10). This chapter focuses on the biology, production, and use of vectors from these two genera, in particular, vectors based on the murine leukaemia virus (MLV, gammaretrovirus) and human immunodeficiency virus (HIV-1, lentivirus). These vectors transduce, or deliver genes to, a wide range of cell types, including non-dividing cells in the case of HIV (11), stably integrate their payload

Charles Coutelle and Simon N. Waddington (eds.), *Prenatal Gene Therapy: Concepts, Methods, and Protocols*,
Methods in Molecular Biology, vol. 891, DOI 10.1007/978-1-61779-873-3_5, © Springer Science+Business Media, LLC 2012

into the host cell chromosomes for long-term gene expression, have cloning capacity sufficient to carry the cDNA from most genes (approximately 10 kb in total), and their tropism can be altered to target different cell populations. These attributes make them ideal vectors for many different applications, including gene therapy.

1.2. Gammaretroviral Vectors

The design of retrovirus vectors has been primarily based on two important criteria. The first is to ensure high-level expression of the transgene from the vector, the second is safety of the vectors developed. Careful consideration has been given to designing vectors that balance both of these factors.

The basic principles of retroviral vector design are based on producing a non-replicating delivery system by deleting pathogenic and non-essential viral accessory proteins and splitting the remainder of the viral genome into separate transcriptional units: *cis* acting (on the same molecule) for vector backbone encoding the transgene construct with a packaging signal and *trans* (supplied on different plasmids) complementing for expression of viral structural and regulatory components.

Removal of everything from viral genome except the sequences that are absolutely necessary for packaging the gene of interest into a functional virus produces minimal *cis*-acting sequences. This still allows efficient packaging, reverse transcription, integration, and expression of the transgene but with a greatly reduced risk of producing a replication competent virus. The regions essential for the correct function of the vector genome are the two long terminal repeats (LTRs, which flank the ends of the viral genome) and the packaging signal, which extends into the 5′ end of a gene called *gag*.

It was discovered that the entire envelope (*env*) sequence could be removed from the vector backbone without affecting viral titre, thereby decreasing the chance of recombination with the *trans* helper constructs. In addition, the elimination of endogenous envelope sequences allows the vector to be pseudotyped by envelope proteins from different families of retroviruses or other enveloped viruses.

Inclusion of part of the *gag* 5′ region is necessary to incorporate the packaging signal on the vector backbone. This raised the possibility that an unwanted viral protein could be produced in the transduced cells. Therefore, a mutation at the ATG start codon for gag translation was engineered to prevent any viral synthesis (12).

Owing to the relative simplicity of the genomic organisation of gammaretroviral vectors, stable gammaretroviral producing cell lines have been developed. In many cases, all *trans* complementing retroviral gene sequences (gag–pol) and envelope (*env*) sequences could be generated stably in 293T (human)- or NIH 3T3 (mouse)-based cell lines. The *gag* genes provide structural proteins to make the capsid or shell of the virus, while *pol* produces the enzymes required for reverse transcription and integration of the genome.

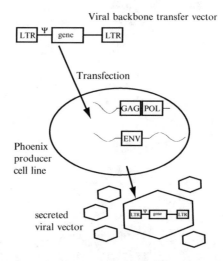

Fig. 1. Production of gammaretroviral vectors. Gammaretroviruses are packaged in Phoenix 293T cells. The vector backbone, with the transgene of interest (gene) flanked by long terminal repeats (LTR), is transfected into the cells and packaged into a particle by necessary structural and functional proteins, gag, pol, and env, which are supplied in trans from stably integrated, selectable constructs. If a self-inactivating vector is used, an internal promoter is required to express the transgene. Only the backbone is packaged into infectious virions because it is flanked by LTRs and contains a packaging signal (Ψ).

The vector backbone encoding the gene of interest/reporter can then be expressed transiently by transfection into a cell line expressing gag/pol/env or established as a clonal stable producer cell line along with the other required factors, thereby ensuring reliable and continuous production of viral vector particles (Fig. 1).

This method of virus production increases the safety of the vectors. The gag/pol/env genes are not flanked by LTRs nor contain a packaging signal which prevents them from being packaged in the vector. Consequently, only the *trans*-supplied structural and enzymatic proteins contribute to the virus, ensuring the vectors only deliver genes and cannot replicate.

1.3. Lentiviral Vectors

HIV-1-based lentiviral vectors have a similar genome, structure, and lifecycle to the gammaretroviruses, so similar principles of separating *cis*-acting and *trans*-acting functions on distinct transcriptional units are applied when producing these vectors. However, as the genome of HIV-based lentiviral vector is more complex than that of gammaretroviruses, certain regulatory genes are necessary components for packaging lentiviral vectors and therefore present in *trans* complementing transcriptional units. Because of the additional complexity, in addition to the LTRs and packaging signal, lentiviruses also require a rev-response element (RRE) to be included in *cis* for correct processing and export of the RNA genome.

To produce vectors in a packaging cell line, the required elements have been split into separate transcriptional units; four

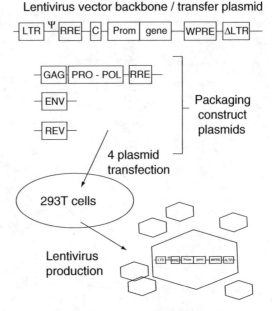

Fig. 2. Production of lentiviral vectors. Lentiviral backbones are flanked by self-inactivating long terminal repeats (ΔLTR) and encode transgenes (gene) from an internal promoter (Prom) and contain a rev-response element (RRE) to ensure correct export of full-length RNA. A central polypurine tract (C) improves transduction of cells and many constructs contain a post-transcriptional regulatory element from the Woodchuck hepatitis virus (WPRE) to enhance viral titres. When transfected into 293T producer cells, the backbone is packaged into a viral particle because it contains a packaging signal (Ψ). All necessary structural, enzymatic, envelope, and accessory genes (gag, pro, pol, env, rev) required to make the viral particle are supplied in trans on three other co-transfected plasmids. Functional vector is secreted into the tissue culture medium for collection and concentration.

plasmids for third-generation lentiviruses (13). Only the desired transgene, encoded on the viral backbone, is flanked by LTRs and has a packaging signal. The other vital components are expressed in *trans* from the three other plasmids, lacking a packaging signal and with LTRs removed, and so will not be inserted into a viral particle. These factors are: plasmid 1; *gag* (structural proteins to make the capsid) and *pol* (the reverse transcriptase polymerase and integrase proteins) (14); Plasmid 2; *env* (the envelope glycoproteins to coat the virus). Lentiviral vectors also require the expression of *rev*, encoded on Plasmid three, for correct processing and export of full-length RNA genomes (Fig. 2).

The risk of recombination between the four plasmids is extremely low. Coupled with the deletion of other virulence factors, this almost abrogates the risk of creating replication competent HIV.

Most lentiviral vectors contain a central polypurine tract (cPPT) element. This is thought to assist in reverse transcription, nuclear entry, and transduction of cells (15, 16) and has been shown to improve transduction and therefore expression levels, irrespective of the internal promoter used (17).

1.4. Self-Inactivating Modifications

To further increase the safety of gammaretro- and lentiviral vectors, the promoter and enhancer regulatory sequences in the U3 region of the 3′ LTR, including the TATA and CAAT boxes, have been deleted to produce self-inactivating vectors (SIN) (18, 19).

Following reverse transcription, this modification generates proviruses with the 5′ LTR lacking the transcriptional activity required to produce full-length vector RNA in target cells. Consequently, no full-length viral backbone can be expressed and then packaged in target cells to make replicating virus. The additional benefit of altering the LTRs in this way is that a heterologous internal promoter is required to express the inserted transgene. This reduces any risk of recombination with wild-type viruses creating replication competent vectors and provides more control over expression of the gene of interest.

1.5. Gene Expression

1.5.1. Multiple Genes

The need for expression of two or three genes from a vector has become apparent. In the case of vectors containing only one transgene, expression can be driven by the 5′ LTR in non-SIN vectors, as the promoter obtained by MLV LTR is very active in different tissues of many species. Alternatively, a heterologous promoter can be inserted to express the transgene. This can be compromised by the heterologous promoter interfering with the promoter in the 5′ LTR, an effect called promoter suppression (which is not a concern in SIN vectors). However, a heterologous promoter can be used to express transgenes in the reverse orientation in which case there is reduced risk of promoter interference, and genetic sequences including introns can be correctly maintained without risk of splicing and removal during virus production. It is important that polyadenylation sites are not present within the gene of interest as this disrupts production of full-length vector genomes for packaging. A polyadenylation site in the 3′ LTR is sufficient to add polyA tails to deliver genes expressed from retroviral vectors. However, if a gene is cloned into the vector in reverse orientation to maintain intron structure, a polyadenylation site should be used.

If two or more transgenes need to be expressed in the vector, a more complex vector design is necessary. Two genes can be expressed from a single transcript by mimicking the splicing ability that allows *env* to be expressed in the wild-type viral genome. As the upstream gene, flanked by a splice donor and a splice acceptor site, is spliced the downstream gene can be translated from the spliced message, although sequence differences can lead to inefficient splicing and therefore unpredictable expression of the second gene (20). Fusion genes can also be used or alternatively, a heterologous promoter can be placed between the two genes to drive the expression of the downstream gene. The transcripts can be expressed from genes in the same orientation or from bidirectional promoter configurations (21).

1.5.2. IRES and 2a Elements

The use of an internal ribosomal entry site (IRES) in the development of retrovirus vectors is becoming increasingly popular (22). The use of IRES allows cap-independent translation of a downstream gene from a single transcript. IRES sequences, derived from picornaviruses and swine vesicular disease virus have been successfully used in the retroviral vector design (23). Similarly, 2A sequences from picornaviruses (most commonly the foot and mouth disease aphthovirus) have been used to express two genes from a single promoter (24). 2A is a short sequence that links two genes to create a polyprotein that is post-translationally separated at the 2A site to produce two distinct proteins. IRES, 2A and heterologous promoters can be used in the same construct to allow the efficient expression of up to three transgenes (25).

1.5.3. WPRE

Many retroviral vectors also contain a post-translational regulatory element (PRE), usually taken from the Woodchuck Hepatitis virus (WPRE) (26, 27). This was initially included to assist RNA processing, maturation, and nuclear export, and was found to improve titres of the vector during production. Additionally, it has also been found that as it is placed following the gene expression cassette, it can also stabilise the expressed RNA from integrated genomes and produce higher levels of transgene expression.

1.6. Retargeting of Retroviral Vectors

Frequently in pre- and postnatal gene therapy, it will be necessary to target gene transfer specifically to certain tissues and cell populations and to avoid others. One way to achieve this is to perform gene delivery to isolated cell populations from patients or animals ex vivo. In this approach, the target tissue is removed and transduced in vitro, before re-implantation of the genetically corrected cells. These procedures can be invasive and restrict the application of gene therapy to accessible tissue such as bone marrow, skin, or blood. For many diseases, in vivo transduction of target cells would be preferable and in some cases is the only strategy available. Thus, targeting is important if direct gene delivery in vivo is to be considered for gene therapy protocols where target cells cannot be treated in isolation, and several different approaches to targeting are discussed in the following sections. With the exception of envelope exchange (pseudotyping) to enhance transduction to certain populations, cell targeting is currently experimental and has not been used clinically.

1.6.1. Pseudotyping: Envelope Exchange

The envelope proteins of retroviruses primarily determine the specificity of the virus–cell interaction and thereby influence the viral tropism. The potential for manipulating the host range of certain viruses, and hence redirecting them to selected targets was first indicated by the phenomenon of pseudotyping. Pseudotyping involves the exchange of envelope proteins between different species of viruses in packaging cells. It was initially demonstrated by dual infection of a host cell by two unrelated enveloped viruses, resulting in the production of virus progeny coated in its own envelope protein and the envelope protein of the second virus, or the genome of one virus was packaged by just the envelope protein of the second unrelated virus (28, 29). This ability to exchange envelope proteins therefore provided a basis for retrovirus vector retargeting.

Gammaretroviral vectors commonly used for gene transfer are derived from the Moloney-Murine leukaemia virus (MoMLV) backbone and can be classified as amphotropic, ecotropic, xenotropic, or polytropic (30) according to their host range. Host range is determined through interaction of the virus with cellular receptors. These function both as viral binding sites and as specific entry ports into the cell. The interaction between virus and receptor is mediated through high affinity binding to the envelope glycoproteins of the virus. The amphotropic retroviruses infect most mammalian cells including human cells whilst ecotropic retroviruses only infect mouse and rat cells. The xenotropic retroviruses infect a wide variety of species with the exception of mouse/rodent cells. Polytropic viruses are the product of recombination events between ecotropic retroviruses and endogenous mouse proviral sequences. The resultant polytropic virus is able to infect cells from a range of species in addition to mice, although the host range is not as wide as that of xenotropic retroviruses.

As such, the genome of MoMLV-based retroviral vectors is easily pseudotyped with different envelope proteins; thus altering the original host range abilities. However, although many recombinant retroviruses show species specificity in the cells they infect, neither amphotropic nor ecotropic vectors exhibit tissue specificity because of the widespread distribution of the cellular receptors they use among various tissue types.

Whilst pseudotyping does not restrict the delivery of genes to very specific populations, it has been used as a technique to improve vector performance in gene therapy clinical trials to treat an immune disorder(7). In this study, CD34[+] haematopoietic stem cells (HSC) were transduced ex vivo with a gammaretrovirus carrying the therapeutic gene. This vector was pseudotyped with a gibbon ape leukaemia virus (GALV) envelope because CD34[+] cells have more of the receptors that GALV interacts with, than the normal amphotropic envelope, so transduction of the target cells was enhanced.

More recently, HIV-based lentiviral vectors have been pseudotyped successfully by expressing envelope proteins from different viruses in packaging cell lines. The wild-type HIV envelope is most commonly replaced with the G glycoprotein from the vesicular stomatitis virus (VSV-g) envelope. This allows almost universal entry into cells in vitro and into most cell types in vivo, allowing one vector to be used in many different targets. It is also very stable and allows easier concentration and storage of vector.

Lentiviral vectors are relatively simple to pseudotype and envelopes from rabies, baculoviruses, ebola, hanta, and Ross River virus have been used to direct transduction to certain cell or tissue types (31). Pseudotyping does not completely retarget the lentivirus host range, but does help to enrich transduction of particular populations, often following the pattern produced by the parent virus. For example, pseudotyping lentivirus with rabies G envelope enhances transduction of neurons and can enable retrograde transport up the axon following peripheral injection (32).

1.6.2. Transductional Targeting: Manipulation of Retroviral and Non-retroviral Envelopes

The fully matured retrovirus envelope comprises heavily glycosylated extra-viral surface (SU) subunits that are non-covalently linked to membrane-anchored transmembrane (TM) subunits. When a mature viral particle encounters a permissive target cell, the SU glycoprotein binds with high affinity to its receptor. Binding of the SU on target cells is thought to trigger a conformational change in the SU–receptor complex that allows the TM subunit to activate fusion (33, 34).

Current retrovirus vector targeting studies have been almost exclusively based on the manipulation of retroviral envelope sequences. The basis of these retroviral targeting strategies has been to modify the host range by directing binding away from the retroviral receptor towards a cell surface molecule with a tissue specific pattern of expression.

Several types of modifications to the viral envelope proteins have been made in attempts to redirect viral binding with the aim of obtaining targeted transduction of particular populations. For example, N terminal extensions to the amphotropic envelope SU glycoproteins, with single chain antibodies against target cell antigens and growth factors, have been used to promote interaction of viral envelope proteins with new cell surface molecules. This method has shown some success in redirecting virus binding to new cellular receptors. However, these modifications involve the fusion of large polypeptides to N-terminally truncated retroviral envelope glycoproteins (35). The resultant chimaeric envelope proteins have a severely reduced ability to form infectious particles due to suboptimal folding, assembly, and/or transport to the cell surface.

In general, these chimaeric enveloped vectors demonstrated efficient binding to target cells but were unable to provide tissue-specific transduction as binding to the novel receptor did not lead to successful fusion and retroviral entry. In order to enhance incorporation of envelope proteins and make retrovirus particles infectious, it has proven necessary to co-express unmodified envelope proteins along with the chimaeric proteins.

To avoid the need for co-expression of chimaeric and unmodified envelope proteins, envelope chimaeras were made by fusing a flexible linker peptide between the foreign polypeptide (or ligand) and the N terminal portion of the SU subunit. This was based on the finding that inter-domain spacing is crucial not only for optimum processing and display of the ligand but also to render the chimaeric envelope able to bind not only to the redirected receptor but also to its natural retroviral receptor on the cells, thereby facilitating the fusogenic potential of the chimaeric envelope (36). Targeted attachment has generally been associated only with low efficiencies.

An alternate strategy for altering the host range of infection of retroviruses is to manipulate an ecotropic retroviral envelope, which normally infects murine but not human cells, so that it becomes able to target and enter human cells as well. One example is the replacement of the SU portion of MoMLV ecotropic envelope, with the polypeptide hormone erythropoietin (epo), resulting in a chimaeric envelope construct directed towards epo receptors on human erythroid cells. This strategy also requires un-manipulated ecotropic envelope sequences to be present to make infectious retrovirus particles. In this way, retroviruses could be targeted to only bind and transduce cells expressing an epo receptor. While this approach successfully achieved specific targeting, transduction was extremely inefficient (37).

More recent envelope modification strategies have now focused on ablating the natural binding properties to its usual receptor utilised on target cells. Such strategies have mainly focused on the use of non-retroviral envelopes whilst maintaining its fusion functions, provided by a fusogenic molecule. These non-retroviral envelopes include VSV-g (38, 39), Sindbis virus (39–41), Measles virus (42), and Influenza virus (43). Re-targeted binding of viruses to target cells can be achieved by co-expression of single chain antibodies or membrane-anchored ligands. Both these binding-redirected and fusogenic molecules can be expressed on the surface of lentiviral vectors and used as targeted transduction viral vectors.

1.6.3. Manipulation of Retroviral Producer Cells to Alter Viral Specificity

It has been shown that modification of retroviral packaging cells with membrane-anchored ligands can be effectively used to target retroviral transduction. When retroviral budding occurs from packaging cells, virtually all host (packaging cell)-associated membrane proteins including retroviral envelope proteins are incorporated into the mature retroviral particle (44). As such the binding of particles can then be redirected to specific target cells by stable expression of membrane-anchored proteins on the packaging cells. Fusion of such retroviral particles can be complemented with the use of a retroviral envelope protein, which is naturally non-infectious (owing to the natural lack of binding abilities) on target cells. Therefore, for example, retroviral particles can be targeted to bind to a specific cell type by the incorporation of membrane-bound stem cell factor (SCF), in the retroviral producing cell, that will recognise a specific cellular receptor, c-kit, which is expressed by HSC. The concurrent use of an ecotropic virus envelope protein on the SCF expressing producer cell, which is not infectious on human cells, would then essentially provide the fusion function for these vectors, allowing the efficient generation of HSC-targeted retroviral particles (45).

1.6.4. Transcriptional Targeting

When self-inactivating vectors are used, the LTR promoter/enhancer function is removed, so internal promoters are required. This allows flexibility to utilise the vectors for different purposes. Initially, strong viral promoters such as those from cytomegalovirus (CMV) or the spleen focus-forming virus (SFFV) were inserted to achieve maximum gene expression, but since efficiency of gene delivery has gradually increased, promoters that give more physiological levels of expression have become increasingly common. Examples include promoters from housekeeping genes, such as PGK and EF1α (46), that have some tissue specificity to enhance targeting or to use the endogenous promoter from the gene being delivered (47). The latter has the advantage of improving gene expression in the required tissue, at the correct time, and limiting influence in irrelevant cell types.

Key elements can also be introduced into the LTR region to provide a measure of cell-specific expression. An example of this strategy is the addition of the muscle creatine kinase enhancer (48) between the viral enhancer and promoter, which resulted in differentiation specific expression in myogenic cells. Additionally, replacement of the viral enhancer with tyrosinase enhancer (49) and promoter enabled specific expression in melanoma cells (50). However, retrovirus titres were relatively low.

More recently, vectors have been "de-targeted" through addition of cell-specific microRNA target sequences to the 3′ end of transgenes to degrade transcripts in chosen cell lineages. For example, in liver gene transfer, despite the vector delivering a gene to several cell types, expression can be limited to hepatocytes. This prevents expression in the liver's Kupffer cells, which could cause an immune response and limit effectiveness of gene delivery (51).

2. Conclusion

1. Gammaretroviral and lentiviral vectors are simple to produce, concentrate, and store, and can deliver the cDNA from most genes.

2. Modifications to the viral genomes and methods of production have greatly reduced any risk or producing replication competent virus and so have a good safety profile for gene transfer experiments.

3. They efficiently infect a wide range or cell types, or can be further engineered to restrict them to particular target populations. Both gammaretroviral and lentiviral vectors work well in most situations, but lentiviruses have the added advantage of being able to transduce non-dividing cells.

4. These vectors can be manufactured in large enough quantities and to sufficient quality for clinical applications and their effectiveness has been demonstrated in several trials.

5. Prenatal gene therapy offers a promising approach in treating many monogenic inherited disorders. For lifelong, stable correction from birth, retroviral vectors can be used to deliver genes to stem or progenitor cells. Retroviral vectors are well suited gene delivery vehicles for in utero delivery and research into developing such vectors for safe, specific targeting with in vivo injections is ongoing.

3. Materials

3.1. General Materials Used to Produce Retroviral Vectors

1. 37°C, 5% CO_2 tissue culture incubator.
2. Centrifuges: bench top and ultracentrifuge.
3. Complete D-MEM (D-MEM with GlutaMAX™ (Invitrogen), containing 10% heat-inactivated fetal bovine serum (FBS) and antibiotics).
4. Trypsin–EDTA.
5. Opti-MEM (Invitrogen).
6. Phosphate-buffered saline (PBS).
7. Tissue culture flasks: T175 for lentivirus, 10-cm dishes for gammaretrovirus.

3.2. Third-Generation Lentiviral Vector Production

1. 10 mM Branched PEI (Sigma Aldrich, cat no: 408727).
2. Plasmids:
   (a) Lentiviral backbone. Examples include: pRRL, pCCL, pLV, pCS.
   (b) VSV-g envelope plasmid. Examples include: pMD2.G, pCMV–VSV-g.
   (c) Third-generation packaging constructs: pMDLg/pRRE (packaging gag–pol plasmid); pRSV–Rev (rev plasmid).

   Plasmids are available at www.addgene.org or from several commercial sources. To pseudotype with different envelopes, exchange VSV-g for the envelope of choice. Contact relevant groups who have published results using them.
3. Cell lines
   Cell lines based on the human embryonic kidney cell line 293, stably expressing SV40 large T antigen, are used for maximum virus production. Examples include:
   (a) 293T cells (DMSZ; cell line number ACC 635, Genhunter cat# Q401, or the National Gene Vector Biorepository (https://www.ngvbcc.org/)).

(b) 293FT cells (a fast growing variant, Invitrogen; R700-07).

293 cells grow in complete D-MEM (D-MEM supplemented with 10% FBS) at 37°C in a 5% CO_2 atmosphere.

3.3. Gammaretroviral Vector Production

1. Calcium phosphate transfection kit (Invitrogen).
2. 50 mM Chloroquine.
3. 20 mg/ml Polybrene.
4. 20 mg/ml Chondroitin sulphate.
5. Plasmids (available from Addgene).

    (a) pBabe (puro/hygro/neomycin).

    (b) PINCO.

      PINCO plasmids are maintained as an episome in the retrovirus producers by virtue of an EBV origin of replication (OriP) and EBNA1 DNA sequences. The retroviral genome also encodes the enhanced green fluorescent protein (EGFP), driven from an internal immediate early CMV promoter.

6. Cell lines.

    Cell lines are available from the National Gene Vector Biorepository (https://www.ngvbcc.org/) or Allele Biotechnology (http://www.allelebiotech.com/allele3/index.php).

    (a) Phoenix Eco (stably expressing the gag–pol and ecotropic envelope).

    (b) Phoenix Ampho (stably expressing the gag–pol and amphotropic envelope).

    (c) Phoenix gag–pol (stably expressing gag–pol only, thereby allowing the flexible use of any other retroviral envelopes such as GALV or RD114 or non-retroviral glycoprotein such as VSV-g).

4. Methods

4.1. General Information for the Generation of All Retroviral Vectors

Production of retroviral vectors is based on providing all accessory, structural, and enzymatic functions in trans within a cell line. The vector backbone carrying the required gene is supplied on a separate plasmid and is the only component flanked by LTRs and a packaging signal. The RNA expressed from this plasmid is therefore assembled into a virion within the packaging cell line and then secreted into the culture medium.

Due to the cytotoxicity of some envelopes, especially VSV-g, and of the viral proteins TAT and REV, lentiviruses are generally made by transient transfection of producer cells, whereas gammaretroviruses are traditionally produced in stable cell lines. However, due to the nature of SIN vectors and the difficulties associated with achieving high expression from randomly integrated plasmid, SIN gammaretroviruses are also often made by transient transfection of producer cell lines. Both methods are described below.

VSV-g pseudotyped HIV-1-based vectors and gammaretroviral vectors are currently classed as safety Level 1, assuming that no sharps or needles are used during and production and application of the vector and that the genes contained on the backbone are not oncogenic or hazardous. All relevant personal protection equipment and safety procedures must be used following a full risk assessment and observing all necessary local and national rules.

4.2. Third-Generation Lentiviral Vector Production

This 5-day protocol is designed for third-generation (four plasmids) HIV-1-based vector production. Similar approaches will be applicable to other lentiviruses (such as EIAV, FIV, SIV), but specifics should be checked before proceeding.

The procedure requires the transient introduction of four plasmids encoding: (1) viral backbone with transgene of interest flanked by LTRs, with possible inclusion of WPRE, (2) envelope, (3) gag–pol, and (4) Rev. HIV-1 is not a lytic virus, and so cells remain relatively healthy throughout the production of virus, which buds into the culture medium for collection.

For advice, background information and a forum on lentiviruses, www.lentiweb.com provides an excellent resource.

4.2.1. Pseudotyping

To pseudotype lentiviral vectors with an alternative envelope, exchange the VSV-g plasmid for that encoded by another virus. The amount of plasmid required will need optimisation if the glycoproteins are toxic. Not all envelopes will effectively pseudotype HIV-1; finding which work can be trial and error, although many have been previously reported.

4.2.2. Protocol

1. Seed 1.5×10^7 293T cells into a 150-cm^2 tissue culture flask the day before the transfection so that the cells were 80–90% confluent (see Note 1).

2. Add the following amounts (see Note 2) of DNA to 5 ml opti-MEM (per flask) and filter through a 0.22-µm filter (see Note 3):

Transfer (backbone) plasmid	40 µg
VSV-g envelope plasmid	8.5 µg
Packaging plasmid	15.5 µg
Rev plasmid	8 µg

3. Per flask, add 1 μl of a 10 mM stock of polyethylenimine (see Note 4) to 5 ml of opti-MEM and filter through a 0.22-μm filter (see Note 3). Solutions from steps 2 and 3 are added together and left at room temperature for at least 20 min (see Note 5).

4. Meanwhile, remove the growth medium from the cells and wash the cells once in opti-MEM (see Note 6) and following the 20-min incubation in step 3, 10 ml of the PEI/DNA complex is added per flask. The cells are incubated at 37°C, 5% CO_2 for 4 h (see Note 7).

5. The medium is then replaced with complete D-MEM (see Note 8).

6. Optional step: 24 h after the transfection the medium is removed and replaced with fresh complete D-MEM (see Note 9).

7. The virus is harvested the following day. The medium is removed from the cells and replaced with fresh medium for repeat harvesting the following day.

8. The medium from the cells (containing the virus) is centrifuged at $5,000 \times g$ for 10 min and then filtered through a 0.22-μm filter to remove any debris from the supernatant. The supernatant is then transferred to a centrifuge tube. The virus is ultra-centrifuged at 4°C, for 2 h at $100,000 \times g$ (see Note 10).

9. Now that the vector has been pelleted to the bottom of the tube, carefully decant off the supernatant and resuspend the virus in the medium/buffered saline of your choice, pipetting up and down three times. Avoid creating bubbles. Seal the tubes with Parafilm and leave on ice for 1 h (see Note 11).

10. Following the incubation on ice, pipette ten times to resuspend the virus before transferring to a microcentrifuge tube. Centrifuge the vector for 10 min at $3,000 \times g$ in a benchtop centrifuge (preferably at 4°C) to pellet any remaining debris. Virus is not concentrated further at this stage and remains in the supernatant, which should be transferred in aliquots to cryovials. Store the vector at −70°C (see Note 12).

4.3. Gammaretroviral Vector Production

4.3.1. Packaging Cell Line Choice

Several models of gammaretroviral producing cell lines are available (AM-12, PG13, Fly) (52, 53) but here we will concentrate on the use of the Phoenix packaging and producer cell lines developed by the Nolan laboratories (54). These packaging cells are suitable for the expression and culture of various gammaretroviral vector backbones such as pBabe and PINCO.

Phoenix ecotropic and Phoenix amphotropic packaging cell lines are second-generation retrovirus-producer lines for the production of helper-free ecotropic and amphotropic retroviruses, respectively. Phoenix gag–pol cells are required for production of retroviruses pseudotyped with VSV-g or other retroviral envelope glycoproteins (such as GALV or RD114).

These cell lines are based on the human embryonic kidney 293T cell line. They contain *gag–pol* genes with hygromycin as a co-selectable marker and envelope proteins (either ecotropic or amphotropic) with diphtheria toxin as a second selectable marker. These cell lines are routinely maintained in complete D-MEM. They should be passaged for 1 week, about every 10–12 weeks, in 300 μg/ml hygromycin and 1 μg/ml diphtheria toxin for reselection.

4.3.2. Protocol

Transient production of viral vector particles.

1. Day 1 (3 P.M.):
   Seed 1.5×10^6–2.0×10^6 Phoenix cells (in complete D-MEM) per 10-cm diameter tissue culture plate (total volume of 15 ml).

2. Day 2 (10 A.M.):
   Cells should be between 40 and 60% confluence (see Note 13). Aspirate media and replenish with 12 ml of fresh complete D-MEM.

3. Day 2 (3 P.M.):
   Perform calcium phosphate transfection. Prepare Vector DNA mixture (15 μg) + water + $CaCl_2$ to a total of 500 μl.

   (a) Add drop wise vector DNA–$CaCl_2$ mixture to 500 μl 2× HBS, while creating bubbles (using a 1-ml pipette) to mix.

   (b) Incubate the DNA–$CaCl_2$–HBS mixture above at room temperature for 30 min.

   (c) Add 1× chloroquine (final 25 μM, stock is 2,000×) (see Note 14) to media in 10-cm plates-packaging cells.

   (d) Add drop wise (b) calcium phosphate–DNA–HBS mixture to packaging cells (c).

   (e) Incubate overnight at 37°C, 5% CO_2.

4. Day 3 (10 A.M.):
   Approximately 18 h post-transfection. Remove media containing the transfection mixture.

5. Wash cells twice in complete D-MEM (see Note 15).

6. Add 8 ml complete D-MEM and incubate for 24 h

7. Day 4:
   Harvest supernatant containing viral vectors and store at –80°C.

8. Repeat step 6 and recollect supernatant and store at –80°C, on Day 5.

9. The vector can be used unconcentrated, but if required, vectors with VSV-g as a pseudotype can be concentrated using the

same method as for a lentiviral vector (as above). Alternately, using any retroviral envelopes, various methods of concentrating and purifying viral preparations are available. These include anion-exchange (55), sucrose density gradients, or precipitation methods (56).

4.3.3. Protocol for the Concentration of Gammaretroviruses

The following protocol describes a method to concentrate vectors, using precipitation, which we have found to be also useful when using lentiviral vectors pseudotyped with VSV-g.

1. From a transient transfection-calcium phosphate of a 10-cm plate, use 8 ml of fresh medium (complete D-MEM) for virus collection/harvest at 37°C overnight.

2. Collect supernatant and centrifuge at $1,500 \times g$, 5 min, at room temperature to remove any debris.

3. Filter the supernatant using 0.45-μm sterile filter unit into 15-ml falcon tube. The sample can be stored at –80°C.

4. Repeat the process the following day.

5. Pool the supernatant collected over 2 days in a 15-ml tube (total volume of about 14 ml). Alternately frozen supernatant can be thawed quickly, pooled, and concentrated.

6. Add 60 μl (20 mg/ml stock) Polybrene (PB) quickly followed by 60 μl (20 mg/ml stock) Chondroitin sulphate (CS), and vortex briefly.

7. Incubate the sample at 37°C in CO_2 incubator for 20 min.

8. Transfer to an appropriate tube and centrifuge at $10,000 \times g$, 16–20°C, for 20 min.

9. A small visible pellet should form at the bottom of the tube (vector PB–CS complex). Carefully discard the supernatant and resuspend the pellet in 1/100th the original volume with fresh complete D-MEM/Opti-MeM/PBS, by gently pipetting up and down until the pellet is dissolved.

10. Aliquot 50 μl into fresh cryovials.

11. Store virus at –80°C.

4.4. Titration of Viral Vectors

There are different methods to calculate the titre or concentration of the prepared virus. Each has advantages and disadvantages and must be selected according to the construct and application. Three of the most common methods are flow cytometry, quantitative PCR, and ELISA (57).

4.4.1. Flow Cytometry

Flow cytometry is useful for obtaining a functional titre for vectors carrying fluorescent marker genes or those which are expressed on the cell surface to which there is an antibody for visualisation. By applying a dilution series of vector to a known number of cells, calculating the percentage of positive cells by flow cytometry will enable the titre to be determined. It is important to choose a dilution that gives between 5% and 30% transduction. Any higher than this and it is likely that cells will have multiple copies of the virus (58), and therefore the real titre will be higher than calculated.

HeLa cells are commonly used as the target cell although it is useful to titre the virus on the cell type which you intend to use the vector on, as HeLa cells may transduce more easily than other cell types. Alternately, when using ecotropic envelopes, murine NIH 3T3 cells are a suitable cell type. Briefly, 50,000 HeLa cells are generally plated into a 12-well plate and allowed to adhere for approximately 3 h. 1 ml amounts of diluted virus (ranging from 1:10,000 to $1:10^7$ of concentrated stock) are then added and left in culture for 3 days. The percentage of transduced cells is then determined with flow cytometry or fluorescence microscopy (if a fluorescent reporter protein is used).

The titre expressed as transducing unit (TU)/ml or as infectious particle (IP/ml) can be calculated with the following formula:

$$\text{Titre} \left(TU / ml \right) = \left(F \times Co / V \right) \times D,$$

where F is the frequency of GFP positive cells; Co is the total number of target cell infected (normally 50,000); V is the volume of the inoculum; D is the viral dilution factor.

4.4.2. Quantitative PCR

Calculation from real-time PCR is similar, but is based on how many viral genomes are delivered, and gives a titre in viral genomes per millilitre. This is normally higher than the functional titre provided by flow cytometry because not all genomes will be intact, or inserted into a region that is transcriptionally active.

The protocol utilised for this approach will depend on different factors, including the vector used and PCR equipment available but several groups have produced an overview of the method to titrate lentiviruses (also applicable to gammaretroviral vectors) (59, 60).

4.4.3. ELISA

ELISA assays can be used to determine the concentration of lentivirus preparations (not gammaretroviruses) by calculating the amount of p24; a protein component of the viral capsid. This method will provide information on the number of viral particles, but not how many of them contain a genome or that are functional. However, a good quality VSV-g pseudotyped lentivirus should produce approximately 100 functional transducing units per picogram of p24. Used in combination with FACS or real-time PCR, it can highlight the quality of the vector preparation as a low quality batch can have a high value for p24 (because of free p24 un-associated with the virus and particles without genomes) but a low titre by other methods. HIV p24 ELISA kits are available from several different companies, such as Roche and Helvetica Health Care.

5. Notes

Third-Generation Lentiviral Vector Production

1. Sufficient cells must be available on the first day (Monday is most convenient). Consider the volume of supernatant that fits into the available centrifuge for concentration when growing cells. For a rotor holding approximately 200 ml of medium, we passage one confluent 150-cm^2 flask of 293Ts 1:10 into ten flasks on Friday. Depending on the growth rate of the cells, this results in 10 flasks, almost confluent, on Monday, which should be ample for setting up 14 flasks for production of sufficient supernatant to concentrate.

   The rate of growth of the packaging cells will depend on a number of factors, including age, passage number, and quality of serum. The resulting confluency is important, as too few cells will not yield much virus. Equally, if the cells are too confluent, they will not divide and transfect well, and will be dying by the end of the procedure. However, we have transfected cells at an estimated confluency ranging from 50% to 100% and generally, high titre virus is obtained.

   The amount of virus produced by the cells will fall over time, so a large stock of low-passage cells should be frozen and a fresh batch defrosted approximately every month, or when the titre of virus produced reduces. When a new batch is defrosted, allow the cells to grow and recover for at least a week before attempting to make virus, otherwise viral titre will be low.

2. Depending on the plasmids used and the size of insert in the transfer (backbone) plasmid, this is an approximate molar ratio of 1:1:1:2 (packaging:rev:env:transfer). The amount of DNA used is in excess for this type of transfection and so can be titrated down or altered for alternative protocols without adversely affecting vector production.

3. A syringe filter or a vacuum-driven Stericup can be used, depending on the volume handled. Make half a flask excess to allow for volume loss during filtration.

4. Polyethylenimine (PEI) bought in bulk and prepared using the method below is very economical, and gives good transfection efficiency in these cells. Other transfection reagents work as well, if not better (those tested include calcium phosphate co-precipitation, Lipofectamine, FuGene 6, or Gene Juice) following the manufacturers instructions. These often give higher transfection efficiencies, but are considerably more expensive and in the case of calcium phosphate, some experience is required to achieve reproducibility. The main advantage of PEI

is its ease of use; for most operators it produces successful results first time. Transfection-ready, optimised PEI is available commercially, but has not been tested in our laboratory.

To make a stock of 10 mM polyethylenimine (PEI, MWT 25,000):

Add 10 ml of water to 10 ml PEI (PEI is viscous and may need to be weighed out. 10 g is adequate.). Mix by vortexing.

Add 12 N HCl, 1 ml at a time and vortex to neutralise (pH 7). This is approximately 12 ml HCl total. Check the pH with litmus paper and top up to 41.2 ml with water. Mix again on a vortex.

Store in 1 ml aliquots in −80°C freezer. 10 mM PEI appears stable at −20°C for over a year and freeze–thaw cycles are not detrimental.

5. Do not mix the DNA and PEI together before filtration as the resulting complex does not pass through a 0.22-μm membrane and so little virus is produced.

6. Take care to deliver the medium down the side of the flask as the cells are only loosely adherent and direct application of fluid onto the monolayer will detach cells. PEI is active in serum, so some investigators feel the wash step is not strictly necessary.

7. Leaving the transfection mixture for longer (up to 6 h has been tested) does not seem to damage the cells, but too long (overnight, for instance) will kill cells.

8. The volume of complete D-MEM growth medium added depends on the final volume from all the flasks that can fit in the centrifuge tubes for the final spin. For example, if a 6-bucket rotor holds 32-ml tubes, 14 plates would be required with approximately 14 ml of medium on each.

9. Omission does not adversely affect virus titre. Include this step if the medium appears yellow or the cells unhealthy. This can happen if too many cells were plated at the beginning.

10. Ultracentrifugation concentrates the virus approximately 500 fold; so if a titre of approximately 10^6 transducing units per millilitre is sufficient, the supernatant can be used at this stage without concentrating.

If required, ultracentrifugation is performed in a swing-out rotor using Sorval Discovery SE or similar. Concentration of the virus appears to be similar when centrifuging at any time between 1 h 45 min and 3 h. These times and forces are suitable for VSV-g pseudotyped lentiviral vectors. Other envelopes are unstable at such forces and lower speeds must be used.

If an ultracentrifuge is unavailable or lower forces are required, similar concentration of the vector can be achieved by centrifuging at $4,000 \times g$ for 20 h at 4°C.

Cell debris and unincorporated VSV-g present in the vector preparation can be cytotoxic to sensitive cells. To further remove contaminants, the virus can be centrifuged through a sterile 20% sucrose cushion: In a 33-ml ultracentrifuge tube, very slowly and carefully overlay 4 ml of 20% sucrose with 26 ml of supernatant and proceed as normal.

We have not found this step to be necessary for in vivo injections of virus, but have observed empirically that reducing the amount of VSV-g plasmid transfected into the producer cells by 25% does not significantly reduce viral titres obtained, but can improve survival rates in recipients.

11. Following centrifugation, a small yellow/brown pellet can be seen. This is probably due to substances contained in the FBS in the growth medium. Resuspend the pellet containing viral vector in medium that is required for the cells in which the vector is going to be applied. For direct injection in vivo, PBS is acceptable. Resuspending in 150 μl of medium/PBS per 30 ml of initial supernatant should give high titre virus. Reducing the volume can concentrate the vector further, but no further concentration is seen below a volume of 50 μl, probably due to insufficient coverage of the pellet. The virus should be resuspended on ice for at least 30 min, preferably 1 h. In our hands, leaving the virus to resuspend for at least 3 h has not been detrimental.

12. Assuming a titre of 10^8–10^9 transducing particles per millilitre will be achieved, consider the smallest volume that will be required for the subsequent experiments and freeze in suitably sized aliquots. Considerable loss in titre occurs with each freeze–thaw cycle, so avoid storing large aliquots and re-freezing. If the virus is going to be used within 2–3 days, place it at 4°C as any loss is likely to be less than a freeze–thawing.

Gammaretroviral Vector Production

13. When using calcium phosphate transfection method, it is important to get the optimum confluence of cells prior to transfection. More than 60% confluent, a cell culture will result in loss/detachment of cells towards the end of the process and less than 40% confluence will result in poor transfection efficiencies. Using the calcium phosphate method, it is worth buying a kit to ensure the pH of the HBS component is optimal.

14. Chloroquine is thought to prevent endosomal degradation of incoming DNA into cells by preventing a change in pH in the lysosome. However, care must be taken to ensure that cells containing chloroquine and transfection mixture are not left for more than 24 h, which may result in cell death.

15. Cells must be washed thoroughly prior to collection of virus supernatant. The presence of chloroquine in viral supernatant and remnants of calcium–phosphate–DNA mix may be toxic to target cells of transduction experiments, especially primary cells.

References

1. Suerth JD, Maetzig T, Galla M et al (2010) Self-inactivating alpharetroviral vectors with a split-packaging design. J Virol 84:6626–6635

2. Guntaka RV, Swamynathan SK (1998) Retroviral vectors for gene therapy. Indian J Exp Biol 36:539–545

3. Palu G, Parolin C, Takeuchi Y, Pizzato M (2000) Progress with retroviral gene vectors. Rev Med Virol 10:185–202

4. Naldini L, Blomer U, Gage FH et al (1996) Efficient transfer, integration, and sustained long-term expression of the transgene in adult rat brains injected with a lentiviral vector. Proc Natl Acad Sci U S A 93:11382–11388

5. Russell DW, Miller AD (1996) Foamy virus vectors. J Virol 70:217–222

6. Hacein-Bey-Abina S, Le DF, Carlier F, Bouneaud C et al (2002) Sustained correction of X-linked severe combined immunodeficiency by ex vivo gene therapy. N Engl J Med 346: 1185–1193

7. Gaspar HB, Parsley KL, Howe S et al (2004) Gene therapy of X-linked severe combined immunodeficiency by use of a pseudotyped gammaretroviral vector. Lancet 364:2181–2187

8. Aiuti A, Cattaneo F, Galimberti S et al (2009) Gene therapy for immunodeficiency due to adenosine deaminase deficiency. N Engl J Med 360:447–458

9. Ott MG, Schmidt M, Schwarzwaelder K et al (2006) Correction of X-linked chronic granulomatous disease by gene therapy, augmented by insertional activation of MDS1-EVI1, PRDM16 or SETBP1. Nat Med 12:401–409

10. Cartier N, Hacein-Bey-Abina S, Bartholomae CC et al (2009) Hematopoietic stem cell gene therapy with a lentiviral vector in X-linked adrenoleukodystrophy. Science 326:818–823

11. Reiser J, Harmison G, Kluepfel-Stahl S et al (1996) Transduction of nondividing cells using pseudotyped defective high-titer HIV type 1 particles. Proc Natl Acad Sci U S A 93: 15266–15271

12. Miller AD, Rosman GJ (1989) Improved retroviral vectors for gene transfer and expression. Biotechniques 7(980–6):989

13. Dull T, Zufferey R, Kelly M et al (1998) A third-generation lentivirus vector with a conditional packaging system. J Virol 72:8463–8471

14. Naldini L, Blomer U, Gallay P et al (1996) In vivo gene delivery and stable transduction of nondividing cells by a lentiviral vector. Science 272:263–267

15. Follenzi A, Ailles LE, Bakovic S et al (2000) Gene transfer by lentiviral vectors is limited by nuclear translocation and rescued by HIV-1 pol sequences. Nat Genet 25:217–222

16. Zennou V, Petit C, Guetard D et al (2000) HIV-1 genome nuclear import is mediated by a central DNA flap. Cell 101:173–185

17. Demaison C, Parsley K, Brouns G et al (2002) High-level transduction and gene expression in hematopoietic repopulating cells using a human immunodeficiency [correction of immunodeficiency] virus type 1-based lentiviral vector containing an internal spleen focus forming virus promoter. Hum Gene Ther 13: 803–813

18. Zufferey R, Nagy D, Mandel RJ et al (1997) Multiply attenuated lentiviral vector achieves efficient gene delivery in vivo. Nat Biotechnol 15:871–875

19. Schwickerath O, Brouns G, Thrasher A et al (2004) Enhancer-deleted retroviral vectors restore high levels of superoxide generation in a mouse model of CGD. J Gene Med 6:603–615

20. Dougherty JP, Temin HM (1986) High mutation rate of a spleen necrosis virus-based retrovirus vector. Mol Cell Biol 6:4387–4395

21. Amendola M, Venneri MA, Biffi A et al (2005) Coordinate dual-gene transgenesis by lentiviral vectors carrying synthetic bidirectional promoters. Nat Biotechnol 23:108–116

22. Fux C, Langer D, Kelm JM (2004) New-generation multicistronic expression platform: pTRIDENT vectors containing size-optimized IRES elements enable homing endonuclease-based cistron swapping into lentiviral expression vectors. Biotechnol Bioeng 86:174–187

23. Pizzato M, Franchin E, Calvi P et al (1998) Production and characterization of a bicistronic Moloney-based retroviral vector expressing

human interleukin 2 and herpes simplex virus thymidine kinase for gene therapy of cancer. Gene Ther 5:1003–1007

24. Chinnasamy D, Milsom MD, Shaffer J et al (2006) Multicistronic lentiviral vectors containing the FMDV 2A cleavage factor demonstrate robust expression of encoded genes at limiting MOI. Virol J 3:14

25. Chinnasamy N, Shaffer J, Chinnasamy D (2009) Production of multicistronic HIV-1 based lentiviral vectors. Methods Mol Biol 515:137–150

26. Zhang XY, La RV, Reiser J (2004) Transduction of bone-marrow-derived mesenchymal stem cells by using lentivirus vectors pseudotyped with modified RD114 envelope glycoproteins. J Virol 78:1219–1229

27. Donello JE, Loeb JE, Hope TJ (1998) Woodchuck hepatitis virus contains a tripartite posttranscriptional regulatory element. J Virol 72:5085–5092

28. Zavada J, Zazadova Z, Malir A, Kocent A (1972) VSV pseudotype produced in cell line derived from human mammary carcinoma. Nat New Biol 240:124–125

29. Zavada J (1976) Viral pseudotypes and phenotypic mixing. Arch Virol 50:1–15

30. Battini JL, Heard JM, Danos O (1992) Receptor choice determinants in the envelope glycoproteins of amphotropic, xenotropic, and polytropic murine leukemia viruses. J Virol 66:1468–1475

31. Rahim AA, Wong AM, Howe SJ et al (2009) Efficient gene delivery to the adult and fetal CNS using pseudotyped non-integrating lentiviral vectors. Gene Ther 16:509–520

32. Mazarakis ND, Azzouz M, Rohll JB et al (2001) Rabies virus glycoprotein pseudotyping of lentiviral vectors enables retrograde axonal transport and access to the nervous system after peripheral delivery. Hum Mol Genet 10:2109–2121

33. Russell SJ, Cosset FL (1999) Modifying the host range properties of retroviral vectors. J Gene Med 1:300–311

34. Morling FJ, Peng KW, Cosset FL, Russell SJ (1997) Masking of retroviral envelope functions by oligomerizing polypeptide adaptors. Virology 234:51–61

35. Cosset FL, Morling FJ, Takeuchi Y et al (1995) Retroviral retargeting by envelopes expressing an N-terminal binding domain. J Virol 69:6314–6322

36. Valsesia-Wittmann S, Morling FJ, Nilson BH et al (1996) Improvement of retroviral retargeting by using amino acid spacers between an additional binding domain and the N terminus of Moloney murine leukemia virus SU. J Virol 70:2059–2064

37. Kasahara N, Dozy AM, Kan YW (1994) Tissue-specific targeting of retroviral vectors through ligand-receptor interactions. Science 266:1373–1376

38. Jeetendra E, Robison CS, Albritton LM, Whitt MA (2002) The membrane-proximal domain of vesicular stomatitis virus G protein functions as a membrane fusion potentiator and can induce hemifusion. J Virol 76:12300–12311

39. Zhang XY, Kutner RH, Bialkowska A et al (2010) Cell-specific targeting of lentiviral vectors mediated by fusion proteins derived from Sindbis virus, vesicular stomatitis virus, or avian sarcoma/leukosis virus. Retrovirology 7:3

40. Morizono K (2005) Lentiviral vector retargeting to P-glycoprotein on metastatic melanoma through intravenous injection. Nat Med 11:346–352

41. Pariente N, Mao SH, Morizono K, Chen IS (2008) Efficient targeted transduction of primary human endothelial cells with dual-targeted lentiviral vectors. J Gene Med 10:242–248

42. Buchholz CJ, Muhlebach MD, Cichutek K (2009) Lentiviral vectors with measles virus glycoproteins—dream team for gene transfer? Trends Biotechnol 27:259–265

43. Szecsi J, Drury R, Josserand V et al (2006) Targeted retroviral vectors displaying a cleavage site-engineered hemagglutinin (HA) through HA-protease interactions. Mol Ther 14:735–744

44. Hammarstedt M, Wallengren K, Pedersen KW et al (2000) Minimal exclusion of plasma membrane proteins during retroviral envelope formation. Proc Natl Acad Sci U S A 97:7527–7532

45. Chandrashekran A, Gordon MY, Casimir C (2004) Targeted retroviral transduction of c-kit+hematopoietic cells using novel ligand display technology. Blood 104:2697–2703

46. Thornhill SI, Schambach A, Howe SJ et al (2008) Self-inactivating gammaretroviral vectors for gene therapy of X-linked severe combined immunodeficiency. Mol Ther 16:590–598

47. Charrier S, Dupre L, Scaramuzza S et al (2007) Lentiviral vectors targeting WASp expression to hematopoietic cells, efficiently transduce and correct cells from WAS patients. Gene Ther 14:415–428

48. Ferrari G, Salvatori G, Rossi C et al (1995) A retroviral vector containing a muscle-specific enhancer drives gene expression only in differentiated muscle fibers. Hum Gene Ther 6:733–742

49. Vile RG, Hart IR (1993) In vitro and in vivo targeting of gene expression to melanoma cells. Cancer Res 53:962–967

50. Vile RG, Hart IR (1994) Targeting of cytokine gene expression to malignant melanoma cells

using tissue specific promoter sequences. Ann Oncol 5(Suppl 4):59–65

51. Brown BD, Venneri MA, Zingale A et al (2006) Endogenous microRNA regulation suppresses transgene expression in hematopoietic lineages and enables stable gene transfer. Nat Med 12:585–591

52. Cosset FL, Takeuchi Y, Battini JL et al (1995) High-titer packaging cells producing recombinant retroviruses resistant to human serum. J Virol 69:7430–7436

53. Loew R, Meyer Y, Kuehlcke K et al (2010) A new PG13-based packaging cell line for stable production of clinical-grade self-inactivating gamma-retroviral vectors using targeted integration. Gene Ther 17:272–280

54. Kinsella TM, Nolan GP (1996) Episomal vectors rapidly and stably produce high-titer recombinant retrovirus. Hum Gene Ther 7: 1405–1413

55. Rodrigues T, Alves A, Lopes A et al (2008) Removal of envelope protein-free retroviral vectors by anion-exchange chromatography to improve product quality. J Sep Sci 31: 3509–3518

56. Landazuri N, Le Doux JM (2006) Complexation with chondroitin sulfate C and Polybrene rapidly purifies retrovirus from inhibitors of transduction and substantially enhances gene transfer. Biotechnol Bioeng 93:146–158

57. Sastry L, Johnson T, Hobson MJ et al (2002) Titering lentiviral vectors: comparison of DNA, RNA and marker expression methods. Gene Ther 9:1155–1162

58. Fehse B, Kustikova OS, Bubenheim M, Baum C (2004) Pois(s)on–it's a question of dose. Gene Ther 11:879–881

59. Lizee G, Aerts JL, Gonzales MI et al (2003) Real-time quantitative reverse transcriptase-polymerase chain reaction as a method for determining lentiviral vector titers and measuring transgene expression. Hum Gene Ther 14:497–507

60. Delenda C (2004) Lentiviral vectors: optimization of packaging, transduction and gene expression. J Gene Med 6(Suppl 1):S125–S138

Chapter 6

Vector Systems for Prenatal Gene Therapy: Principles of Adeno-Associated Virus Vector Design and Production

Christopher J. Binny and Amit C. Nathwani

Abstract

Vectors based on adeno-associated virus (AAV) show great promise for safe, efficacious therapeutic gene transfer in extensive pre-clinical data and, recently, in clinical trials. Careful vector design and choice from a range of natural or synthetic pseudotypes allow targeted, efficient, and sustained expression of therapeutic genes. The efficiency of gene delivery can be further enhanced through the use of drug pre-treatment or co-infection with a suitable helper virus. This chapter describes current best practice for AAV production, including complete methods for: (1) efficient generation of vector without the use of helper viruses, simplifying the transition to GMP-grade production for clinical applications; (2) efficient and easily scalable purification of the virus by affinity chromatography, allowing rapid production of highly concentrated, high titre stocks; (3) reliable quantification and assaying of viral stocks, along with short- and long-term storage considerations.

Key words: Adeno-associated virus, Recombinant AAV, Transfection, Preparation, Purification of AAV, Affinity chromatography, Concentration, Titration, AAV serotype tropism, AAV pseudotype targeting, Transcapsidation, Rational AAV-vector design, Directed evolution of AAV vector, Species and sex dependence of AAV transduction, Proteosomal degradation

1. Introduction

1.1. Background

Adeno-associated virus (AAV), a helper dependent parvovirus, shows great potential as a gene transfer tool (1). It can mediate efficient and stable transduction of a variety of post-mitotic tissues in adults and crucially, it has an excellent safety profile (2, 3). The great potential of this vector for use in novel therapeutic applications is complemented by the ability to generate large quantities of high-titre stocks that have little or no contamination with wild-type AAV or helper viruses.

AAV is a relatively small virus (approx 20 nm diameter), with icosahedral capsids comprised of three proteins VP1, VP2, and VP3, present in the ratio 1:1:20 (4, 5). The genome of a typical

Charles Coutelle and Simon N. Waddington (eds.), *Prenatal Gene Therapy: Concepts, Methods, and Protocols*,
Methods in Molecular Biology, vol. 891, DOI 10.1007/978-1-61779-873-3_6, © Springer Science+Business Media, LLC 2012

AAV is approximately 4.7 kb of single-stranded DNA (ssDNA) (6). The termini of this genome consist of 145 nucleotide long inverted terminal repeats (ITRs), which fold upon themselves to form T-shaped hairpin structures, providing a free 3′-hydroxyl group to prime the necessary step of synthesising a complementary DNA strand before gene expression can begin. Within the ITRs are the terminal resolution sites (*trs*), involved in the regulation of viral gene expression and necessary for packaging of the genome into the capsid (7, 8). These ITRs flank the two viral genes *rep* and *cap*, which encode the AAVs' four regulatory proteins and three capsid proteins, respectively.

Completion of AAVs' lytic replication cycle is dependent on the activity of helper genes from another active virus within the host cell, e.g., Adenovirus E1, E2a, E4, and virus-associated RNA (VA-RNA) (9). In the absence of a helper virus, wild-type AAV enters a latent phase in which maintenance of the virus through cell divisions is accomplished by site-specific integration of head to tail concatemers of the viral genome into the long arm of chromosome 19. This integration event is dependent on the activity of the AAV Rep60 or Rep78 proteins and is inefficient in recombinant AAV (rAAV) vectors lacking this gene, with maintenance of the viral genome being mostly episomal (5, 10).

1.2. Vector Design

As a gene therapy vector, AAV is typically modified to completely remove the *rep* and *cap* genes with their promoters. rAAVs therefore consist of the desired therapeutic gene under control of a chosen promoter, with the flanking ITRs as the only remaining viral DNA. This configuration allows maximum use of the virus' limited packaging capacity.

Complete removal of the viral genes also renders the virus non-replicating, while eliminating the potential for unexpected interactions of viral proteins or RNA with the inserted gene or with the targeted cells. This gives these vectors a favourable safety profile when compared to other viral vector systems such as most Adenovirus-based vectors.

When the AAV has successfully entered the cell, dissociation of the AAV particle leaves the single-stranded genome vulnerable to digestion by cellular ssDNAses. Additionally, gene expression cannot begin until a second strand of DNA is synthesised or until two complementary copies of the genome, each introduced to the cell by a separate AAV particle, meet and bind to each other. Together, these vulnerabilities result in an inefficient transduction of cells and, in those cells that are successfully transduced, relatively slow initiation of expression of the therapeutic gene.

Some controversy exists over the true packaging limit of rAAV vectors. As mentioned above, wild-type AAVs have genomes of approximately 4.7 kb. Successful packaging and subsequent expression of genes up to a length of 8.9 kb has been reported using AAV5 capsid (11). However, recent data suggest that a

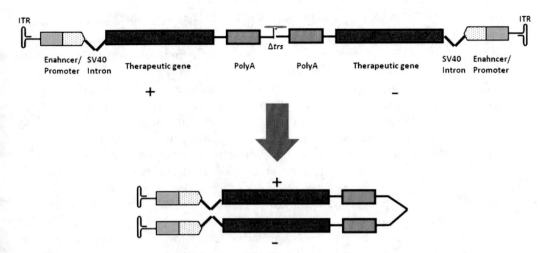

Fig. 1. Genome organisation of scAAV vectors.

mixture of AAV particles are produced and packaged, each containing a maximum of 5.2 kb ssDNA, representing either the 3′ or 5′ half of the genome. In multiply infected cells, these genome fragments are presumed to join together to form complete, over-sized AAV genomes through homologous recombination or non-homologous end joining, resulting in in vivo expression of the oversized genome (6, 11–13).

When the AAV expression cassette is half the size of the wild-type AAV genome, there is a tendency to package two single-stranded proviruses in the same virion. If this short form of the genome is designed with a deleted 3′ *trs*, complementary copies of the AAV genome are able to join as inverted repeats and be pack-aged within the same viral particle. When released into the cell, these inverted repeats of the genome fold together into a hairpin structure to give a dsDNA genome, as illustrated in Fig. 1. The vector is thus protected from cellular ssDNAses and able to begin expression of the therapeutic gene immediately.

Using a self-complementary vector dramatically reduces the maximum size of the expression cassette to around 2.3 kb thus mandating modification/engineering of the promoter and thera-peutic genes. However, where the reduced packaging capacity is not prohibitive, the use of self-complementary vectors can significantly improve the efficiency of transduction and rapidity of onset of gene expression as compared to an identical vector in a single-stranded format since the vector already forms a self-complementary unit.

1.3. Choice of Pseudotype

1.3.1. Tropisms, Transduction Efficiency, Pre-existing Immunity

AAV2 is undoubtedly the best-characterised serotype, and to date most extensively used in clinical trials. This is largely due to the fact that it was the first serotype identified and cloned into a bacterial plasmid, rather than any inherent superiority over other serotypes. Indeed, with approximately 80% of humans already expressing antibodies against AAV2, its potential for use as a vector for gene therapy is limited (8). Other serotypes, whether wild-type, hybrid, or synthetic, are therefore emerging into popular use.

A wide range of naturally occurring AAV serotypes has been isolated from a variety of mammalian species (5, 14), defined on the basis of each serotype being antigenically distinct from the others. This diversity of antigenicity is due to variations in the structural proteins forming the exterior of the capsid, particularly the protrusions responsible for adsorption to target cells and triggering internalisation. As might be expected, this extensive variation in the structure of attachment domains is reflected in a wide variation in tropisms between serotypes. The choice of which serotype to work with must therefore be driven largely by the tissue(s) that the vector is intended to target. Analysis of the tropism of known and newly isolated serotypes is an ongoing process. Screening of AAV serotypes for specific tropisms has, for example, revealed a range of transduction patterns in the mouse brain, allowing researchers control over their vector's targeting of non-neuronal cells and vector transport along neuronal projections (15). Such variation in tropism can be attributed to surprisingly small variations between the serotypes. For example, a single amino-acid difference between AAV1 and AAV6 (K531E) is sufficient to account for a dramatic change in tropism by suppressing the heparin binding ability of AAV6 (16).

The most complete study of vector tropism to date was conducted in mice, following a single tail vein injection of a Luciferase expression cassette packaged in capsids of serotypes 1–9 (17). Broadly speaking, the use of AAV7 and AAV9 leads to the highest overall expression throughout the examined animals. Expression from AAV9 was detectable in all tested tissues except the kidney while AAV7 showed more restricted expression, being reduced in the lungs and absent from the brain, testes, and kidney. Data from these studies should be interpreted with caution, however, as the route of delivery can strongly influence the eventual distribution of the virus through examined tissues.

1.3.2. Modified Capsids

Packaging of AAV genomes into capsids relies on an interaction between the ITRs and VP3 which is, to some extent, serotype-specific. Therefore, efficient packaging of a given vector into your capsid of choice may necessitate modification of the ITR or the use of hybrid structural proteins, designed to maximise the strength of the interaction between the two factors (5). An example of this technique—commonly referred to as "transcapsidation" or "cross-packaging"—is the use of helper plasmids 2–8, which, along with the other viral transcripts, encodes a hybrid VP3 able to efficiently package sequences bearing AAV2 ITRs into the AAV8 capsid.

Efforts to create new serotypes of AAV have included rational design and directed evolution, both with some degree of success. Identifying a key binding/attachment region on the VP3 trimer of

AAV2 and testing a panel of alternative peptide sequences for that region yielded viruses with potentially useful modified tropisms, in particular a variant (AAV2i8) that showed efficient and specific gene transfer to muscle tissues (18). In one example of using directed evolution to generate new serotypes, a combination of genome shuffling and passage through a mouse model of epilepsy generated an AAV serotype capable of selectively crossing the blood:brain barrier at sites of damage to infect oligodendrocytes and neurones (19). A similar technique of DNA family shuffling and recovery of packaged vector after passage through target tissues has also been used to generate a hybrid AAV capsid that showed improved targeting and gene delivery to liver cells while avoiding pre-existing immunity against wild-type capsids (20).

The availability of wild-type capsids targeting a range of tissues and the rapidly developing ability to generate new serotypes selectively targeting specific tissues, together hold the promise that AAV vectors can be used to deliver gene therapy vectors efficiently to a wide range of tissues, and are therefore of great potential use in the development of new therapeutic approaches.

1.4. Influence of Sex and Animal Model

A final consideration when working with AAV vectors is that, in mice, the efficiency of both transduction and gene expression is sex-dependent. Poor liver transduction and gene expression led to 5- to 13-fold lower efficiency of expression in females relative to males. Although this relative inefficiency in females can be partially corrected using pretreatment with bortezomib or androgens, or through the use of self-complementary vectors, it must be borne in mind when designing studies using linear AAV vectors in mice (21, 22).

1.5. Administration of AAV Vectors

Our understanding of the biology underlying AAV vector tropism, cell entry, and gene expression can be exploited to improve the efficiency of gene transfer using AAV vectors. Work in murine bronchial epithelial cells showed that the major obstacles to transduction are efficient endosomal processing and ubiquitination of the viral capsid leading to proteosomal degradation. As might be expected from these findings, treatment with proteosome inhibitors markedly increased transduction efficiency in these cells (23). Related work examining transduction of murine hepatocytes confirmed this data, demonstrating that pre-treatment with bortezomib lead to a twofold increase in expression of the therapeutic gene (22). This increase in expression was sex-dependent, and had the effect of raising expression in females to a similar level to that achieved in males, correcting the poor transduction efficiency in females described above.

It has been suggested that packaging DNA strands longer than the wild-type AAV genome sensitises AAV particles to proteosomal degradation. Administering bortezomib concurrently with AAV2 or AAV8 containing a 5.6-kb genome led to an increase in therapeutic gene expression from this over-large vector of 600% or 300%, respectively (24). Further modulators of proteosome activity such as doxorubicin (adriamycin) and alcarubicin (aclacinomycin A) have been demonstrated to have similar beneficial effects in mouse lungs and in human polarised lung epithelial cells (25).

Co-infection with adenovirus increases the rate at which ssDNA AAV genomes are converted to dsDNA, improving the stability of AAV genomes and shortening the lag between transduction and gene expression (26). Later administration of a helper adenovirus did not increase the number of AAV genomes present or the proportion present as dsDNA, but was associated with an increase in mRNA from the AAV vector and a resulting increase in transgene expression (22).

2. Materials

2.1. Materials for AAV Production

Plasmids should be dissolved in nuclease-free water or TE buffer, and kept sterile.

2.1.1. Plasmids

1. Helper plasmid carrying Adenovirus genes. Examples include: HGTI, XX6.
2. Packaging plasmid carrying AAV rep and cap genes. Examples include:
   XX2, pAAV5-2.
3. Recommended: plasmid unrelated to AAV encoding GFP under constitutive promoter. Examples include:
   pCL10.1 EF2α GFP.

2.1.2. Cell Culture

1. 293 T cells (DMSZ; cell line number ACC 635, Genhunter cat# Q401).
2. DMEM supplemented with 10% fetal bovine serum (FBS).
3. 37°C incubator with a 5% CO_2 atmosphere.
4. Tissue culture-treated 150-mm polystyrene dishes (Corning B.V. 430599).

2.1.3. Transfection

1. Vortexer.
2. 0.22-μm PES membrane vacuum filtration unit (Millipore FDR-125-050E).
3. Vacuum source.
4. 225-ml Polypropylene conical centrifuge tube (BD Biosciences 352075) (NB: calcium phosphate method only).
5. Fluorescent microscope, with UV source and filters to observe emitted light at 509 nm.

Table 1
2× HBS buffer

	Molecular weight	Final concentration (mM)	Mass for 1 l (g)
NaCl	58.44	273	16
KCl	74.56	9.92	0.74
NaH$_2$PO$_4$H$_2$O	137.99	1.45	0.2
Dextrose	180.16	11.1	2.0
HEPES	238.3	42.0	10.0

Adjust pH to 7.05 with 1N NaOH. Filter-sterilise and store in 53 ml aliquots at –20°C

6. 2.5 M CaCl$_2$ Solution (calcium phosphate method only).

7. 277.5 g in 1 l, sterilise by 0.2 μm filtration and keep in aliquots at –20°C.

8. 0.1% PEI (PEI Method only) dissolve 1 g polyethanimine (PEI) in 1,000 ml water, adjusting pH to 7.2 with NaOH. Filter sterilise, store in 10 ml aliquots at –20°C.

9. 2× HBS buffer (for calcium phosphate method only) (Table 1).

Materials for AAV Harvesting and Purification

1. Cell scraper (Greiner Bio-One 541080).

2. 225 ml Polypropylene conical centrifuge tubes (BD Biosciences 352075) (NB: calcium phosphate method only).

3. Centrifuge with swinging-bucket rotor suitable to spin the above 225-ml conical centrifuge tubes at $4,000 \times g$, e.g. Thermo Sorvall Heraeus Swinging Bucket Rotor (7500 6445) with tissue culture bucket 250 ml (6497).

4. 20% Sodium deoxycholate solution.

5. Benzonase nuclease (Sigma, E1040-25KU).

6. 37°C water bath.

7. -80°C freezer.

8. 25 mm Syringe filter, NY, 0.22 μm (Corning 431224), with 20 ml syringes.

9. Recommended: Flow Cytometer for analysis of GFP expression, with suitable buffers and FACS tubes.

10. 1× TD buffer (Table 2).

11. Materials for vector purification by affinity chromatography (e.g. heparin- or mucin-sepharose).

12. HPLC system, e.g. AKTA Explorer (GE Healthcare).

13. AVB Sepharose High Performance packed into 5 ml column (GE Healthcare 28-4112-12).

Table 2
1× TD buffer

	Molecular weight	Final concentration (mM)	Mass for 1 l (g)
NaCl	58.4	140	8.2
KCl	74.6	5	0.37
K₂HPO₄	174.2	0.7	0.12
MgCl₂	203.3	3.5	0.7
Tris	121.4	25	3.0

Adjust pH to 7.5. Autoclave and keep at 4°C

Table 3
Glycine elution buffer for affinity chromatography of AAV

	Molecular weight	Final concentration (mM)	Mass for 1 l (g)
Glycine	75.07	50	3.75

Make up to 1 l, adjusting pH to 2.7. Remove particulates by filtering through a 0.45-μm PES membrane. Store at 4°C

14. PBS, adjusted to pH 7.5 and autoclaved.

15. Dialysis Cassette 3–12 ml capacity, 20 kDa cut-off (Thermo Scientific Pierce 66012).

16. 50 mM Glycine elution buffer for affinity chromatography of AAV pH 2.7 (Table 3).

2.2. Assaying Virus Stocks

2.2.1. Alkaline Gel Electrophoresis

1. Restriction Digest Buffer 3 (New England Biolabs, B7003S).

2. 0.5 M EDTA.

3. Proteinase K (New England Biolabs P8102S).

4. 10% SDS.

5. Glycogen.

6. Electrophoresis grade agarose.

7. Gel casting tray.

8. Horizontal electrophoresis tank and power supply.

9. Ethidium bromide.

10. UV Transilluminator with camera.

11. Buffers for denaturing alkaline gel electrophoresis (Table 4).

2.2.2. Coomassie Staining of Protein Gels

1. Pre-cast 10% acrylamide denaturing gel, with running buffers.

2. 2× Laemmli protein loading buffer.

Table 4
Buffers for denaturing alkaline gel electrophoresis

50× Alkaline buffer		Alkaline sample loading buffer	
Reagent	Final concentration	Reagent	Final concentration
NaOH	2.5 M	NaOH	0.4 M
EDTA	50 mM	EDTA	5 mM
		Ficoll	18% (w/v)
		Xylene cyanol	0.01% (for colour only; mass used need not be accurate)

NB: Alkaline sample loading buffer should be stored at 4°C and kept for no more than 1 week

Table 5
Buffers for coomassie staining of protein gels

Coomassie blue protein gel stain		Destain solution	
Methanol	500 ml	Methanol	500 ml
Water	400 ml	Water	1,360 ml
Acetic acid	100 ml	Acetic acid	140 ml
Coomassie R-250	2.5 g		

3. Vertical electrophoresis tank and power supply.

4. Shallow tray to hold gel.

5. Rocking platform.

6. Buffers for coomassie staining of protein gels (Table 5).

2.2.3. Real-Time PCR

1. SYBR green Master Mix (e.g. Applied Biosystems 4385612).

2. DNA- and nuclease-free water.

3. Custom primers.

3. Methods

3.1. Three-Plasmid Transfection

3.1.1. Plasmids

Successful replication of AAV requires the presence of key genes from a viable "helper" virus, typically adenovirus. The use of adenovirus as a helper in the production of AAV is a well-established and efficient technique. However, ensuring that these AAV stocks are free from adenoviral contamination is more challenging and may present hurdles in the progression of novel AAV vectors from the bench through animal studies and to large-scale GMP production for the clinic. Therefore, a production system that avoids the use of any competent virus is to be preferred.

In the three plasmid transient transfection system, no viable helper viruses or wild-type AAV are involved in the production process, removing the challenge of preventing contamination with the helper virus and reducing the associated risk to the patient. Instead of a viable helper virus, its essential functions are provided by plasmid bearing key adenovirus genes such as XX6 (27) or HGTI (28). Each of these plasmids encodes the Adenoviral VA-RNAs and the E2A and E4 proteins, which together are sufficient to support efficient AAV replication and packaging.

Packaging of the vector is achieved by expressing the two AAV genes *rep* and *cap* from a second plasmid, e.g. XX2 (27) or pAAV5-2 (29). These genes are not flanked by the viral ITRs and therefore cannot be excised from the plasmids for packaging into new AAV particles. These genes provide the necessary AAV polymerase, endonuclease, and other proteins required for preparation of the viral genome for packaging, along with structural proteins to form the AAV capsid.

Finally, a third plasmid contains the genome which is to be packaged into the AAV vector. As the functions of *rep* and *cap* are provided in *trans*, the only viral elements needed in this modified genome are the AAV ITRs for packaging, flanking the researcher's promoter and transgene of choice.

The combination of these three plasmids—Helper, Packaging, and Genome—is sufficient for the production of viable AAV particles to high titre while eliminating the need for, or risk of, creating other contaminating viruses. The removal from the genome of all AAV genes except the ITRs also serves to maximise the space available for the insertion of transgenes.

A useful addition to this technique is the inclusion of a fourth plasmid designed to express a fluorophore such as GFP in the packaging cells, but which cannot itself be packaged into the vector. This plasmid is not involved in the virus production but serves as an indicator of transfection efficiency, useful for refining the transfection technique (see Note 1).

3.1.2. Packaging Cell Line   The 293 HEK cell line was originally derived from human embryonic kidneys, transformed by the insertion of nucleotides 1–4,344 from the adenovirus genome into chromosome 19q13.2 (30). This inserted sequence encodes the adenoviral immediate-early genes E1a, E1b, and Eb1, sufficient for further Ad gene transcription, progression of the cell into S phase and suppression of the apoptotic response.

The variant cell line 293 T is further transformed by the addition of a temperature-sensitive mutant of the SV40 large T antigen, which enhances the replication and maintenance of plasmids containing the SV40 origin of replication when grown at 37°C (31). Plasmids to be used for AAV production in 293 T cells should therefore be designed to include the SV40 origin of replication. This will improve the plasmid copy number in transfected cells and their progeny, and thus increase the yield of virus from the protocol.

3.1.3. Calcium Phosphate Transfection Protocol (See Note 2)

In this protocol, phosphate ions in a HEPES-buffered salt solution are mixed with $CaCl_2$ in the presence of plasmid DNA. Calcium and phosphate ions combine to form an insoluble precipitate with a net positive charge. DNA in the solution binds to these crystals which are then internalised by cells. The mechanism of this internalisation is not known in detail, although it has been suggested that the precipitate/DNA complex is first drawn into acidifying endosomes before DNA is transferred to the nucleus.

293 T cells should be grown in DMEM supplemented with 10% heat-inactivated FBS. Antibiotics may be used if desired: Penicillin and streptomycin at 100 units/ml and 100 μg/ml, respectively. Grow in plates or flasks suitable for adherent cells, at 37°C with 5% CO_2. Cells should be allowed to reach approximately 80% confluence before passage at a ratio of 1:3.

While transfections can in principle be performed on any scale, this protocol is optimised for transfection of 40×15 cm plates. This quantity of plates strikes a good balance between a manageable workflow and virus yield.

1. On the morning of transfection, plate 293 T cells at $6–7 \times 10^6$ cells per 15 cm plate so that they will be approximately 70% confluent.

2. Return to the incubator for 6–8 h, to allow the cells to attach to the plate surface.

3. In a sterile 250 ml conical flask, combine the plasmids (Table 6) and make up to 45 ml with sterile water. To this, add 5 ml $CaCl_2$ solution.

4. Mix using the vortexer. Allow a stable vortex to become established to ensure rapid and thorough mixing in the next steps.

5. While vortexing, add 25 ml of thawed 2× HBS dropwise at a fairly quick pace (approx 2 drops per second). Add the final 25 ml in a rapid stream.

6. While continuing to vortex the mixture, immediately and rapidly add 100 ml of DMEM with 10% FBS.

7. After 30 s to 1 min, remove the transfection mixture from the vortexer. Allow to stand for approx 2 min, to allow bubbles in the mixture to rise to the top.

Table 6
Plasmids for three-plasmid transfection of 40 plates

Plasmid	Amount per plate (μg)	Amount for 40 plates (μg)
HGTI	45	1,800
Packaging plasmid (e.g. pAAV2-8)	15	600
Genome	15	600
Plasmid encoding GFP	1	40

8. Gently add 5 ml of transfection mixture to each plate, being careful not to disturb the cell layer.

9. Swirl plates to mix transfection mixture evenly through the medium.

10. Return plates to incubator for approx 72 h.

3.1.4. PEI Transfection Protocol

Rather than calcium phosphate crystals, polyethyenimine (PEI) can be used to condense DNA out of solution and act as a carrier into targeted cells. Often used as an aid to cell attachment, this polymer adheres well to the cell surface and, when in aggregates, is internalised along with the bound DNA. Its relatively strong positive charge combined with the acidification of the endosome leads to an influx of anions. The resulting osmotic gradient across the endosome's membrane leads to an influx of water, bursting the endosome and allowing the PEI and bound DNA to escape to the cytoplasm. Further details of the DNA's transit to the nucleus are not well established, but extensive use has shown PEI to be an efficient transfection reagent, yielding transfection efficiencies and viral titres similar to the calcium phosphate method of transfection.

The protocol for PEI transfection is very similar to that for calcium phosphate transfection, as described above.

1. The day before transfection, plate 293 T cells at 6–7×10^6 cells per 15 cm plate so that they will be approximately 70% confluent.

2. On the day or transfection, combine 54 ml serum-free DMEM with 6 ml 0.1% PEI in a sterile bottle.

3. In a separate tube, combine the plasmids (Table 6) and make up to 62 ml with sterile water.

4. Filter (using a 0.2-μm PES filter) the plasmid mixture into the DMEM:PEI mixture and mix.

5. Incubate for 15 min.

6. Gently add 3 ml of transfection mixture to each plate, being careful not to disturb the cell layer.

7. Swirl plates to mix transfection mixture evenly through the medium.

8. Return plates to incubator for approx 72 h.

3.1.5. Harvesting Virus

1. Three days after transfection, the cells should appear confluent on the plate possibly with extensive cell rounding. The precipitate formed in the transfection mixture may be visible under the microscope, looking similar to clumps of cellular debris. The cell medium is typically an orange-pink colour, as may be expected from confluent cell culture plates.

2. Thoroughly scrape the cells from the surface of each transfected plate, collecting cells and media together into 4×250 ml conical tubes.

3. Also scrape cells and medium from one untransfected plate into a 50-ml falcon tube.

4. Pellet the scraped cells by centrifugation at $2,000 \times g$ for 10 min at 18°C.

5. If harvesting cells are transfected using the calcium phosphate precipitation method, aspirate and discard the supernatant. If cells have been transfected using PEI method, decant the media into a sterile vessel and store at 4°C until ready to purify virus. In either case, retain the pellets and proceed to step 6.

6. Resuspend the four pellets of transfected cells in 50 ml each of cold TD buffer, combining into one tube for a total of 200 ml.

7. Also resuspend the pellet of untransfected cells in 5 ml TD buffer.

8. If a reporter plasmid was included to track transfection efficiency, transfer 3 ml from each cell suspension—transfected and untransfected—into separate FACS tubes on ice. Assess reporter gene expression by flow cytometry when convenient, within 2 h of harvest.

9. Centrifuge the remaining transfected cell suspension at $2,000 \times g$ for 10 min at 18°C.

10. Resuspend the pellet in 40 ml TD buffer, in a 50-ml centrifuge tube.

11. To destroy the cell membranes and thus release the virus, perform five complete freeze–thaw cycles: [30 min at –80°C, 30 min at +37°C, vortex] $\times$ 5 (see Note 3).

12. To 40 ml of cell lysate, add 1 ml of 20% sodium deoxycholate to give a final concentration of 0.5%, and Bezonase nuclease to a final concentration of 50 units/ml. Incubate at 37°C for 30 min to allow degradation of cellular DNA and membranes to complete (see Note 4).

13. To clarify the lysate, spin at $4,000 \times g$ for 30 min at 4°C, then remove any remaining particulates by filtration through a 0.45-μm PES membrane. Virus may be purified from this lysate immediately, or the lysate may be stored at 4°C for up to 16 h.

14. If cells were transfected with the PEI method and medium retained, clarify the medium by centrifugation at $4,000 \times g$ for 30 min, followed by filtration through a 0.45-μm PES membrane. Virus may be purified from this medium immediately, or the medium may be stored at 4°C for up to 16 h.

3.2. Virus Purification

3.2.1. Gradient Centrifugation

Purification of AAV particles has traditionally been via CsCl gradient centrifugation. The difference in density between a packaged AAV particle (1.41 g/ml) and an adenovirus (1.34 g/ml) allows the efficient separation of AAV from contaminating or "helper" viruses used in production (32). Due to the difference in density between empty and properly packaged capsids, this process also allows the purification of AAV with a high ratio of packaged:empty vectors. However, achieving a high degree of purity required multiple rounds of purification. This lengthy exposure to high salt concentrations has been suggested to result in gradual deactivation of the vector, with a resulting decrease in the yield of infectious units per particle, although other groups report successful CsCl purification without this loss in infectious yield (33). Additionally, the limited capacities of laboratory ultracentrifuges make large-scale production with this method either impractically time consuming or—if more ultracentrifuges are bought and dedicated to the project—very expensive.

In order to avoid the possible loss of infectivity associated with high salt concentrations, an alternative protocol was developed based on iso-osmotic idoxinol medium density gradient centrifugation. The non-ionic and inert nature of idoxinol evades the risk of chemical damage to the virus particles and is amenable to direct use of the purified virus in additional purification steps, e.g. chromatography-based protocols such as heparin–agarose-based affinity chromatography (34), However, as with CsCl ultracentrifugation, this system has limited load capacity and is not easily scalable, making it inappropriate for production of virus in quantities necessary for use in the clinic (33) .

3.2.2. Purification by Ion Exchange Chromatography (See Notes 5 and 6)

The net surface charge of AAV particles causes the particles to bind to various ion exchange media in a pH- and salt-dependant manner. Examples of suitable exchange media include quaternary ammonium, diethylaminoethyl, or diethylaminopropyl substituted anion-exchange chromatography media equilibrated with low-salt, bis–Tris buffer at pH 6.0 (35). The precise concentration of salt required to disrupt this interaction—and therefore to elute the virus from the column—is unique to each combination of virus capsid and column medium. Applying a linear concentration gradient of NaCl to the column of 10–500 nM (in bis–Tris binding buffer) over 2.5 column volumes allows elution of the virus at the lowest possible salt concentration in an acceptably small fraction of the eluate. The rate of elution of virus from the column may be observed by monitoring the absorbance of the eluate at 280 and 260 nm, at which wavelengths an increase in absorbance can be taken to indicate an increase in viral capsids (protein) and genomes (DNA), respectively (35, 36).

While this technique is relatively simple and easily scalable by increasing column volumes, virus purified using this technique often requires additional purification steps, due to concerns regarding contamination of the eluate by cellular proteins which happen to have the same isoelectric point of the virus particles surface domains. One such step commonly used is gel filtration chromatography in which a gel filtration medium separates proteins by size. This second stage of purification, often referred to as a "polishing" stage, permits the isolation of packaged AAV particles free from contamination with cellular proteins (35).

3.2.3. Affinity Chromatography (See Note 6)

Based on an improving the understanding of the binding activities of various AAV serotypes, affinity media can be designed to specifically select for AAV particles in a one-step process, rather than involving multiple steps of centrifugation and/or selection by isoelectric point and physical size.

Based on AAV2's interaction with heparin sulphate proteoglycan, Heparin affinity chromatography was developed for purification of AAV2 particles, and has been demonstrated as suitable for use as a purification step following idoxinol centrifugation or as a one-step purification technique (37–39). For purification of vectors based on AAV5, which does not interact with heparin, an affinity medium based on the sialic acid-rich protein mucin bound to CNBr-activated Sepharose has been shown to be suitable for single-step affinity purification (32, 40).

AVB Sepharose High Performance (GE Healthcare) is another commercially available affinity medium, suitable for one-step purification of AAV, AAV2, and AAV5 (34, 41). Extensive experience in our laboratory has also confirmed its suitability for production of AAV8. This protocol is optimised for approximately 5 ml of packed AVB Sepharose, but is readily scalable (see Note 7).

1. Equilibrate the affinity medium using five column volumes of filtered PBS, pH 7.5 at room temperature. Monitor the pH of the flow-through emerging from the column, which should stabilise at pH 7.5. Also monitor the absorbance of the flow-through at 260 and 280 nm; once the pH and absorbance readings have stabilised, zero the spectrophotometer on these readings.

2. If purifying virus from clarified cell lysate (from "0 Harvesting virus"), dilute with two volumes of filtered PBS.

3. To load virus onto column, flow lysate through packed affinity medium at a rate of 5 ml/min.

4. Monitor the flow-through. The pH should remain within the range of 7–8. Absorbance at 260 and 280 nm should increase significantly as unbound protein and DNA exit the column. No detectable virus is expected in the flow-through, but consider collecting aliquots throughout this step to assist with troubleshooting.

5. Wash off unbound protein and DNA by flowing three column volumes of filtered PBS through the affinity medium at 5 ml/min.

6. Monitor the flow-through. The pH should remain within the range of 7–8. Absorbance at 260 and 280 nm should decrease and stabilise at approximately zero as the last of the unbound protein and DNA are washed from the column. No detectable virus is expected in the flow-through, but consider collecting aliquots throughout this step to assist with troubleshooting.

7. Elute the bound virus from the column by flowing two column volumes of filtered glycine pH 2.7 through the affinity medium at 5 ml/min. Collect the flow-through in 1 ml fractions and monitor the pH and absorbance at 280 and 260 nm of the flow-through.

8. Examine the pH and absorbance data to identify the fractions containing purified virus: a decrease in pH from approx 7.5 to approx 2.7 should be concomitant with simultaneous sharp peaks in the absorbance at 260 nm (indicating the presence of viral genomes) and 280 nm (indicating presence of viral capsids). Typical data are shown in Fig. 2, in which the pH (pale grey line) can be seen falling to approximately 2.9, accompanied by peaks of approximately equal height in absorbance at 260 and 280 nm.

9. Pool the fractions containing virus. To restore the pH of the eluate and remove glycine, dialyse overnight in ≥1 l filtered PBS at 4°C using a 10-kDa cut-off membrane (see Note 8).

3.3. Storing Virus Stocks

Virus stocks should be sterilised by filtration through 0.2-μm membrane filters, to prepare them for use in vitro or in vivo, but also to prevent the growth of moulds or bacterial cultures in the stock when stored above freezing point.

When sterile, virus stocks may be stored at +4°C for long periods (up to a year), or aliquoted and stored at –80°C for longer periods.

3.4. Confirming Self-complementarity of sc Vectors (See Note 9)

After the production of an scAAV vector, it is advisable to confirm that the genome packaged into the vector is, as intended, a short double-stranded hairpin structure rather than a short ssDNA or a long, non-folded ssDNA.

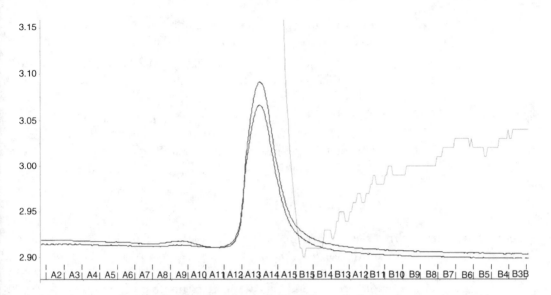

Fig. 2. Example pH (shown in *grey*) and absorbance (280 nm shown in *blue* and 260 nm shown in *red*) data during elution of virus from AVB Sepharose affinity chromatography column.

This confirmation can be achieved by extracting the viral genome from purified virus particles and running them on two agarose gels: one under denaturing alkaline conditions, one under non-denaturing conditions. Under alkaline conditions, the hairpin structure of the scAAV genome is forced to unfold and therefore runs with an apparent size of its full length. Under non-denaturing conditions, the DNA maintains its folded structure and therefore runs with an apparent size of half its full length. Packaged viral DNA (viral genome copy number, vg) must first be isolated from AAV particles, then run under these two conditions.

1. To 1×10^{11} vg of virus, add 4 μl DNAseI in a suitable buffer (e.g. restriction digest buffer 3, NEB) and add nuclease free water to a final volume of 200 μl.

2. Incubate at 37°C for 1 h.

3. To deactivate DNAse, add 2 μL 0.5 M EDTA and heat at 65°C for 15 min.

4. To degrade the viral capsid, add 4 μl Proteinase K, 2 μL 10%SDS, and 0.5 μl Glycogen.

5. Incubate at 37°C for 2 h.

6. Divide mixture into two 100 μl aliquots and store on ice until ready to run gels.

7. Melt 1 g agarose into 98 ml water.

8. Allow to cool to 30–40°C, without setting.

9. Prepare the running buffer (10 ml of 50× alkaline running buffer in 490 ml water) and use with the alkaline gel from step 5 to set up an electrophoresis tank as normal.

10. In a separate gel tank, prepare a non-denaturing 1% agarose gel according to local protocol.

11. To one aliquot of freed viral DNA, add 20 μl 5× alkaline loading buffer, mix, and then load into the alkaline gel. Run this gel at 50 V.

12. To the second aliquot of freed viral DNA, add 20 μl 5× non-denaturing DNA loading buffer, mix, and then load into the non-denaturing gel. Run this gel at 200 V.

13. When each gel has run to completion, incubate in 200 ml TBE with 5 μg/ml ethidium bromide and visualise DNA by illumination with UV.

3.5. Quantifying Virus Stocks

3.5.1. By qPCR

In order to account for the possibility of having purified large quantities of free DNA or empty capsids, quantification of viral stocks should be based on the measurement of both DNA (viral genome copy number) and protein (number of capsids).

Viral genome copy number can be measured using qPCR. A small sample of the purified virus is serially diluted (10-, 100-, 1,000-fold) and quantified using primers designed for the vector in question. The standard curve should be serial dilutions of the plasmid encoding the genome, based on a copy number derived from the concentration of the plasmid solution and the mass of a single copy of the plasmid in question. The usual considerations when designing qPCR reactions, including assessing efficiency of the primers and providing suitable controls, should be observed.

3.5.2. By Coomassie Staining

To assess the quantity and integrity of viral proteins present in the purified stock, samples should be run on a denaturing agarose gel and visualised using coomassie staining. The three structural proteins VP1, VP2, and VP3 should be easily revealed and observed to be present at the correct sizes and relative intensities.

1. In a microcentrifuge tube, mix 20 μl of virus stock with 20 μl 2× protein loading buffer.

2. Heat to 95°C for 10 min, then briefly centrifuge to collect sample at the bottom of the tube.

3. Assemble a 10% acrylamide denaturing gel with wells >30 μl, in laemmli running buffer.

4. Load 30 μl of this denatured protein mixture onto a 10% acrylamide gel, alongside a protein ladder.

5. Run the gel under conditions suitable for the equipment and gel chemistry being used. Typical conditions for a 10% acrylamide mini-gel are 100 V for 45 min.

6. When the loading buffer has run to the bottom of the gel, end the run and transfer the gel to a plastic tray.

7. Cover the gel with approx 1 cm of coomassie blue staining buffer, cover the plate with cling film or similar, and incubate at room temperature on a rocker for 2–12 h.

8. Discard the blue stain buffer and gently rinse the gel twice with deionised water to wash away the remaining buffer.

9. Place a piece of rolled-up paper towel at the edge of the tray, then pour in destain buffer to a depth of approx 1 cm over the gel.

10. Cover with cling film or similar and incubate at room temperature on a rocker until bands are easily distinguishable from the background (4–12 h). Note that the tissue is acting as a sink for coomassie blue as it is freed from the gel; therefore, the tissue should be periodically replaced as it becomes saturated with dye.

4. Notes

1. *Monitoring transfection efficiency.* Given an efficient promoter controlling the reporter gene, transfected cells should begin to fluoresce within 24 h, allowing an early semiquantitative check of transfection efficiency by fluorescent microscopy. Fluorescing cells should be counted accurately by flow cytometry immediately after being harvested, to give a more accurate reading of transfection efficiency and thus help to trouble-shoot transfection technique. This count should weakly correlate with the viral yield, although the influence of other factors (principally age, health, and confluence of producer cells along with quality and purity of the plasmid stocks) prevents it from being a reliable predictive factor.

2. *Calcium phosphate precipitation.* The size of the precipitate formed during this reaction has a large effect on the transfection efficiency, and therefore on the eventual virus yield. The reaction conditions must therefore be controlled carefully. Experience in our laboratory suggests that:

   • Reagents (excluding the complete medium) should be kept on ice before use

   • The speed at which 2× HBS is added to the plasmid + calcium mixture should be kept as described

   • Establishing a stable vortex—as opposed to simply agitating the mixture—is important to ensure rapid and thorough mixing of the reagents

   • Addition of complete medium to the mixture to stop the precipitation reaction should be as immediate and rapid as possible.

3. *Freeze–thawing harvested cell pellets.* Pellets may be resuspended into a smaller volume of TD buffer if desired. This will speed the freeze/thaw cycles, allowing a faster progression to the purification steps. However, after free/thawing, TD buffer should be added to bring the total volume of the mixture to 40 ml before adding sodium deoxycholate and Benzonase.

4. *Sodium deoxycholate.* This solution should be clear and straw-yellow in colour. It should be protected from light; loss of its colour indicates that the sodium deoxycholate has begun to break down and that a fresh batch should be made. Before use, check whether it has begun to precipitate out of solution and re-dissolve if necessary by warming to 37°C.

5. *Ion exchange chromatography.* While a full protocol for purifying AAV by this method is outside the scope of this chapter, it should be noted that the strongly polar nature of sodium deoxycholate interferes with ion exchange chromatography methods for purification of AAV, as does Benzonase nuclease. When preparing clarified lysate for purification by ion exchange chromatography, octylglucopyranoside (OGP) should be used instead of DOC, at a final concentration of 1.5%, and Benzonase nuclease should not be used (42).

6. *Affinity purification of virus from cell medium.* Purification of AAV from the media of transfected cells in addition to the cell pellet is desirable to maximise virus yield, as virus shed into the medium of transfected cells has been reported to comprise 30–70% of the total virus produced. However, experience in our laboratory suggests that transfecting cells using calcium phosphate precipitation markedly increases the viscosity of the medium, even after filtration through a 0.45 μm PES membrane. Within the pressure limits imposed by the chosen HPLC system and chromatography columns, this viscous liquid may only be filtered at a flow rate so low as to be impractical for most purposes. In contrast, medium from cells transfected using the PEI method does not have increased viscosity, making purification of AAV from this medium markedly more practical.

7. *Identifying virus in eluate.* As seen in Fig. 2, the drop in pH is accompanied by a sharp peak in absorbance at 260 and 280 nm, indicating high concentrations of DNA and protein being released from the column. Ideally, the two peaks will be of approximately equal magnitude, indicating a high proportion of correctly packaged vectors. A disproportionately high absorbance at 280 nm suggests a relatively high proportion of empty capsids.

In the example shown in Fig. 2, tubes A13–A15 would be collected for dialysis and purification. While the fractions to be collected remain fairly consistent between purification runs, minor changes do occur, presumably due to minor variation in factors such as pH and temperature of reagents, temperature, state of the chromatography medium, and viral capsid being used. Therefore, it is advisable to consult the absorbance data before collecting fractions rather than establishing a perfectly repetitive collection system.

8. *Care for chromatography columns.* Once packed into chromatography columns, affinity chromatography media such as Mucin or AVB Sepharose should not be allowed to dry out. Take care to avoid introducing bubbles into the column and, when the column is not in use, to prevent evaporation of liquid from the column medium by sealing the ends. The surface of the medium visible through the column wall should appear smooth; a cracked, pitted, or gritty surface indicates trapped air, which will significantly degrade column performance.

    To prevent contamination between runs, degradation of the chromatography medium and microbial growth, the affinity medium should be washed with five volumes of filtered PBS after each use and stored at 4°C. Check the medium for discolouration before each run (Mucin and AVB Sepharose should both be pure white) and consider replacing it if noticeable discolouration occurs.

9. *Alkaline gel electrophoresis.* After disrupting the viral capsid, an attempt may be made to purify the viral DNA from the denatured protein using DNA chromatography columns such as the QIAGEN PCR cleanup kit. However, experience in the authors' laboratory suggests that, for this protocol only, recovery of DNA from this step is unreliable, presumably due to reagents used elsewhere in this protocol conflicting with the chemistry of the chromatography system. Allowing the digested DNA and denatured proteins to remain in the mixture to be run on the gels adds a small but acceptable amount of background smearing to the resulting images. Special attention must be paid to the temperature of the agarose solution before adding the alkaline buffer. If the solution is too hot, the agarose will be degraded by alkaline hydrolysis, resulting in disruption of the pore structure of the gel and therefore interfering with the running properties of the gel. Ensure that cooling of the gel is uniform, to avoid pockets of alkaline hydrolysis in hot spots and to avoid premature setting of the gel in cool spots.

    The alkaline gel is typically more brittle than an ordinary non-denaturing gel and should be handled with care. Also, samples will also run considerably more slowly than on a non-denaturing gel; consider running the alkaline gel overnight at 30 V.

References

1. Mueller C, Flotte TR (2008) Clinical gene therapy using recombinant adeno-associated virus vectors. Gene Ther 15(11):858–63

2. Heilbronn R, Weger S (2010) Viral vectors for gene transfer: current status of gene therapeutics. Handb Exp Pharmacol 197:143–70

3. Walther W, Stein U (2000) Viral vectors for gene transfer: a review of their use in the treatment of human diseases. Drugs 60(2):249–71

4. Vandenberghe LH, Wilson JM, Gao G (2009) Tailoring the AAV vector capsid for gene therapy. Gene Ther 16(3):311–9

5. Choi VW, McCarty DM, Samulski RJ (2005) AAV hybrid serotypes: improved vectors for gene delivery. Curr Gene Ther 5(3):299–310

6. Wu Z, Yang H, Colosi P (2010) Effect of genome size on AAV vector packaging. Mol Ther 18(1):80–6

7. Koczot FJ, Carter BJ, Garon CF, Rose JA (1973) Self-complementarity of terminal sequences within plus or minus strands of adenovirus-associated virus DNA. Proc Natl Acad Sci USA 70(1):215–9

8. Goncalves MA (2005) Adeno-associated virus: from defective virus to effective vector. Virol J 2:43

9. Zhang H, Xie J, Xie Q et al (2009) Adenovirus-adeno-associated virus hybrid for large-scale recombinant adeno-associated virus production. Hum Gene Ther 20(9):922–9

10. Smith RH (2008) Adeno-associated virus integration: virus versus vector. Gene Ther 15(11):817–22

11. Allocca M, Doria M, Petrillo M et al (2008) Serotype-dependent packaging of large genes in adeno-associated viral vectors results in effective gene delivery in mice. J Clin Invest 118(5):1955–64

12. Lai Y, Yue Y, Duan D (2010) Evidence for the failure of adeno-associated virus serotype 5 to package a viral genome > or = 8.2 kb. Mol Ther 18(1):75–9

13. Dong B, Nakai H, Xiao W (2010) Characterization of genome integrity for oversized recombinant AAV vector. Mol Ther 18(1):87–92

14. Schmidt M, Voutetakis A, Afione S et al (2008) Adeno-associated virus type 12 (AAV12): a novel AAV serotype with sialic acid- and heparan sulfate proteoglycan-independent transduction activity. J Virol 82(3):1399–406

15. Cearley CN, Vandenberghe LH, Parente MK et al (2008) Expanded repertoire of AAV vector serotypes mediate unique patterns of transduction in mouse brain. Mol Ther 16(10):1710–8

16. Wu Z, Asokan A, Grieger JC et al (2006) Single amino acid changes can influence titer, heparin binding, and tissue tropism in different adeno-associated virus serotypes. J Virol 80(22):11393–7

17. Zincarelli C, Soltys S, Rengo G, Rabinowitz JE (2008) Analysis of AAV serotypes 1-9 mediated gene expression and tropism in mice after systemic injection. Mol Ther 16(6):1073–80

18. Asokan A, Conway JC, Phillips JL et al (2010) Reengineering a receptor footprint of adeno-associated virus enables selective and systemic gene transfer to muscle. Nat Biotechnol 28(1):79–82

19. Gray SJ, Blake BL, Criswell HE et al (2010) Directed evolution of a novel adeno-associated virus (AAV) vector that crosses the seizure-compromised blood-brain barrier (BBB). Mol Ther 18(3):570–8

20. Grimm D, Lee JS, Wang L et al (2008) In vitro and in vivo gene therapy vector evolution via multispecies interbreeding and retargeting of adeno-associated viruses. J Virol 82(12):5887–911

21. Davidoff AM, Ng CY, Zhou J et al (2003) Sex significantly influences transduction of murine liver by recombinant adeno-associated viral vectors through an androgen-dependent pathway. Blood 102(2):480–8

22. Nathwani AC, Cochrane M, McIntosh J et al (2009) Enhancing transduction of the liver by adeno-associated viral vectors. Gene Ther 16(1):60–9

23. Duan D, Yue Y, Yan Z et al (2000) Endosomal processing limits gene transfer to polarized airway epithelia by adeno-associated virus. J Clin Invest 105(11):1573–87

24. Monahan PE, Lothrop CD, Sun J et al (2010) Proteasome inhibitors enhance gene delivery by AAV virus vectors expressing large genomes in hemophilia mouse and dog models: a strategy for broad clinical application. Mol Ther 18(11):1907–16

25. Yan Z, Zak R, Zhang Y et al (2004) Distinct classes of proteasome-modulating agents cooperatively augment recombinant adeno-associated virus type 2 and type 5-mediated transduction from the apical surfaces of human airway epithelia. J Virol 78(6):2863–74

26. Douar AM, Poulard K, Stockholm D et al (2001) Intracellular trafficking of adeno-associated virus vectors: routing to the late endosomal compartment and proteasome degradation. J Virol 75(4):1824–33

27. Xiao X, Li J, Samulski RJ (1998) Production of high-titer recombinant adeno-associated

virus vectors in the absence of helper adenovirus. J Virol 72(3):2224–32

28. Streck CJ, Dickson PV, Ng CY et al (2005) Adeno-associated virus vector-mediated systemic delivery of IFN-beta combined with low-dose cyclophosphamide affects tumor regression in murine neuroblastoma models. Clin Cancer Res 11(16):6020–9

29. Chiorini JA, Kim F, Yang L, Kotin RM (1999) Cloning and characterization of adeno-associated virus type 5. J Virol 73(2):1309–19

30. Louis N, Evelegh C, Graham FL (1997) Cloning and sequencing of the cellular-viral junctions from the human adenovirus type 5 transformed 293 cell line. Virology 233(2):423–9

31. Rio DC, Clark SG, Tjian R (1985) A mammalian host-vector system that regulates expression and amplification of transfected genes by temperature induction. Science 227(4682):23–8

32. Burova E, Ioffe E (2005) Chromatographic purification of recombinant adenoviral and adeno-associated viral vectors: methods and implications. Gene Ther 12(Suppl 1):S5–17

33. Kohlbrenner E, Aslanidi G, Nash K et al (2005) Successful production of pseudotyped rAAV vectors using a modified baculovirus expression system. Mol Ther 12(6):1217–25

34. Zolotukhin S, Potter M, Zolotukhin I et al (2002) Production and purification of serotype 1, 2, and 5 recombinant adeno-associated viral vectors. Methods 28(2):158–67

35. Smith RH, Ding C, Kotin RM (2003) Serum-free production and column purification of adeno-associated virus type 5. J Virol Methods 114(2):115–24

36. Anderson R, Macdonald I, Corbett T et al (2000) A method for the preparation of highly purified adeno-associated virus using affinity column chromatography, protease digestion and solvent extraction. J Virol Methods 85(1–2):23–34

37. Zolotukhin S (2005) Production of recombinant adeno-associated virus vectors. Hum Gene Ther 16(5):551–7

38. Clark KR, Liu X, McGrath JP, Johnson PR (1999) Highly purified recombinant adeno-associated virus vectors are biologically active and free of detectable helper and wild-type viruses. Hum Gene Ther 10(6):1031–9

39. Zolotukhin S, Byrne BJ, Mason E et al (1999) Recombinant adeno-associated virus purification using novel methods improves infectious titer and yield. Gene Ther 6(6):973–85

40. Auricchio A, O'Connor E, Hildinger M, Wilson JM (2001) A single-step affinity column for purification of serotype-5 based adeno-associated viral vectors. Mol Ther 4(4):372–4

41. Smith RH, Levy JR, Kotin RM (2009) A simplified baculovirus-AAV expression vector system coupled with one-step affinity purification yields high-titer rAAV stocks from insect cells. Mol Ther 17(11):1888–96

42. Kaludov N, Handelman B, Chiorini JA (2002) Scalable purification of adeno-associated virus type 2, 4, or 5 using ion-exchange chromatography. Hum Gene Ther 13(10):1235–43

Chapter 7

Vector Systems for Prenatal Gene Therapy: Principles of Non-viral Vector Design and Production

Suet Ping Wong, Orestis Argyros, and Richard P. Harbottle

Abstract

Gene therapy vectors based on viruses are the most effective gene delivery systems in use today and although efficient at gene transfer their potential toxicity (Hacein-Bey-Abina et al., Science 302:415–419, 2003) provides impetus for the development of safer non-viral alternatives. An ideal vector for human gene therapy should deliver sustainable therapeutic levels of gene expression without affecting the viability of the host at either the cellular or somatic level. Vectors, which comprise entirely human elements, may provide the most suitable method of achieving this. Non-viral vectors are attractive alternatives to viral gene delivery systems because of their low toxicity, relatively easy production, and great versatility. The development of more efficient, economically prepared, and safer gene delivery vectors is a crucial prerequisite for their successful clinical application and remains a primary strategic task of gene therapy research.

Key words: Non-viral, Minicircle, Plasmid, S/MAR, PEI, Chitosan, Gene delivery, In utero, Gene expression

1. Introduction

Gene delivery vectors comprising non-viral components have been utilized and developed alongside their viral counterparts (1). In addition to reduced pathogenicity and the lower risk of insertional mutagenesis, non-viral vectors are relatively easy to produce at a large scale and low cost. Non-viral vectors also provide flexibility of capacity allowing any size of gene and regulatory sequence to be incorporated without restriction.

However, given their relative inefficient gene transfer, many studies focus on the development of techniques that can be applied to improve the efficiency of gene expression and the targeting of cells in biological environments. Non-viral vectors must be designed

Charles Coutelle and Simon N. Waddington (eds.), *Prenatal Gene Therapy: Concepts, Methods, and Protocols*,
Methods in Molecular Biology, vol. 891, DOI 10.1007/978-1-61779-873-3_7, © Springer Science+Business Media, LLC 2012

to overcome a range of physiological barriers. They must be sufficiently stable after administration to permit bulk delivery into a tissue of interest. This requires protection from potential bloodstream shear-forces, enzymatic degradation, and the ability to gain access to cells after exposure to host interstitial components, such as albumin and antibodies. Once inside the target cell, additional requirements for transfection include resistance to cytoplasmic degradation and passage through the double-membrane structure of the nuclear envelope, which encapsulates the nucleus.

Here, we describe the principles of non-viral vector technology and provide detailed protocols for the generation of minimally sized, persistently expressing, episomally maintained DNA vectors. Additionally, we describe a variety of procedures for their delivery in vivo.

1.1. Principles of Non-viral Transfection—Gene Delivery Vectors

1.1.1. Use of Naked DNA

The simplest form of non-viral vector is unformulated or naked genetic material. Wolff et al. demonstrated that a direct injection of plasmid DNA (pDNA) into skeletal muscle could result in strong marker gene expression (2). Low levels of gene expression were also detected following direct pDNA injection intraportally for liver gene transfer (3, 4), direct cutaneous injection for skin transduction (5), or airway instillation for lung expression (6). Furthermore, transfer of naked pDNA to the heart by intramyocardial injection (7) and to the brain by intracerebral injection (8) appears promising. However, a broad application of unformulated pDNA may not be feasible as expression is limited only to cells adjacent to the site of injection, and due to its large size and hydrophilic nature, DNA is efficiently kept out of cells by physical barriers.

1.1.2. Gene Transfer by Physical Methods

Physical approaches of gene transfer induce transient permeabilization of the cell membrane so that DNA can enter the cells by diffusion, electrical current, or pressure. Gene delivery methods currently employed use mechanical methods (by particle bombardment or gene gun delivery), electric methods (by electroporation), ultrasound, or hydrodynamic transfer.

Gene Gun

Ballistic transfection can be used for gene transfer to superficial cell layers of the skin and to tissues such as liver and skeletal muscle that are easy to expose surgically. In this procedure, pDNA is adsorbed onto the surface of gold or tungsten particles, which are accelerated to a high velocity to penetrate cell membranes. This method allows relatively efficient transfer, as it bypasses the barriers associated with cell entry. However, only shallow penetration of the particles is achievable (less than 0.5 mm in mouse skeletal tissue) (9). Further improvements include chemical modification of the surface of the particles to allow higher capacity and better consistency for DNA coating and fine tuning of the acceleration imparted by the gene gun to control the final DNA distribution (10).

Ultrasound-Mediated Transfection

Another approach for plasma membrane permeabilization is the application of focused ultrasound pulses at the site where gene transfer is desired, which is found to facilitate gene transfer of DNA (11). Exposure to low-intensity ultrasound increases cell membrane permeabilization and studies have shown increased gene expression in skeletal muscle and the carotid artery when used in conjunction with naked DNA (12, 13). When combined with ultrasound exposure, microbubble-mediated gene delivery of DNA achieved a 300-fold greater expression in vitro compared to that of naked DNA alone (14). Although most studies have used this technique mainly in vitro or ex vivo, the general flexibility, safety, and repeatability of ultrasound delivery make this a promising technique for future clinical application.

Electroporation

Electroporation is a versatile method that involves plasmid-based gene transfer accompanied by the application of a series of electric pulses. In vivo electroporation was first demonstrated in skin and muscle, and readily exposed tissue like the liver. Muscle, in particular, has proven to be a particularly suitable tissue for this application due to long-term maintenance of pDNA in episomal form, as demonstrated by successful transfer of a dystrophin gene to the mdx mouse model of Duchenne muscular dystrophy (15, 16).

Hydrodynamic Delivery

Hydrodynamic gene delivery uses a hydrodynamic force generated by a pressurized injection of a large volume of DNA solution into the blood vessel to drive DNA molecules out of the blood circulation into the tissue by temporarily forcing enlargement of endothelial fenestrae. The overall efficiency of this technique depends on the capillary structure, the architecture of the cells surrounding the capillary, and the hydrodynamic force applied to the interior of the vasculature. This method has been used primarily to deliver transgenes to the liver (17, 18). It has also been applied to muscle tissue by rapid injection of pDNA into the femoral artery (19). The standard procedure involves a tail vein injection within 5–7 s of physiological solution, equivalent to 10% of body weight. The high volume of DNA solution enters the inferior vena cava stretching myocardial fibres beyond their capability, causing a momentary cardiac congestion and driving the solution back into the hepatic vein (20).

Animals survive the hydrodynamic procedure well, and its utility has been explored for gene delivery into various other tissues (such as the kidney, myocardium, and tumour cells) and on other species (such as fish, chicken, monkey, pigs, and even humans).

1.1.3. Gene Transfer by Chemical Carriers

Prior to the development of physical methods of gene transfer, systemic delivery of naked DNA tended to result in its rapid degradation by nucleases and clearance by the mononuclear phagocyte system (21). Additionally, the phosphate group on the deoxyribose groups of DNA confer it a net negative charge, limiting the potential for electrostatic interactions with anionic lipids in the cell membrane. An injection of pDNA into the tail vein of mice without hydrodynamic pressure results in the complete absence of detectable gene transfer (22). Localized administration of the vector may significantly increase the concentration of the vector in proximity of the target cell, such as pulmonary administration for lung gene transfer and hepatic vein injection for liver gene transfer, although this is not always possible due to difficulty with access for certain tissues.

To protect DNA and permit successful gene administration, it can be complexed with positively charged lipids or polymers. DNA is compacted into particles formed by the neutralization of its atomic charge and becomes surrounded by a chemical moiety that serves to protect it from degradation by nucleases, increases its in vivo stability, facilitates cell entry, and assists in intracellular trafficking. Compacted DNA can be subdivided into two categories called lipoplexes or polyplexes depending on whether a lipid or polymer was used in its formulation.

Liposomes

The earliest description of a chemical carrier that effectively binds and delivers DNA to cultured cells was a cationic liposome formulation, called Lipofectin™, reported in 1987 (23). Since then, hundreds of new cationic lipids have been developed which differ in the number of charges in their hydrophilic head group and by the detailed structure of their hydrophobic moiety (24). Transfection efficiency of lipoplexes is affected by the chemical structure of the cationic lipid, the charge ratio between the cationic lipid and DNA, and the structure and proportion of the helper lipid in the complexes. In addition to determining the structure, the complex charge, and transfection activity, these factors also determine the overall toxicity to target cells and the susceptibility of the cells to a particular lipid-based transfection reagent.

Lipoplexes form spontaneously when cationic liposomes are mixed with DNA. The process involves an initial association between the polycationic liposomes and the polyanionic DNA through electrostatic interactions followed by a slower lipid rearrangement process. Once delivered, the structure then undergoes a process of "lipofection", where it fuses with the cell membrane and is internalized by endocytosis (25).

Lipoplexes do, however, appear to be immunogenic, although at a lesser extent than viral vectors. Nonetheless, they can elicit an immune response by the toll-like receptor 9 (TLR-9), which causes rapid induction of proinflammatory cytokines (26, 27). This has been reported to lead to vector-associated toxicity and rapid elimination of the transfected cell and gene expression (28). For these reasons, cationic lipids are predominantly used in in vitro application.

Polyplexes

Another class of synthetic vectors that have been actively studied are cationic-polymer-based gene delivery systems, which are particularly efficient in condensing and protecting DNA. Many different types of polymer have been evaluated as gene delivery vehicles, such as polyethylenimine (PEI), poly-L-lysine (PLL), and chitosan. Among them, PEI appears to be the most promising and has been used for in vivo gene transfer via many routes, such as lung instillation, kidney perfusion, intracerebral injection, and intravenous administration (25). In addition to its ability to compact DNA into easily delivered particles, prevent degradation, and enhance uptake to target cells, PEI exhibits relatively high gene transfer activity due to its intrinsic endosome-buffering property, which facilitates release of complexes into the cytoplasm. Furthermore, it has also been proposed that the multiple positive charges on PEI mimic nuclear localizing signals, which enhance nuclear entry. In support of this, studies have shown PEI complexes to be less dependent on cell division to achieve gene expression compared to lipid- or PPL-based complexes (29). There are two different forms of PEI polyplexes, linear and branched, which differ in their structure and molecular weight. Linear PEI has been shown to be less inflammatory than branched PEI following systemic administration, and the lack of detectable neutralizing antibodies suggests that repeated administrations are possible (30). One such commercially available linear PEI polymer is jetPEI™, which has been shown to transduce a variety of tissues successfully in vivo (31–33).

Additionally, PEI can be readily targeted with a conjugated ligand leading to an increased level of uptake into receptor-positive cells. Therefore, PEI represents a versatile and promising approach to delivering DNA. A summary of the advantages and limitations of the currently available non-viral vector system is shown in Table 1.

1.2. Principles of Non-viral Transfection—The Genetic Component

Once the genetic component of a non-viral vector is successfully transfected into a target cell, the sustenance and efficiency of transgene expression is dependent on the composition of its DNA.

1.2.1. Gene Expression: Role of Promoters and Enhancers

The promoter is perhaps the key regulatory element that determines the efficiency of transgene expression from a pDNA vector. Within the promoter are binding sites for RNA polymerase and its associated factors that initiate transcription. An enhancer is a separate DNA sequence, which binds proteins other than RNA polymerase, such as transcription factors and chromatin-remodelling complexes, and acts upon promoters to enhance their transcription efficiency. Depending on the presence of associated upstream or downstream enhancer elements, the promoter can range from 100 to 2,000 bp in length.

Table 1
Advantages and limitations of current non-viral delivery systems
(Gao, 2007 #465 (11))

Method	Route of gene delivery	Advantages	Limitations
Direct injection	Intratissue	Simple and safe	Low efficiency
Gene Gun	Topical	Good efficiency	Tissue damage in some applications and limited working range of up to 100uM deep
Ultrasound	Topical Systemic	Good site-specific delivery	Low efficiency in vivo, a limited working range, and requires surgical procedure for internal applications
Electroporation	Topical Intratissue	High efficiency	Limited working range and requires surgical procedure for internal applications
Hydrodynamic delivery	Systemic Intravascular	Simple with high efficiency for liver gene delivery	Localized delivery in the organ of interest is required for large animals
Cationic lipids	Topical Intratissue Systemic Airway	High efficiency in vitro with low to medium efficiency for local and systemic delivery	Acute immune responses and limited activity in vivo
Cationic polymers	Topical Intratissue Systemic Airway	High efficiency in vitro with low to medium efficiency for local and systemic delivery	Toxic to cells and acute immune responses

In a pDNA vector, the promoter is not subject to the same controls as endogenous genes, making it difficult to predict the performance of the promoter simply by sequences. Both promoters and enhancers can be ubiquitously active or stringently tissue specific, depending on the expression profile of the binding factors required for their activity and the variety of both types of elements in vectors. The same promoter and enhancer that confer long-term expression in a viral vector may provide only transient expression in a plasmid. Many promoters are only able to confer short-term expression when contained in non-viral vectors, most likely due to promoter-independent loss of vector DNA from transduced tissue or because of transcriptional inactivation of the promoter.

The CMV Promoter

The promoter from the immediate early gene of human cytomegalovirus (CMV) is one of the most widely used promoters as it confers robust expression in most cell types tested (34). The strong activity of this promoter is due to the presence of multiple repeats of several transcription binding sites within the promoter and an upstream enhancer region (35). The expression profile of this promoter typically peaks 1–2 days following administration of the vector. Several reasons have been suggested to be behind this rapid inactivation, such as cytokine inhibition (36) and activation by repressor proteins that cause methylation (36, 37).

Constitutive Promoters

Promoters from genes that encode abundant cellular proteins, such as β-actin and elongation factor 1α (EF1α), provide better sustenance of transgene expression, although their expression is generally lower than that driven by the CMV promoter. For example, the EF1α promoter, from the gene encoding EF1α which is an abundant, widely expressed protein responsible for catalyzing GTP-dependent binding of tRNA to ribosomes, resulted in a tenfold lower marker gene expression compared to that conferred by CMV but declined much more gradually unlike CMV and was still detectable 4 weeks post instillation in the lung (38).

The UbC Promoter

More promising is the UbC promoter, which drives expression from three known cellular human ubiquitin genes UbA, UbB, and UbC. Ubiquitin is abundantly expressed in all eukaryotic cells and attaches covalently to abnormal, misfolded, or short-lived proteins marking them for destruction. The UbC promoter is shown to provide high-level ubiquitous expression when inserted into transgenic mice (39). When included in pDNA vectors, equally strong levels of reporter gene expression were detected in the lung for 2 months after instillation, but gradually declined and were lost 6 months after administration (38). A hybrid of the CMV-UbC promoter was able to provide sustained expression for up to 84 days after lung instillation and for 42 days following hydrodynamic delivery to the liver (40).

Enhancers

Different promoter–enhancer combinations have been incorporated into gene therapy vectors to improve their expression. Most native promoters contain one or two enhancer elements fused to a heterologous promoter sequences. To produce higher promoter activity, studies have investigated the possibility of combining the enhancer elements endogenously present in control regions of genes to promoters and the effect of enhancer/promoter combinations on transgene expression.

1.2.2. Post-transcriptional Regulation

While selection of a strong promoter is the first step to ensure high-level transcription of mRNA, the post-transcriptional events concerning mRNA processing can also be enhanced, either by increasing stability of this message after transcription, promoting its export from the nucleus, or by increasing the efficiency of translation. One approach is the inclusion of an intron sequence at the 5′ or 3′ end of the RNA of interest. Most naturally occurring eukaryotic genes are interrupted by introns and expression from these intron-containing transcripts is often higher than from the equivalent intron-deficient cDNA sequences (41). In some cases, expression is entirely dependent on the presence of an intron, for example β-globin (42).

Post-transcriptional Regulatory Elements

Another element capable of stimulating gene expression post-transcriptionally is called a post-transcriptional regulatory element (PRE). The hepadnavirus hepatitis B virus (HBV) encodes several intronless mRNAs and employs a *cis*-acting PRE RNA element for cytoplasmic accumulation of viral RNAs (43). This element called the HPRE contains two stem loops that confer its function. It compensates for the absence of introns by directing the viral transcripts into a Crm1-dependent cellular nuclear export pathway. The related hepadnavirus woodchuck hepatitis virus (WHV) possesses a similar element called the WPRE. In addition to the double stem loop in the HPRE, the WPRE contains a third functional domain (the γ element), which provides an increased enhancement of transgene levels, above that conferred by the HPRE (44). The WPRE has widely been used as a *cis*-acting regulatory module in various types of plasmid or viral gene vectors to increase transgene expression from gene therapy vectors.

1.2.3. Effect of CpG Depletion

The innate immune system is able to recognize pDNA principally by its higher frequency of unmethylated CpGs compared to mammalian DNA (45). Chen et al. showed that both the persistence and silencing of transgene expression were associated with specific increases in heterochromatin-associated histone modifications and a subsequent decrease in modifications associated with euchromatin (46). Unmethylated CpG motifs are a major contributor to the acute inflammatory response that occurs following administration of cationic-lipid pDNA complexes (47).

Typically, pDNA-mediated transgene expression peaks at 12–48 h after hydrodynamic delivery to the liver or after instillation to the lungs, but then drops rapidly lasting rarely beyond 1 week, even though vector DNA is retained in cells of these tissues (48–50).

Modifying non-viral vectors by reducing their unmethylated CpG content, e.g. by methylating or removing CpGs from the pDNA, can avoid immune responses. It has been reported that certain sequence motifs specifically inhibit the recognition and signal transduction of CpG oligo deoxynucleotides (for example replacing a GCGTT or ACGTT motif with GCGGG or ACGGG, respectively) converted a stimulatory CpG oligo deoxynucleotide to an inhibitory one that prevents apoptosis induced by stimulatory CpG oligo deoxynucleotides (51).

1.2.4. Gene Maintenance

A major limitation of non-viral vectors is the short duration of expression observed in many animal models. There may be several reasons for the loss of expression and a primary one is the reduction in the copy number of vector DNA itself. This can occur as a consequence of the removal of cells that have been lethally damaged during transfection, immune responses to cells expressing the transgene, and removal of pDNA that has been recognized as foreign by the cell in the nucleus.

1.2.5. Scaffold/Matrix Attachment Region Vectors: Structure and Function

Following gene delivery once the genetic material is introduced into a target cell, the efficiency and sustenance of gene expression are determined by the composition of the DNA. Ideally, a vector should be retained episomally, replicate once per cell cycle, and provide persistent transgene expression. Particular DNA motifs, which provide such properties, are the Scaffold/Matrix Attachment Region (S/MAR) elements.

S/MARs are DNA elements that mediate the attachment of chromatin to the nuclear scaffold, forming looped domains that participate in various matrix-based processes involving DNA replication and transcription, RNA processing, as well as transport and signal transduction (52).

A prototype S/MAR vector was originally described by the group of Professor Hans Lipps; the functional element on this plasmid is the S/MAR from the human β-interferon gene cluster called pEPI (see Fig. 1) (53).

The utility that this element provides to plasmid molecules is threefold:

1. Plasmids harbouring an S/MAR motif are rendered resistant to integration.

2. Their expression cassettes are not subject to epigenetic silencing (54).

3. They exhibit extrachromosomal, mitotic stability (55).

These properties are conferred without the need for virally encoded proteins or selective pressure (53, 56).

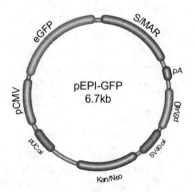

Fig. 1. Schematic representation of pEPI. Plasmid pEPI is shown with all the necessary elements for replication and maintenance.

Although S/MARs are evolutionary conserved, they have no consensus sequence but contain several recognizable motifs within and between species (57).

Insulator Function

The interaction between S/MARs and the nuclear matrix is important for the organization of chromosomal loops that define the boundaries of independent chromatin domains (55), thereby insulating coding regions from the surroundings and establishing local access of transcription factors to promoters and enhancers. The function of S/MARs as insulators has been shown in studies demonstrating that S/MARs enhance expression of a reporter gene following integration in a cellular chromosome only in stably transfected cell lines, but not in transient transfection assays (58, 59). This idea was reinforced by the "one gene-one S/MAR" hypothesis that predicts that each active gene has its own S/MAR element (60) and that S/MAR elements could effectively shield transgenes from position effects by creating independent domains. In addition, the presence of S/MARs was shown to establish independent domains in transgenic mice completely restoring correct hormonal regulation (61).

The idea that S/MARs could insulate from methylation effects that cause gene silencing emerged from immunoglobulin (Ig) gene studies. In most pre-B cells, the κ chain gene is methylated and transcriptionally inert, whereas in B and plasma cells the κ chain is hypomethylated and transcriptionally active. Studies showed that B cells transfected with Ig κ transgenes were only demethylated when S/MAR was present (62). Indeed, other studies found that the S/MARs are able to initiate transcription from CpG-methylated genes (63). Furthermore, S/MAR binds to chromatin-remodelling proteins such as Bright, SAF-A, and p300, which in turn affect histone acetylation and remodelling of nucleosomes (64).

Transcription Augmentation

S/MARs may also act in *cis* to increase transcription initiation rates, even in the absence of an enhancer. However, how these S/MARs switch on gene expression remains unclear. One suggestion is that the single chromatin loop formed by the attachment of S/MAR to chromatin is not immobile and switching on gene expression involves changing the attachment points of the loops to the nuclear matrix (65). This suggests that the S/MAR is able to regulate gene expression by mediating changes in the structure of chromatin, increasing the likelihood of establishing an active locus (66). Harraghy et al. reported that the inclusion of an S/MAR increases the probability of acquiring a permissive state while decreasing the occurrence of silencing events associated with transgene integration in chromosomes of mammalian cells (67).

In addition to their chromatin-remodelling activities, S/MAR-binding transcription factors may also contribute to transgene regulation by directly interacting with components of the general transcription machinery. S/MAR binds to ubiquitous nuclear matrix proteins such as SAF-B and SATB1, which in turn associate with RNA polymerases, thereby increasing the accessibility of the plasmid to a "transcription factory". Consistent with this are FISH analyses showing plasmid pEPI clusters in active sites located at the periphery of the chromosome, where the chromatin remains open with easy access to transcription complexes (68, 69).

Mitotic Stability

The molecular mechanisms behind the mitotic stability of pEPI in vitro remain unclear, but Stehle et al. showed that once the episome is established, it is efficiently maintained in active chromatin and associates with early replicating chromosomal sequences (70). In contrast, vectors without the S/MAR or in which transcription terminates upstream of the S/MAR can become integrated. Following integration, the loss of gene expression correlates with epigenetic silencing by promoter methylation (54). Jenke et al. suggested how the plasmid is segregated to daughter cells by showing that S/MAR binds to nuclear matrix proteins, such as topoisomerase II, lamin B1, SATB1, and histone H1. In particular, S/MAR was found to associate preferentially to SAF-A, a principal component of cellular chromatin and chromosomes. This suggests that during mitosis, pEPI interacts with mitotic chromosomes, mediated by the binding of S/MAR to SAF-A, enabling its co-segregation with the chromosomes during mitotic division (55). Furthermore, it is suggested that S/MAR recruits nuclear components in a manner similar to viral proteins like Tag and EBNA-1 that lead to helix destabilization forming an open chromatin domain at the origin of replication, which is necessary for assembly of the replicating machinery such as the origin of replication complex (ORC).

Random distribution of the pEPI in chromosomes prior to mitosis would suggest that during each division a significant percentage of cells would lose the episome. This is not observed and plasmid pEPI was shown to stably replicate at a low copy number (five to ten copies per cell), indicating that the vector molecules localize in specific regions on the host chromosome. This is supported by an almost equal distribution of pEPI in daughter nuclei after mitosis shown by FISH analysis, providing a mechanistic insight to how the S/MAR is able to achieve propagation in proliferating cells (70).

1.3. Minicircle Vectors—Principles of Design and Production

An alternative method to reduce unmethylated CpG content is to produce plasmids comprising only the expression cassette (promoter-transgene-polyA signal), thereby removing any elements not required for transcription of the transgene of interest. These constructs are called "minicircles" and contain neither the bacterial *ori* nor antibiotic resistance genes (the bacterial backbone).

Several groups have observed that by removing the extraneous bacterial sequences from a pDNA vector the minicircle vectors exhibited persistent expression and had a pattern of histone modifications consistent with euchromatin and sites of active transcription (48, 71–75).

The superiority of minicircles over that of standard pDNA vectors has been shown in vivo using reporter proteins (48, 76), expression of human manganese superoxide dismutase that protects from radiation (77), and other human proteins such as AAT (72) and vascular endothelial growth factors (78).

Several strategies have been developed to create minicircles, mainly utilizing site-specific recombination of integrase systems, such as φC31(72), λ integrase (73), Flippase (Flp) recombinase (74), or Cre recombinase (71). Essentially, the bacterial backbone of the pDNA is removed using a recombinase enzyme and two recombination sites, creating a closed covalently circular minicircle.

Initial pioneering experiments by Darquet et al. (73) and Bigger et al. (71) produced minicircle DNA vectors by utilizing λ integrase and Cre recombinase, respectively. This was followed by production of minicircle DNA using φC31 integrase by Chen et al. (72) and finally the use of Flp recombinase (74).

Here, we describe the three common methods used (those using Cre, φC31, and Flp recombinases). Readers interested in the λ integrase method are referred to ref. 73. A brief outline of the three production methods is provided in Fig. 2.

In principle, φC31 and Flp are more efficient compared to other recombinases such as Cre because they mediate a unidirectional reaction. Consequently, the intramolecular recombination can proceed to completion, resulting in higher yields of a unique population of circular expression cassette monomers.

The λ recombinase can mediate a unidirectional reaction in the absence of the Xis protein. The Xis protein is responsible for the eXicISion of a DNA piece, which is replaced by λ recombinase (equivalent to the Cre recombinase) and promotes recombination (79). However, small amounts of dimer were always seen in the preparation of minicircle DNA using λ integrase in the absence of Xis protein in the studies of Darquet et al. (73), and the mechanism underlying the presence of this contamination is unclear.

1.3.1. Cre/lox Recombination

This method is based on the homologous recombination between the lox recombination sites and is catalyzed by Cre recombinase. Cre mediates a bidirectional reaction, resulting in relatively small yields of minicircular DNA. Moreover, the product of the bidirectional reaction is a mix of circular molecules composed of a different number of repeated sequences, making it difficult to produce pure monomers. Despite these problems, several reports utilizing this recombinase have been published (48, 71, 76).

1.3.2. Construction of Mother Plasmid

Removal of bacterial sequences needs to be efficient, using the smallest possible excision site while creating supercoiled DNA

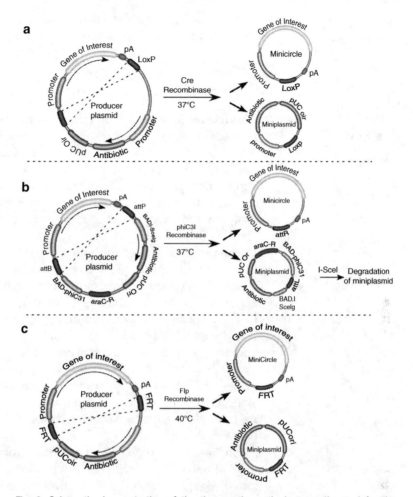

Fig. 2. Schematic demonstration of the three main methods currently used for the generation of minicircles. (a) Generation of minicircles using the Cre/Lox recombination method, where Cre recombinase catalyzes specific recombination of DNA between loxP sites. (b) Generation of minicircles using the φC31 recombinase, which catalyzes the site-specific recombination of two recognition sites that differ in sequence, typically known as attachment sites attB and attP. (c) Generation of minicircles using the Flp recombinase, which catalyzes recombination between the 34-bp-long Flippase Recognition Target (FRT) sites.

minicircles, consisting solely of gene expression elements under appropriate mammalian control regions.

This can be achieved by the use of Cre recombinase, a bacteriophage P1-derived integrase (80, 81), catalyzing site-specific recombination between direct repeats of 34-bp loxP sites. In the case of a supercoiled plasmid containing DNA flanked by two loxP sites in the same orientation, Cre recombination produces two DNA molecules that are topologically unlinked, circular, and mainly supercoiled (80), each containing a single 34-bp1 loxP site (see Fig. 2a). Efficient minicircle production requires the use of a stable bacterial based Cre expression system for efficient production of supercoiled

DNA. Bigger et al. (71) described such a system utilizing the tightly controlled arabinose expression system to create a Cre-expressing bacterial strain (called MM219Cre), which is both stable and easily controllable by altering the carbon source available for metabolism by these bacteria.

1.3.3. Minicircle Production Using φC31

Construction of Mother Plasmid

The recombinase φC31 is derived from Streptomyces temperate phage. It belongs to the resolvase/invertases family, and its intra-molecular recombination activity has been characterized thoroughly by Thorpe et al. (82). Groth and colleagues (83) have determined the minimal functional sizes of the attB and attP recombination sites as 34 and 39 bp, respectively. In the most common method used, Chen et al. placed the desired expression cassette between the attB and attP recombination sites and then they placed the φC31 under the control of arabinose promoter, outside the attB and attP sites (72). To prepare the attB and attP sites, they are amplified using the corresponding DNA oligonucleotides (83) and then inserted into the target sites of plasmid DNA, to flank the desired expression cassette, using restriction cut endonucleases.

The φC31-mediated intracellular recombination between attB and attP is induced in *Escherichia coli* Top 10 by the BAD promoter inducer L-arabinose. This results in two circular DNA molecules: one is the transgene expression cassette with a 36-bp hybrid attR, termed the minicircle, and the second is the circular miniplasmid, containing the bacterial backbone and the 37-bp hybrid attL (see Fig. 2b).

In order to simplify the production method and increase the yield of minicircle, Chen et al. placed the intron-encoded endonuclease gene, I-SceI, under the control of the BAD promoter as they did with the φC31 gene, and added an I-SceI recognition sequence outside the attB and attP sites of their plasmid (84). I-SceI was chosen because it cuts an 18-bp recognition sequence that is not present in the *E. coli* genome. When the bacteria are incubated in the presence of the inducer, 1% L-arabinose, two reactions in the minicircle-producing plasmid occur. The first reaction is the φC31-mediated intramolecular recombination between attB and attP, which results in the formation of two DNA circles, the minicircle and plasmid bacterial DNA circle. The second is the linearization of the plasmid bacterial DNA circle by I-SceI cleavage of the I-SceI site, followed by degradation of the linearized DNA by bacterial exonucleases (85) leaving the minicircle as the only episomal DNA in the bacteria. This episome can then be purified from bacteria by affinity column chromatography, as for any routine plasmid DNA.

Yield and purity of the minicircle DNA preparation may be optimized by use of plasmids encoding multiple copies of φC31 integrase or using *E. coli* strains that encode L-arabinose-inducible I-SceI as part of the bacterial genome (e.g. *E. coli* strain ZYCY10P3S2T).

**1.3.4. Preparation
of Minicircle Using Flippase
Recombinase**

The recombinase Flippase has also been used for the generation of minicircles under the control of a heat-inducible promoter (86). For Flp recombination to occur, the desired expression cassette is flanked by two identical 34-bp Flippase Recognition Target (FRT) sites, and Flp recombinase is induced in *E. coli* strain MM294-Flp by the temperature shift cycle (30°C–40°C–35°C). The 34-bp-long FRT site sequence is 5′-GAAGTTCCTATTC tctagaaaGTATAGGAACTTC-3′.

Flippase binds to the 13-bp 5′-GAAGTTCCTATTC-3′ and the reverse complement of 5′-GTATAGGAACTTC-3′ (5′-GAAG TTCCTATAC-3′). The FRT site is cleaved just before 5′-tctagaaa-3′, the 8-bp asymmetric core region, on the top strand and behind this sequence on the bottom strand. This results in two circular DNA molecules: one is the minicircle vector containing the transgene expression cassette and the second is the circular plasmid bacterial backbone, called miniplasmid (see Fig. 2c).

**1.4. Formulating
Polymer/DNA
Complexes for
In Vivo Use**

Currently, non-viral research is focused around developing and optimizing current transfection agents such as polymer- and lipid-based amphiphiles that belong to a diverse class of macromolecules capable of forming stable complexes with DNA. At optimal formulations, these agents condense DNA into positively charged particles capable of interacting with anionic proteoglycans at the cell surface and entering cells by endocytosis (87).

Comparison of different PEI/DNA complexes for prenatal gene delivery has been accessed where plasmid DNA was complexed with both branched and linear forms of PEI and generated under various conditions in salt-free or salt-containing solutions and at different N/P ratios (88). In accordance with this, it was demonstrated that the smallest complexes were formed with linear 25 kDa PEI in salt-free solutions. For effective in vivo transfection, a commercially available linear PEI (25 kDa) called in vivo-jetPEI has been developed and was demonstrated to mediate gene delivery to various tissues via many routes, such as lung instillation, kidney perfusion, intracerebral injection, and intravenous administration. As mentioned earlier, PEI exhibits a relatively high gene transfer activity due to its "proton sponge" effect that facilitates its escape from the intracellular endocytotic pathway that buffers the endosomal pH and protects pDNA from degradation. Cytoplasmic injection of PEI/DNA polyplexes has shown up to 50% of cells expressing the transgene compared to 1 and 5% of cells with lipid/DNA and polylysine/DNA complexes, respectively (89). Delivery with PEI/DNA is reportedly less toxic when compared to branched PEI formulations following systemic administration, and the lack of detectable neutralizing antibodies suggests that repeated administrations are possible (30).

Effective uptake into cells depends on the cationic charges on the particles. The ionic balance of in vivo-jetPEI and DNA anions should, therefore, be highly cationic once complexed. The N/P ratio describes the number of nitrogen residues of PEI per DNA phosphate. As not every nitrogen atom of PEI is a cation, a neutral charged of in vivo-jetPEI/DNA complexes is reached for N/P = 2–3. Good transfection results have been published with N/P = 6–10 and optimal ratios for each application can be easily determined experimentally. As the formulation of small and stable in vivo-jetPEI/DNA complexes is only possible in salt-free solutions, ionic solutions such as PBS and cell culture medium are not recommended. A sterile isotonic 5 or 10% glucose (w/v) is, therefore, provided to dilute PEI and DNA to obtain a final concentration of 5% glucose. To avoid precipitation of PEI/DNA complexes, the final concentration of DNA in the total volume should not exceed 0.5 µg/µL.

The volume of in vivo-jetPEI solution to be mixed with concentration of DNA in order to obtain the desired N/P ratio is calculated using the following equation (adapted from the manufacturer's instructions):

$$\text{Volume}(\mu L) \text{ of in vivo jet PEI}$$
$$= \frac{(\mu g \text{ of DNA} \times 3) \times \text{N/P ratio required}}{150}$$

While systemic injection of the PEI-complexed pDNA is reported to achieve the strongest expression in lung tissue, in our hands systemic delivery of PEI/DNA appears to achieve significantly higher levels of transfection in the liver (90).

1.4.1. Chitosan–DNA Delivery

Chitosan is derived from the exoskeletons of shellfish by the alkaline deacetylation of chitin (91). In addition to being a non-toxic, non-immunogenic polymer, it has several biophysical characteristics that allow it to be an attractive gene delivery agent. In addition to having positively charged amine groups for DNA affinity, similar to polymers and lipids, it is mucoadhesive permitting enhanced delivery across mucosal surfaces. Investigations into chitosan as a gene transfer vehicle have been extensively researched in vitro and in the adult in vivo, but gene transfer in the fetus using chitosan is relatively new.

For in vivo studies, ultrapure chitosan is recommended to minimize the effects of unknown contaminants or adjuvants when interpreting the results. Optimized chitosan–DNA complexes are stabilized by strong hydrophobic forces, where excessive amounts of neither salt, detergent, nor other polyanion can dissociate the polyplexes. Therefore, the DNA is only released after incubation with a chitosan-degrading enzyme, such as lysozymes in animal cells. The high physical stability allows the DNA to be protected from degradation by nuclease and serum.

1.5. Application of Non-viral Vectors In Utero

Delivery in utero using non-viral vectors remains relatively uncommon when compared to viral vector delivery. However, with the development of new formulations and techniques, many recent studies have shown the feasibility of gene delivery to many target tissues in fetuses.

1.5.1. Electroporation In Utero

Delivery in utero by electroporation has been performed targeting neural tissue of mice at embryonic days 12–17 (92, 93). For this method of gene delivery, a procedure known as "intraamniotic gene transfer and subsequent in vivo electroporation" (IAGTE) is utilized. The efficacy of electroporation is highly dependent on DNA concentration. The higher the concentration of DNA used, the higher the level of expression is obtained. The location of DNA solution is usually monitored by the addition of 0.1% Fast Green or trypan blue solution to the DNA solution to be injected. IAGTE should be completed within 20–30 min for each dam treated.

1.5.2. Gene Delivery to Fetal Neural Tissue

For targeting neural tissue, expression of the transgene introduced particularly depends on the promoter in the expression cassette, even if suggested to be ubiquitously active. In the embryonic stage, expression led by the CMV promoter is generally low and does not persist throughout development following birth. EF1α generates a modest level of expression while the CAG promoter is suggested to produce strong expression. However, both EF1α- and CAG-mediated expression is lost at least 3 weeks after birth (92). However, gene transfer to the brain by this method results in 65–80% of fetuses expressing the transgene.

1.5.3. Gene Delivery to Fetal Skin

Due to its visibility and access, the surface epithelium of embryos is an attractive target for gene therapy. Sato et al. (94) demonstrated that delivery of a non-viral lacZ expression vector into the intraamniotic cavity of a normal fetus followed immediately by electroporation (IAGTE) targeted the skin surface. In addition, his group introduced a Cre-lox-expressing vectors by IAGTE and when expressed by fetal cells, the loxP-flanked EGFP sequence would be excised and a downstream lacZ gene would become active. Once recombination had occurred, lacZ expression persists throughout fetal development enabling lineage analysis of trans-duced cells. The injection involves 1–0.5 μg of plasmid DNA in a 1–2-μL final volume.

The efficacy of electroporation is highly dependent on DNA concentration. The results following gene delivery to the skin in utero using non-viral vectors differ from those using adenoviral vectors. For example, lacZ expression was found in the fetal epidermal layers including the periderm and the basal layer proximal to the periderm, which differs from transgene expression occurring over the entire fetal surface after injection with an adenoviral lacZ construct. Optimization of various conditions for in vivo electroporation would no doubt increase gene transfer and enable gene delivery to deeper portions of the fetal skin and possibly even into muscle tissue.

1.5.4. Gene Delivery to Fetal Retinal Cells

Retinal ganglion cells can be targeted for gene expression by electroporation of the eye of the mouse embryo (95). During eye development, neurogenesis occurs in the mouse retina between E14 and P10. Among the many different cell types in the retina, ganglion cells (RGCs) are the only cell type of which axons leave the retina, transmitting visual information to the brain. RGCs differentiate with their axons leaving through the optic disc in the centre of the eye and form the optic nerve. Therefore, gene transfer to RGCs during early stages of embryonic development would prove useful to visualize the shape and location of RGCs at different stages of retinal differentiation.

Garcia-Frigola et al. (95) describe the delivery of GFP-expressing plasmid DNA driven by a CAG promoter resulting in strong expression in the retinal layer. This procedure was reported to have a high survival rate with approximately 90% of injected embryos surviving the electroporation with over 70% successfully expressing GFP in the retina. No damage was evident in the electroporated retinas. The total volume of DNA that can be applied with this procedure is 1 µl. The number of GFP-expressing cells varied depending on the concentration of the injected DNA solution (0.2–1 µg/µl) with no expression being detected at the lowest level.

2. Materials

2.1. General Chemicals and Reagents

1. Agarose electrophoresis grade.
2. Ampicillin.
3. Bovine serum albumin (BSA).
4. Carbon tetrachloride (CCL_4).
5. Chloramphenicol.
6. Dimethyl sulfoxide (DMSO).
7. 2′-desoxynucleotide 5′-triphosphate mix (dNTPs).
8. Ethylenediaminetetraacetic acid (EDTA).
9. Ethanol.
10. Ethidium bromide.
11. Formaldehyde.
12. Glucose.
13. Glutaraldehyde.

14. Glycerol.

15. Isopropanol.

16. Kanamycin.

17. L-arabinose.

18. Luria–Bertani (LB) broth.

19. Lithium chloride.

20. D-Luciferin.

21. Molecular weight DNA markers.

22. Paraformaldehyde.

23. Potassium acetate.

24. Potassium chloride.

25. Potassium dihydrogen orthophosphate.

26. Sodium dodecyl sulphate (SDS).

27. Super Optimal broth with Catabolite repression (S.O.C) medium.

28. Sodium chloride.

29. Sodium tricitrate.

2.2. Production of Standard Plasmids

1. LB broth (1% sodium chloride, 1% tryptone, 0.5% yeast extract, pH adjusted to 7.0 with sodium hydroxide).

2. 10 mg/mL kanamycin sulphate.

3. Commercial kits (Endotoxin-free maxiprep kit QIAGEN).

2.3. Production of Minicircle Vectors

2.3.1. Culture and Induction

1. LB broth (1% sodium chloride, 1% tryptone, 0.5% yeast extract, pH adjusted to 7.0 with sodium hydroxide).

2. LB agar (LB medium supplemented with 1.5% agar).

3. S.O.C medium (2% (v/v) tryptone, 0.5% (v/v) yeast extract, 10 mM glucose, 10 mM sodium chloride, 2.5 mM potassium chloride).

4. 5× M9 medium salts (64 g sodium phosphate, 15 g potassium hydrogen phosphate, 2.5 g sodium chloride, 5.0 g ammonium chloride (made up 1 L M9 medium with sterile deionized water)). 20% glucose (added after autoclaving).

2.3.2. Extraction of Producer Plasmid DNA

1. Solution 1 (50 mM glucose, 25 mM Tris–HCl pH 8.0, 10 mM EDTA).

2. Lysis solution (0.2 M sodium hydroxide, 1% SDS).

3. Isopropanol.

4. Distilled water.

5. 12 M lithium chloride.

6. 40 mg/mL RNase solution.

7. 10 M ammonium acetate.

8. Ethanol.

9. Tris–EDTA (TE) buffer (10 mM Tris pH 7.5, 1 mM EDTA).

10. Chloroform.

11. Phenol.

2.3.3. φC31 Recombinase

1. 1% L-arabinose.

2. I-SceI Endonuclease.

3. *E. coli* strain ZYCY10P3S2T.

2.3.4. Flp Recombinase

1. Any mother plasmid with an expression cassette flanked by the FRT recombination sites.

2. MM294Flp *E. coli* strain (F- λ- supE44 endA1 thi-1 hsdR17 lacZ:cI857-FLP).

3. LB Powder.

4. LB-Agar Plates for kanamycin-resistant *E. coli* strains (LB medium supplemented with 1.5% agar).

5. 10 mg/mL Kanamycin Sulfate Solution (Invitrogen, cat. no. 15160).

6. Plasmid-Safe™ ATP-dependent DNase (Epicentre/Biozym Scientific GmbH).

2.3.5. Production of Minicircle Using φC31

1. Any mother plasmid with an expression cassette flanked by the attB–attP recombination sites.

2. ZYCY10P3S2T *E. coli* strain (96) or TOP10 Competent cells.

3. Terrific Broth (TB) Powder (Invitrogen).

4. LB Powder.

5. LB-Agar Plates for kanamycin-resistant *E. coli* strains.

6. 1 N NaOH solution.

7. L-arabinose.

8. 10 mg/mL Kanamycin sulphate solution.

2.3.6. Production of Minicircle Using Flp

1. Any mother plasmid with an expression cassette flanked by the FRT recombination sites.

2. MM294Flp *E. coli* strain (F- λ- supE44 endA1 thi-1 hsdR17 lacZ:cI857-FLP).

3. LB Powder.

4. LB-Agar Plates for kanamycin-resistant *E. coli* strains.

5. 10 mg/mL Kanamycin sulphate solution.

6. Plasmid-SafeTM ATP-dependent DNase (Epicentre/Biozym Scientific GmbH).

2.4. Formulating Polymer/DNA Complexes for In Vivo Use

1. In vivo-jetPEI™ (Polyplus Transfection, UK).
2. 5% Glucose.

2.5. Delivery of PEI/DNA

2.5.1. Intrauterine Application of PEI/DNA

1. In vivo-jetPEI™ (Polyplus Transfection, UK).
2. 5% Glucose 0.5× HBS (75 mM NaCl, 20 mM HEPES pH 7.4+5% glucose) or HBG (20 mM HEPES pH 7.4+5% glucose).

2.5.2. Systemic Injection via the Fetal Yolk Sac Vessels

1. In vivo-jetPEI™ (Polyplus Transfection, UK).
2. 5% Glucose 0.5× HBS (75 mM NaCl, 20 mM HEPES pH 7.4+5% glucose) or HBG (20 mM HEPES pH 7.4+5% glucose).

2.5.3. Chitosan–DNA Delivery

1. 0.2 mg/mL Chitosan stock solutions.
2. Sterile water at pH 6.1.
3. 28-gauge needle.
4. Sterile normal saline solution.

2.5.4. Electroporation In Utero

1. Electroporator.
2. 0.1% Fast green or trypan blue.

2.5.5. Gene Delivery to Fetal Neural Tissue

1. Micropipettes made with a puller and 1-mm-diameter glass capillary tube attached to an aspirator tube.
2. Sterile gauze.
3. Fast Green solution (0.1%).
4. Phosphate-buffered saline (PBS).
5. Nylon and silk sutures.

2.5.6. Gene Delivery to Fetal Skin

1. Sterile gauze.
2. 0.1% Trypan blue.
3. Glass micropipette made from a microcapillary tube.
4. Caliper electrodes with a 5-mm-diameter disc at the tip.

2.5.7. Gene Delivery to Fetal Retinal Cells

1. Sterile gauze.
2. 0.1% trypan blue.
3. Glass micropipette made from a microcapillary tube.
4. Caliper electrodes with a 5-mm-diameter disc at the tip.

3. Methods

3.1. Preparation of Endotoxin-Free Plasmids

For production of large quantities of DNA, the Endotoxin-free Plasmid Maxiprep kits are used according to the manufacturer's instructions.

1. Inoculate 500 mL fresh LB broth with 5-mL starter cultures in a 2-L flask containing the appropriate antibiotic.

2. Pellet the cells for 15 min at $6,000 \times g$ and resuspend by vortexing in resuspension buffer containing RNase. The cells are then lysed by addition of lysis buffer at room temperature for 5 min. The lysed cells are neutralized with neutralizing buffer and the lysate is incubated in endotoxin-removal buffer prior to binding to the column.

3. Pass the lysate over a column with an ion-exchange resin, which binds DNA at a pH of 7 and ionic strength of 750 mM NaCl.

4. Following elution, precipitate the DNA with 0.7 volumes isopropanol, pellet by centrifugation at $15,000 \times g$ for 30 min, and wash with 70% ethanol before resuspending in an endotoxin-free elution buffer. The theoretical yield is approximately 1–1.5 mg of endotoxin-free plasmid DNA.

3.2. Preparation of Minicircle Vectors

3.2.1. Culture and Induction

1. Autoclave flasks with the nominal volume 2 L. They should be closed with at least four layers of foil. Autoclave bottles with 0.5 L of LB medium each (see Note 1).

2. After autoclaving, add sterile glucose solution to produce a final glucose concentration of 0.5%. Thus, add 5 mL of 50% glucose for 0.5-L culture. Add appropriate antibiotic (e.g. chloramphenicol to a final concentration of 30 mg/mL) (see Note 2).

3. Seed the cultures with the stock of MM219Cre strain harbouring a minicircle mother plasmid. 0.5 mL of the seeding culture per 0.5-L culture is a good example of the seeding density (see Note 3).

4. Incubate the cultures at 37°C overnight. The shaker should run at no less than 250 rpm (preferably 300 rpm or more, if it is safe).

5. Autoclave another couple of flasks to be used at the next induction step. The nominal volume of these flasks is 2 L.

6. Prepare M9 Medium and if necessary, supplement the M9 medium with stock solutions of the appropriate amino acids.

    $5 \times$ M9 salt solution is divided into 200-mL aliquots and sterilized by autoclaving for 15 min at 15 lb/sq. in. on liquid cycle.

7. Next day, centrifuge the cells down (10 min at $5,000 \times g$). Use sterile centrifuge tubes. Resuspend the cells from a 500-mL culture in 250 mL of sterile M9 medium. Centrifuge down

again (10 min at $5,000 \times g$). Resuspend the cells in 250 mL of sterile M9 medium. Centrifuge down again (10 min at $5,000 \times g$). Resuspend the cells in 500 mL of sterile M9 medium. Fill in new sterile flasks. Add thiamin to a final concentration of $1 \mu g/mL$. Add sterile $MgSO_4$ to a final concentration of 10 mM. Add sterile L-arabinose to the final concentration of 0.5% (e.g. add 12.5 mL of 20% arabinose stock to 500 mL of cells). Incubate the cells at 37°C with vigorous shaking (at least 250 rpm, preferably 300 rpm or more, if it is safe) for 6 h or overnight. Collect the cells and extract DNA by lithium chloride protocol (see below) or other DNA extraction protocol (see Note 4).

3.2.2. Extraction of Producer Plasmid DNA (See Note 5)

1. Centrifuge down 500 mL of induced culture (10 min, $5,000 \times g$). Resuspend the cells in 10 mL of "Solution 1". Add 10 mL of 10 mg/mL lysozyme prepared in "Solution 1". Mix the suspension and leave for 4–5 min on the bench (see Note 6).

2. Add 40 mL of 0.2 M NaOH, and 1% SDS (this solution should be prepared immediately before addition from 2 M NaOH and 10% SDS stocks). Gently mix and leave on the bench until completely clear (normally 15 min) (see Note 7).

3. Add 30 mL of 3 M potassium acetate pH 4.8. Mix gently but thoroughly by inverting the bucket or the tube several times.

4. Cool the tubes in an ice bath for at least 20 min (see Note 8).

5. Centrifuge them down ($5,000 \times g$ or better $10,000 \times g$, 10 min), and collect the supernatant in a clean tube by careful use of a pipette. It is often necessary to repeat this step to avoid unwanted bits of bacterial debris. The debris on the top can be removed by disposable inoculation loop. Try to get rid of this debris as far as possible, but do not be fussy if some very small bits do come along.

6. Add 1 volume of isopropanol (under the hood), and leave on the bench for 20–30 min. Centrifuge at $5,000 \times g$ or better $10,000 \times g$, 10 min. Carefully remove the supernatant, spin down, and again remove the remaining supernatant.

7. After brief drying under the hood (normally 10 min), add 5 mL of water, mix, let the precipitate dissolve (normally 10–20 min), and add 5 mL of 12 M lithium chloride. Mix. Put at –20°C for at least 30 min (see Note 8).

8. Centrifuge the tubes to remove the precipitate of protein and RNA ($5,000 \times g$ or better $10,000 \times g$, 10 min). Transfer the supernatant to new tubes, add 10 mL of isopropanol (under the hood), mix, and leave at –20°C for at least 30 min.

9. Centrifuge the tubes ($5,000 \times g$ or better $10,000 \times g$, 10 min), and remove the supernatant. Repeat to remove the remaining supernatant. Briefly dry precipitate in the Speedvac system (5–10 min

without heating) or under the hood, add 5 mL of 40 mg/mL RNase solution, mix, and leave for 30 min on the bench mixing again from time to time. DNA should be dissolved.

10. Do one phenol–chloroform extraction in Eppendorf or Falcon tubes followed by one chloroform extraction. Precipitate DNA using appropriate acetate salt and 2 volumes of ethanol (e.g. 5 mL DNA + 2.5 mL 10 M ammonium acetate + 15 mL of ethanol). After brief drying, DNA should be dissolved in 5 mL of water or TE buffer (see Note 8).

11. Then, the DNA can be digested by restriction enzymes to check for purity and correct size.

3.3. Preparation of Minicircle Using φC31

3.3.1. Minicircle Induction Broth

First, prepare a stock of 20% L-arabinose by dissolving 2 g L-arabinose in 100 mL of ddH$_2$O. Freeze stock at –20°C for up to 6 months. To prepare induction broth, mix 384 mL LB, 16 mL 1 N NaOH, and 0.4 mL 20% L-arabinose (final concentration of L-arabinose is 0.01%). Make fresh just prior to use.

3.3.2. Minicircle DNA Production

1. Day 1: Streak a kanamycin-containing LB-agar plate with *E. coli* transformed with mother plasmid. Incubate overnight for 12–16 h at 37°C.

2. Day 2: Inoculate 5 mL of LB Broth containing kanamycin (50 µg/mL) with a colony from the freshly streaked plate. Incubate this starter culture at 37°C with agitation at 250 rpm for 8 h.

3. Autoclave 400 mL of TB in a 2-L flask. When flask is cool enough to touch (~55°C), add 0.2 mL of 10 mg/mL kanamycin sulphate solution to a final concentration of 50 µg/mL.

4. Inoculate 100 µL of the starter culture of step 2 into 400 mL of TB prepared in step 4 and incubate overnight for 16–18 h at 37°C with shaking at 250 rpm.

5. Day 3: Take a 0.5 mL sample of the overnight bacterial culture and confirm that the OD$_{600}$ reading is between 4 and 5 and the pH is 6.5.

6. Combine 400 mL of the Minicircle Induction Broth (see Subheading 2) with 400 mL of the overnight culture into a 2-L flask. Incubate at 32°C with shaking at 250 rpm for greater than 5 h (see Note 9).

7. Pellet bacterial cells by centrifugation at $6,000 \times g$ for 15 min at 4°C.

8. Resuspend the bacterial pellet in 100 mL of Qiagen Buffer Resuspension buffer. Follow the maxi prep protocol as per the manufacturer's instructions, but use double the volume of resuspension, lysis, and neutralization buffers (100 mL each).

9. Measure DNA concentration and purity using a spectrophotometer (see Note 10).

3.4. Preparation of Minicircle Using Flp Recombinase

1. A single colony of *E. coli* MM294Flp is transformed with the mother plasmid and grown overnight in a shaking incubator at 220 rpm and 30°C in LB media containing 25 mg/mL kanamycin.

2. Pellet cells at $4,000 \times g$ before resuspension in 4:1 (v/v) LB.

3. After washing, cells are re-pelleted at $4,000 \times g$ and resuspended in 2:1 (v/v) fresh LB.

4. Initiate Flp recombinase expression by incubation at 40°C for 20 min.

5. Incubate the bacteria for 2.5 h at 35°C in a shaking incubator at 180 rpm.

6. Initiate Flp recombinase for a second time by incubating cells at 40°C for 20 min.

7. Incubate for an additional 1.5 h at 35°C in a shaking incubator at 180 rpm.

8. Apply a restriction endonuclease that digests the mother plasmid and the mini-plasmid, but not the minicircle, in the pool of DNA products.

9. Undigested supercoiled minicircle can then be separated from the linearized mother plasmid and the bacterial mini-plasmid by agarose gel electrophoresis.

10. The respective minicircle band is excised from the gel and the DNA is then extracted using the Qiagen Gel Purification Kit, according to manufacturer's instructions.

11. A further purification step, the application of ATP-dependent nuclease, can be applied to free the gel-extracted minicircle from nicked or linear contaminants.

    For this purification, 42 μL of the gel extract are mixed with 5 μL 10× Plasmid Safe™ reaction buffer, 2 μL of 25 mM ATP, and 1 μL Plasmid-Safe™ ATP-dependent DNase.

12. After shaking at 225 rpm, 37°C, for 1 h, supercoiled DNA can be recovered by the QIAquick PCR purification kit according to the manufacturer's instructions.

3.5. Formulating Polymer/DNA Complexes for In Vivo Use

A basic protocol for formulating in vivo-jetPEI/DNA is as follows.

1. Plasmid solution is prepared by conventional methods, ensuring endotoxin-free yields. If precipitation forms in the plasmid solution, further purification with phenol/chloroform is required.

2. Allow the in vivo-jetPEI solution to thaw at room temperature prior to use.

3. Under sterile conditions, dilute DNA in required amount of 5% glucose (w/v). Vortex gently and spin down.

4. Dilute in vivo-jetPEI in the required volume of 5% glucose (w/v). Vortex and spin down gently.

5. Add the diluted PEI to diluted DNA in one go into the diluted DNA (important: do not mix in the reverse order), vortex briefly, and allow to complex for 30 min before injection.

6. Particle size is measured by Photon Correlation Spectroscopy and should be between 100 and 200 nm.

3.6. Delivery of PEI/DNA

The suggested amounts of DNA to be injected are provided in the table below, adapted from manufacturer's instructions. The concentration of DNA used should be adjusted according to the size of the animal (Table 2). Usually, in vivo-jetPEI at an N/P ratio = 8 is recommended. Strong transgene expression may be observed 24–48 h post transfection (see Note 11).

3.7. Intrauterine Application of PEI/DNA

The feasibility of direct intrauterine injection of PEI/DNA into fetal livers has been successfully demonstrated (88). Accordingly, the following procedure should be followed.

1. Anaesthetize pregnant mice at E17.5 and open abdomen by ventral midline laparotomy under sterile conditions.

2. Expose one uterine horn and inject solution of PEI/DNA transuterine with an aluminosilicate capillary directly into the fetal livers.

3. Following this procedure, return the uterine horn into the abdominal cavity and close the abdominal wall with two layers of surgical sutures.

4. Keep animals on a warming blanket as the body temperature of the mouse decreases as a result of uterus exposure and the effect of anaesthesia. The mouse is kept on the warmer until active and then housed in an undisturbed environment under observation.

5. Allow the animals to give birth naturally and investigate organs of treated animal to confirm gene transfer (see Note 12).

Table 2
General guidelines for the preparation of complexes of DNA and jetPEI

| Amount of DNA (µg) | Volume (µl) of in vivo-jetPEI at | | | | |
	N/P = 4	N/P = 5	N/P = 6	N/P = 8	N/P = 10
5	0.4	0.5	0.6	0.8	1.0
10	0.8	1	1.2	1.6	2

All fetal injections are performed at days 15–16 of gestation. A volume of 20 µL PEI/DNA is injected into the yolk sac vessels of two to three fetuses per dam. The following protocol describes PEI/DNA formation for a single fetal injection at N/P = 8 using 5 µg DNA and administration to fetal mice in utero.

1. Dilute 5 µg of DNA in 5% isotonic glucose solution to a final volume of 10 µL while also diluting 1.6 µL of thawed in vivo-jetPEI in another aliquot of 5% isotonic glucose solution to a final volume of 10 µL.

2. Add diluted PEI to the diluted DNA all at once, vortex briefly, and allow to complex for at least half an hour prior to use.

3. Anaesthetize time-mated pregnant females by inhalation of isoflurane. After disinfecting the abdominal skin with 70% ethanol, make a 1.5–2-cm midline skin incision in the abdominal wall along the linea alba. Place piece of sterile gauze with a hole cut in the centre over the incision and draw out both horns of the gravid uterus through the hole in the gauze (see Note 13).

4. After determining the direction of the blood flow, inject the fetal yolk sac vessel in the direction of the blood flow with a 33-gauge microlitre syringe (see Note 11). A maximum of four fetuses are injected per dam (see Note 14).

5. For identification, injected fetuses are marked with a 5 µL subcutaneous injection of colloidal carbon marker dye to allow identification after birth.

6. Following injection, return the uterus to the abdominal cavity and close the abdominal wall in two layers with nylon and silk sutures, respectively.

7. Keep animals on a warming blanket as the body temperature of the mouse decreases as a result of uterus exposure and the effect of anaesthetization. Keep the mouse on the warmer until active and then house in an undisturbed environment under observation.

8. After 3 days (approximately 12 h prepartum), if natural birth has not occurred, the fetuses are removed by caesarean section and the neonates are fostered on MF1 dams with litters while retaining the original MF1 litter sizes.

The basic protocol for formulating chitosans for in vivo gene delivery is as follows (97).

1. Chitosan stock solution (0.2 mg/mL) is prepared by dissolving sterile chitosan in sterile water at pH 6.1 followed by sterile filtration.

2. Chitosan polyplexes are formulated by adding chitosan and then plasmid DNA stock solution to sterile water under intense stirring on a vortex mixer.

3. To prepare chitosan polyplexes at a charge ratio of 1:1 (+/-) using ultrapure chitosan C (15;190), 0.70 µg of chitosan per µg of DNA are used.

4. Polyplexes are then concentrated by mild evaporation under vacuum at $330 \times g$ for approximately 90 min to obtain DNA concentrations of around 500 µg/mL.

3.7.3. Gene Delivery to Fetal Lung and Intestinal Tissue Using Chitosan–DNA

Yang et al. (98) report successful gene transfer of GFP-expressing plasmid DNA to the luminally exposed cells of lung and intestinal tissue following injection of vector into amniotic fluid, which the fetus breathes and swallows (99). Accordingly, the protocol is as follows.

1. Anaesthetize pregnant mice at E16–17 and make a midline laparotomy under sterile conditions. Expose the horns of the gravid uterus (see Note 13).

2. Using a 28-gauge needle, inject chitosan–plasmid DNA in a 30-µl volume containing 12.5 µg of transgene DNA into each fetal amniotic sac (see Note 14).

3. Close the maternal abdomen after injecting 10 mL of pre-warmed sterile normal saline solution into the peritoneal cavity.

4. Keep the treated mother warm and leave undisturbed following the procedure to recover.

5. After natural parturition, perform tail skin biopsies for reporter gene analysis and only pups with transgene DNA present in tail skin are presumed to have been successfully transfected in vivo.

3.7.4. Gene Delivery to Fetal Neural Tissue

Neural tissue electroporation in utero was shown by Tabata et al. (92). Accordingly, the following protocol is performed.

1. Make micropipettes for injection with a puller and 1-mm-diameter glass capillary tubes with the tip cut obliquely. Attach the micropipettes to an aspirator tube assembly.

2. Anaesthetize a pregnant mouse at the desired gestation time and make a 1.5–2-cm midline incision along the linea alba. Place a piece of sterile gauze with a hole cut in the centre placed over the incision and carefully draw out one uterine horn through the hole in the gauze (see Note 13).

3. After determining the orientation of the embryos through the wall of the uterine horn, insert a micropipette into the lateral ventricle and inject 1–2 µL of plasmid by expiratory pressure using the aspirator tube assembly. Use Fast Green solution (0.1%) to enable the visualization of the DNA solution in the lateral ventricle to be observed through the uterine wall (see Note 14).

Table 3
Electroporation conditions for gene delivery to fetal neural tissue

Age	Electrode diameter (mm)	Voltage (V)	Pulse-on (ms)	Pulse-off (ms)	Number of pulses
E12.5	3	33	30	970	4
E13	5	30–35	50	950	4

4. After soaking the uterine horn with PBS, pinch the head of the embryo with a forceps-type electrode and apply an electronic pulse. The plasmid can be introduced to a larger region of the cortex with the cathode placed on the chin and the anode is placed on the centre of the injected hemisphere. The electroporation conditions are summarized in the table below (Table 3) (92).

5. After the electroporation is completed on the first uterine horn, replace the uterine horn into the uterine cavity and expose the second uterine horn to be subjected to the same conditions.

6. Following the completion of the procedure, close the abdominal wall and skin with nylon and silk sutures. The animal is kept on a warming blanket until active and housed in an undisturbed environment. The mice should give birth to the treated pups normally and expression should be visible on postnatal day 1.

3.7.5. Gene Delivery to Fetal Skin

The protocol is based on previous methods developed by Sato et al. (94) and is as follows.

1. Anaesthetize a pregnant female and make a 1.5–2-cm vertical abdominal incision along the linea alba. Expose both gravid uterine horns and place on sterile gauze moistened with sterile saline (see Note 13).

2. Inject 1–2 μl of plasmid DNA (0.5–1 μg) together with 0.1% trypan blue with a glass micropipette made from a microcapillary tube into the amniotic cavities of fetuses near a target region (see Notes 14 and 15).

3. Immediately following DNA transfer, place caliper electrodes with a 5-mm-diameter disc at the tip encompassing the target region of the embryos and administer four square pulses of 40 V amplitude for duration of 50 ms.

4. Apply a similar procedure to the other fetuses. After gene transfer, return the uterine horns into the uterine cavity and close the abdominal wall with nylon and silk sutures. Confirm the viability of the fetuses with the presence of a visibly beating heart and normal gross morphology (Note 16).

3.7.6. Gene Delivery
to Fetal Retinal Cells

For delivery into RGCs, the method for IAGTE was adapted by Garcia-Frigola et al. (95) as follows.

1. Anaesthetize pregnant females at E13 and using adhesive tape, immobilize the anterior and posterior limbs to facilitate rotation and targeted injection of DNA into the eyes of the embryos.

2. Open the abdominal cavity by a midline incision along the linea alba and expose the uterine horns under sterile conditions (see Note 13).

3. Using a graduated pulled glass mouth-micropippette and a portable stereoscope, inject plasmids monocularly through the uterus wall into each embryo.

4. Immediately following plasmid injection, pass an electric current through the head of each embryo using a caliper electrode. Treat all the embryos in each litter to eliminate the need for identification by subcutaneous marking (see Note 17).

4. Notes

1. Ensure sterility of flasks by supplementation with foil and cotton plugs. It is absolutely important to grow no more than 0.5 L of culture in each of the 2-L flasks—otherwise, poor aeration will reduce DNA yield.

2. It is crucial to add glucose after autoclaving. The purpose of glucose is to guarantee the absence of any premature dissociation of a minicircle mother plasmid. The purpose of antibiotic is to prevent loss of the plasmid by providing selection pressure.

3. While preparing the seeding stock, do not forget—it also requires glucose in the medium to block premature Cre activation and dissociation of minicircle and mini-plasmid.

4. Never wash the cells in water because some of them will die of osmotic shock. Chloramphenicol is not important at the induction step. Thiamin is not crucial but desirable. Magnesium is needed as a cofactor of Cre-recombinase.

5. All procedures are done at room temperature on the bench until the potassium acetate step. This ensures proper lysozyme treatment and good lysis.

6. Sometimes, it is difficult to resuspend the bacterial pellet directly in the lysozyme solution because of foam formation. Accordingly, this protocol recommends addition of lysozyme after resuspension.

7. This is a crucial step for good DNA yield. The resultant solution should be as clear as possible.

8. Chloroform extraction is used to remove the remaining protein and also dissolved phenol. It is important to cool ethanol mixture at least for 30 min at −20°C. It can take more than 1 h to redissolve the precipitated DNA.

9. Longer incubation periods of up to 9 h will yield minicircle DNA of greater purity.

10. Minicircle DNA preparations can be frozen in 5-μg aliquots at −20°C for long periods. Avoid repeated freeze–thaw cycles.

11. In our hands, Hamilton syringes (1705 RN/1710 RN) fitted with 33-gauge needles have proven best for intrauterine injection at all sites.

12. Expression of reporter genes luciferase and lacZ was demonstrated by Gharwan et al. (88) in the mouse livers after parturition (usually 48 h after injection) of 3 μg of plasmid DNA conjugated with PEI at different N/P ratios (4.8, 6.0, or 7.2) to a final volume of 250 μL in either 0.5× HBS (75 mM NaCl, 20 mM HEPES pH 7.4 + 5% glucose) or HBG (20 mM HEPES pH 7.4 + 5% glucose). Approximately 10-20 times lower levels of transgene expression were detected in the lungs, the next organ in the route of circulation. However, the authors state that this method yields gene expression levels far below therapeutic relevance. Furthermore, maintenance of gene expression after initial successful transfection was not further investigated.

13. Cover the exposed uterine horn or embryos (except those to be injected) with sterile moistened gauze to prevent drying.

14. Embryos close to the vaginal duct are not injected to avoid any spontaneous abortions caused by any damage to the embryos.

15. Successful skin-targeted gene transfer appears if the solution remains in the target site even after the removal of the micropipette for a certain time. If the solution disperses through the yolk sac cavity immediately following injection, the efficacy is found to be much reduced.

16. The survival for this procedure appears relatively low at 55–60% following inspection at 1 day after the procedure, when compared to IAGTE to the fetal brain (>80% survival) (92, 93), but the reasons for this are unclear.

17. At 1 day after electroporation (E14), Garcia–Frigola report that GFP expression could be observed in the ventricular zone of the neural retina. However, at E16, the positive cells were positioned in a more internal layer than at E14 and by E18 the majority of GFP cells were located in the innermost retina layers, where the RGCs are located. Furthermore, it was reported that electroporation of the retinae at E13 resulted in GFP-expressing cells surrounding the optic disc, whereas when the retinae were electroporated 1 day later, at E14, GFP expression was observed in various regions of the retina depending

on the position of the electrodes. The authors suggest that this is due to the increase in size of retinae between days E13 and 14. At E13, the retina was sufficiently small enough, so the plasmid DNA solution extends over the entire retina following injection while at E14 plasmid DNA was accessible to a proportionally smaller area of the retina and the location of GFP expression was therefore dependent on the position of the electrodes.

References

1. Hacein-Bey-Abina S, Von Kalle C, Schmidt M et al (2003) LMO2-associated clonal T cell proliferation in two patients after gene therapy for SCID-X1. Science 302:415–419

2. Wolff JA, Malone RW, Williams P et al (1990) Direct gene transfer into mouse muscle in vivo. Science 247:1465–1468

3. Hickman MA, Malone RW, Lehmann-Bruinsma K et al (1994) Gene expression following direct injection of DNA into liver. Hum Gene Ther 5:1477–1483

4. Budker V, Zhang G, Knechtle S et al (1996) Naked DNA delivered intraportally expresses efficiently in hepatocytes. Gene Ther 3:593–598

5. Choate KA, Khavari PA (1997) Direct cutaneous gene delivery in a human genetic skin disease. Hum Gene Ther 8:1659–1665

6. Meyer KB, Thompson MM, Levy MY et al (1995) Intratracheal gene delivery to the mouse airway: characterization of plasmid DNA expression and pharmacokinetics. Gene Ther 2:450–460

7. Reilly JP, Grise MA, Fortuin FD et al (2005) Long-term (2-year) clinical events following transthoracic intramyocardial gene transfer of VEGF-2 in no-option patients. J Interv Cardiol 18:27–31

8. Schwartz B, Benoist C, Abdallah B et al (1996) Gene transfer by naked DNA into adult mouse brain. Gene Ther 3:405–411

9. Zelenin AV, Kolesnikov VA, Tarasenko OA et al (1997) Bacterial beta-galactosidase and human dystrophin genes are expressed in mouse skeletal muscle fibers after ballistic transfection. FEBS Lett 414:319–322

10. Mehier-Humbert S, Guy RH (2005) Physical methods for gene transfer: improving the kinetics of gene delivery into cells. Adv Drug Deliv Rev 57:733–753

11. Gao X, Kim KS, Liu D (2007) Nonviral gene delivery: what we know and what is next. AAPS J 9:E92–104

12. Taniyama Y, Tachibana K, Hiraoka K et al (2002) Local delivery of plasmid DNA into rat carotid artery using ultrasound. Circulation 105:1233–1239

13. Taniyama Y, Tachibana K, Hiraoka K et al (2002) Development of safe and efficient novel nonviral gene transfer using ultrasound: enhancement of transfection efficiency of naked plasmid DNA in skeletal muscle. Gene Ther 9:372–380

14. Lawrie A, Brisken AF, Francis SE et al (2000) Microbubble-enhanced ultrasound for vascular gene delivery. Gene Ther 7:2023–2027

15. Wells DJ (2004) Gene therapy progress and prospects: electroporation and other physical methods. Gene Ther 11:1363–1369

16. Wolff JA, Williams P, Acsadi G et al (1991) Conditions affecting direct gene transfer into rodent muscle in vivo. Biotechniques 11:474–485

17. Liu F, Song Y, Liu D (1999) Hydrodynamics-based transfection in animals by systemic administration of plasmid DNA. Gene Ther 6:1258–1266

18. Zhang G, Budker V, Wolff JA (1999) High levels of foreign gene expression in hepatocytes after tail vein injections of naked plasmid DNA. Hum Gene Ther 10:1735–1737

19. Budker V, Zhang G, Danko I et al (1998) The efficient expression of intravascularly delivered DNA in rat muscle. Gene Ther 5:272–276

20. Zhang G, Gao X, Song YK et al (2004) Hydroporation as the mechanism of hydrodynamic delivery. Gene Ther 11:675–682

21. Mahato RI, Takakura Y, Hashida M (1997) Nonviral vectors for in vivo gene delivery: physicochemical and pharmacokinetic considerations. Crit Rev Ther Drug Carrier Syst 14:133–172

22. Mahato RI, Kawabata K, Nomura T et al (1995) Physicochemical and pharmacokinetic characteristics of plasmid DNA/cationic liposome complexes. J Pharm Sci 84:1267–1271

23. Felgner PL, Gadek TR, Holm M et al (1987) Lipofection: a highly efficient, lipid-mediated DNA-transfection procedure. Proc Natl Acad Sci U S A 84:7413–7417

24. Liu D, Ren T, Gao X (2003) Cationic transfection lipids. Curr Med Chem 10:1307–1315

25. Li S, Huang L (2000) Nonviral gene therapy: promises and challenges. Gene Ther 7:31–34

26. Hemmi H, Takeuchi O, Kawai T et al (2000) A Toll-like receptor recognizes bacterial DNA. Nature 408:740–745

27. Ito Y, Kawakami S, Charoensit P et al (2009) Evaluation of proinflammatory cytokine production and liver injury induced by plasmid DNA/cationic liposome complexes with various mixing ratios in mice. Eur J Pharm Biopharm 71:303–309

28. Niidome T, Huang L (2002) Gene therapy progress and prospects: nonviral vectors. Gene Ther 9:1647–1652

29. Brunner S, Furtbauer E, Sauer T et al (2002) Overcoming the nuclear barrier: cell cycle independent nonviral gene transfer with linear polyethylenimine or electroporation. Mol Ther 5: 80–86

30. Kawakami S, Ito Y, Charoensit P et al (2006) Evaluation of proinflammatory cytokine production induced by linear and branched polyethylenimine/plasmid DNA complexes in mice. J Pharmacol Exp Ther 317:1382–1390

31. Boussif O, Lezoualc'h F, Zanta MA et al (1995) A versatile vector for gene and oligonucleotide transfer into cells in culture and in vivo: polyethylenimine. Proc Natl Acad Sci U S A 92:7297–7301

32. Hackett PB, Podetz-Petersen KM, Bell JB et al (2010) Gene expression in lung and liver after intravenous infusion of polyethylenimine complexes and hydrodynamic delivery of sleeping beauty transposons. Hum Gene Ther 21(2): 210–20

33. Oh YK, Kim JP, Yoon H et al (2001) Prolonged organ retention and safety of plasmid DNA administered in polyethylenimine complexes. Gene Ther 8:1587–1592

34. Guo ZS, Wang LH, Eisensmith RC et al (1996) Evaluation of promoter strength for hepatic gene expression in vivo following adenovirus-mediated gene transfer. Gene Ther 3:802–810

35. Boshart M, Weber F, Jahn G et al (1985) A very strong enhancer is located upstream of an immediate early gene of human cytomegalovirus. Cell 41:521–530

36. Zhang XY, Ni YS, Saifudeen Z et al (1995) Increasing binding of a transcription factor immediately downstream of the cap site of a cytomegalovirus gene represses expression. Nucleic Acids Res 23:3026–3033

37. Sinclair JH, Baillie J, Bryant LA et al (1992) Repression of human cytomegalovirus major immediate early gene expression in a monocytic cell line. J Gen Virol 73(Pt 2):433–435

38. Gill DR, Smyth SE, Goddard CA et al (2001) Increased persistence of lung gene expression using plasmids containing the ubiquitin C or elongation factor 1alpha promoter. Gene Ther 8:1539–1546

39. Schorpp M, Jager R, Schellander K et al (1996) The human ubiquitin C promoter directs high ubiquitous expression of transgenes in mice. Nucleic Acids Res 24:1787–1788

40. Yew NS, Przybylska M, Ziegler RJ et al (2001) High and sustained transgene expression in vivo from plasmid vectors containing a hybrid ubiquitin promoter. Mol Ther 4:75–82

41. Cullen BR (2003) Nuclear RNA export. J Cell Sci 116:587–597

42. Buchman AR, Berg P (1988) Comparison of intron-dependent and intron-independent gene expression. Mol Cell Biol 8:4395–4405

43. Huang J, Liang TJ (1993) A novel hepatitis B virus (HBV) genetic element with Rev response element-like properties that is essential for expression of HBV gene products. Mol Cell Biol 13:7476–7486

44. Donello JE, Loeb JE, Hope TJ (1998) Woodchuck hepatitis virus contains a tripartite posttranscriptional regulatory element. J Virol 72:5085–5092

45. Krieg AM (2000) The role of CpG motifs in innate immunity. Curr Opin Immunol 12:35–43

46. Chen ZY, Riu E, He CY et al (2008) Silencing of episomal transgene expression in liver by plasmid bacterial backbone DNA is independent of CpG methylation. Mol Ther 16: 548–556

47. Yew NS, Wang KX, Przybylska M et al (1999) Contribution of plasmid DNA to inflammation in the lung after administration of cationic lipid: pDNA complexes. Hum Gene Ther 10: 223–234

48. Argyros O, Wong SP, Fedonidis C et al (2011) Development of S/MAR minicircles for enhanced and persistent transgene expression in the mouse liver. J Mol Med 89:515–29

49. Gill D, Pringle I, Hyde SC (2009) Progress and prospects: the design and production of plasmid vectors. Gene Ther 16:165–171

50. Wong SP, Argyros O, Coutelle C et al (2009) Strategies for the episomal modification of cells. Curr Opin Mol Ther 11:433–441

51. Rothenfusser S, Tuma E, Wagner M et al (2003) Recent advances in immunostimulatory CpG oligonucleotides. Curr Opin Mol Ther 5: 98–106

52. Jackson DA, Cook PR (1995) The structural basis of nuclear function. Int Rev Cytol 162A: 125–149

53. Piechaczek C, Fetzer C, Baiker A et al (1999) A vector based on the SV40 origin of replication

and chromosomal S/MARs replicates episomally in CHO cells. Nucleic Acids Res 27:426–428

54. Jenke AC, Scinteie MF, Stehle IM et al (2004) Expression of a transgene encoded on a non-viral episomal vector is not subject to epigenetic silencing by cytosine methylation. Mol Biol Rep 31:85–90

55. Jenke BH, Fetzer CP, Stehle IM et al (2002) An episomally replicating vector binds to the nuclear matrix protein SAF-A in vivo. EMBO Rep 3:349–354

56. Papapetrou EP, Ziros PG, Micheva ID et al (2006) Gene transfer into human hematopoietic progenitor cells with an episomal vector carrying an S/MAR element. Gene Ther 13:40–51

57. Girod PA, Nguyen DQ, Calabrese D et al (2007) Genome-wide prediction of matrix attachment regions that increase gene expression in mammalian cells. Nat Methods 4:747–753

58. Stief A, Winter DM, Stratling WH et al (1989) A nuclear DNA attachment element mediates elevated and position-independent gene activity. Nature 341:343–345

59. Klehr D, Schlake T, Maass K et al (1992) Scaffold-attached regions (SAR elements) mediate transcriptional effects due to butyrate. Biochemistry 31:3222–3229

60. Bonifer C, Vidal M, Grosveld F et al (1990) Tissue specific and position independent expression of the complete gene domain for chicken lysozyme in transgenic mice. EMBO J 9:2843–2848

61. McKnight RA, Shamay A, Sankaran L et al (1992) Matrix-attachment regions can impart position-independent regulation of a tissue-specific gene in transgenic mice. Proc Natl Acad Sci U S A 89:6943–6947

62. Lichtenstein M, Keini G, Cedar H et al (1994) B cell-specific demethylation: a novel role for the intronic kappa chain enhancer sequence. Cell 76:913–923

63. Forrester WC, Fernandez LA, Grosschedl R (1999) Nuclear matrix attachment regions antagonize methylation-dependent repression of long-range enhancer-promoter interactions. Genes Dev 13:3003–3014

64. Girod PA, Mermod N (2003) Use of scaffold/matrix-attachment regions for protein production. Elsevier Science B.V. Makrides SC (Ed.) Gene Transfer and Expression in Mammalian Cells, Chapter 10

65. Kalos M, Fournier R (1995) Position-independant transgene expression mediated by boundary elements from the apoliprotein B chromatin domain. Mol Cell Biol 15:198–207

66. Ottaviani D, Lever E, Takousis P et al (2008) Anchoring the genome. Genome Biol 9:201

67. Harraghy N, Gaussin A, Mermod N (2008) Sustained transgene expression using MAR elements. Curr Gene Ther 8:353–366

68. Mielke C, Kohwi Y, Kohwi-Shigematsu T et al (1990) Hierarchical binding of DNA fragments derived from scaffold-attached regions: correlation of properties in vitro and function in vivo. Biochemistry 29:7475–7485

69. Allen GC, Hall G Jr, Michalowski S et al (1996) High-level transgene expression in plant cells: effects of a strong scaffold attachment region from tobacco. Plant Cell 8:899–913

70. Stehle IM, Postberg J, Rupprecht S et al (2007) Establishment and mitotic stability of an extra-chromosomal mammalian replicon. BMC Cell Biol 8:33

71. Bigger BW, Tolmachov O, Collombet JM et al (2001) An araC-controlled bacterial cre expression system to produce DNA minicircle vectors for nuclear and mitochondrial gene therapy. J Biol Chem 276:23018–23027

72. Chen ZY, He CY, Ehrhardt A et al (2003) Minicircle DNA vectors devoid of bacterial DNA result in persistent and high-level transgene expression in vivo. Mol Ther 8:495–500

73. Darquet AM, Rangara R, Kreiss P et al (1999) Minicircle: an improved DNA molecule for in vitro and in vivo gene transfer. Gene Ther 6:209–218

74. Nehlsen K, Broll S, Bode J (2006) Replicating minicircles: generation of nonviral episomes for the efficient modification of dividing cells. Gene Ther Mol Biol 10:233–244

75. Riu E, Chen ZY, Xu H et al (2007) Histone modifications are associated with the persistence or silencing of vector-mediated transgene expression in vivo. Mol Ther 15:1348–1355

76. Vaysse L, Gregory LG, Harbottle RP et al (2006) Nuclear-targeted minicircle to enhance gene transfer with non-viral vectors in vitro and in vivo. J Gene Med 8:754–763

77. Zhang X, Epperly MW, Kay MA et al (2008) Radioprotection in vitro and in vivo by minicircle plasmid carrying the human manganese superoxide dismutase transgene. Hum Gene Ther 19:820–826

78. Chang CW, Christensen LV, Lee M et al (2008) Efficient expression of vascular endothelial growth factor using minicircle DNA for angiogenic gene therapy. J Control Release 125:155–163

79. Kim S, Landy A (1992) Lambda Int protein bridges between higher order complexes at two distant chromosomal foci attL and attR. Science 256:198–203

80. Abremski K, Hoess R (1984) Bacteriophage P1 site-specific recombination. Purification and properties of the Cre recombinase protein. J Biol Chem 259:1509–1514

81. Sternberg N, Sauer B, Hoess R et al (1986) Bacteriophage P1 cre gene and its regulatory region. Evidence for multiple promoters and for regulation by DNA methylation. J Mol Biol 187:197–212

82. Thorpe HM, Wilson SE, Smith MC (2000) Control of directionality in the site-specific recombination system of the Streptomyces phage phiC31. Mol Microbiol 38:232–241

83. Groth AC, Olivares EC, Thyagarajan B et al (2000) A phage integrase directs efficient site-specific integration in human cells. Proc Natl Acad Sci U S A 97:5995–6000

84. Chen L, Woo SL (2005) Complete and persistent phenotypic correction of phenylketonuria in mice by site-specific genome integration of murine phenylalanine hydroxylase cDNA. Proc Natl Acad Sci U S A 102:15581–15586

85. Benzinger R, Enquist LW, Skalka A (1975) Transfection of Escherichia coli spheroplasts. V. Activity of recBC nuclease in rec + and rec minus spheroplasts measured with different forms of bacteriophage DNA. J Virol 15: 861–871

86. Buchholz F, Ringrose L, Angrand PO et al (1996) Different thermostabilities of FLP and Cre recombinases: implications for applied site-specific recombination. Nucl Acids Res 24: 4256–4262

87. Mislick KA, Baldeschwieler JD (1996) Evidence for the role of proteoglycans in cation-mediated gene transfer. Proc Natl Acad Sci U S A 93:12349–12354

88. Gharwan H, Wightman L, Kircheis R et al (2003) Nonviral gene transfer into fetal mouse livers (a comparison between the cationic polymer PEI and naked DNA). Gene Ther 10: 810–817

89. Pollard H, Remy JS, Loussouarn G et al (1998) Polyethylenimine but not cationic lipids promotes transgene delivery to the nucleus in mammalian cells. J Biol Chem 273: 7507–7511

90. Wong SP, Argyros O, Howe SJ et al (2010) Systemic gene transfer of polyethylenimine (PEI)-plasmid DNA complexes to neonatal mice. J Control Release 150:298–306

91. Felt O, Buri P, Gurny R (1998) Chitosan: a unique polysaccharide for drug delivery. Drug Dev Ind Pharm 24:979–993

92. Tabata H, Nakajima K (2001) Efficient in utero gene transfer system to the developing mouse brain using electroporation: visualization of neuronal migration in the developing cortex. Neuroscience 103:865–872

93. Saito T, Nakatsuji N (2001) Efficient gene transfer into the embryonic mouse brain using in vivo electroporation. Dev Biol 240: 237–246

94. Sato M, Tanigawa M, Kikuchi N (2004) Nonviral gene transfer to surface skin of mid-gestational murine embryos by intraamniotic injection and subsequent electroporation. Mol Reprod Dev 69:268–277

95. Garcia-Frigola C, Carreres MI, Vegar C et al (2007) Gene delivery into mouse retinal ganglion cells by in utero electroporation. BMC Dev Biol 7:103

96. Kay MA, He CY, Chen ZY (2010) A robust system for production of minicircle DNA vectors. Nat Biotechnol 28:1287–1289

97. Koping-Hoggard M, Tubulekas I, Guan H et al (2001) Chitosan as a nonviral gene delivery system. Structure-property relationships and characteristics compared with polyethylenimine in vitro and after lung administration in vivo. Gene Ther 8:1108–1121

98. Yang PT, Hoang L, Jia WW et al (2011) In utero gene delivery using Chitosan-DNA nanoparticles in mice. J Surg Res 171(2):691–9

99. Sase M, Miwa I, Sumie M et al (2005) Gastric emptying cycles in the human fetus. Am J Obstet Gynecol 193:1000–1004

Chapter 8

Use of Manipulated Stem Cells for Prenatal Therapy

Jessica L. Roybal, Pablo Laje, Jesse D. Vrecenak, and Alan W. Flake

Abstract

Prenatal stem cell therapy has broad potential for therapeutic application. "Stem cells" of interest include multipotent adult-derived stem cells, cord blood, amniotic fluid, or fetal stem cells, and embryonic or induced pluripotent stem cells. Potential manipulations of stem cells prior to their administration may include harvest, processing, enrichment, expansion, and genetic transduction. A complete description of the methodology related to all of the above is well beyond the scope of this chapter. In the interest of practical application and proven efficacy, we limit our description to adult-derived hematopoietic stem cells (HSCs) and their application to in utero transplantation with or without HSC-targeted gene transfer.

Key words: In utero transplantation, Hematopoietic stem cell, Lentiviral vector, Gene transfer

1. Introduction

The hematopoietic stem cell (HSC) remains the most well-characterized and clinically useful stem cell identified thus far. The paradigm by which stem cells are defined and most of the technical advances related to the unique identification, isolation, enrichment, and validation of stem cell populations, were developed in the race to characterize and purify the HSC (1). Although this chapter must by necessity limit itself to representative examples of stem cell manipulations for the purpose of gene transfer and in utero transplantation, there are many overriding considerations presented that apply to the manipulation of any stem cell population from any tissue.

Generally speaking, all stem cells are characterized by a set of surface molecules, defined within the "cluster of differentiation" (CD) protocol, which alone or in combination have specificity for a cell population that satisfies the biological activity of the stem cell

Charles Coutelle and Simon N. Waddington (eds.), *Prenatal Gene Therapy: Concepts, Methods, and Protocols*,
Methods in Molecular Biology, vol. 891, DOI 10.1007/978-1-61779-873-3_8, © Springer Science+Business Media, LLC 2012

in question. Thus, HSCs are defined by CD surface markers that have been validated to identify cells that functionally exhibit the biological activity of hematopoietic reconstitution. In general, stem cells function as a network within a tissue, residing in a "niche" that determines the functional state of the stem cell (2, 3). Most stem cell populations in adults are relatively quiescent, and exist at very low frequencies, proliferating in response to homeostatic, regenerative, or repair signals. This presents one of the major challenges for the experimental or therapeutic utilization of stem cells and is the primary reason for the need for extensive processing prior to utilization. In the case of adult bone marrow, the HSC has a frequency of 1 in 25,000–100,000 BM cells. The utility of these cells in various experimental and clinical protocols requires an adequate HSC dose, varying degrees of HSC purity, and may be hindered by the presence of associated cells that normally reside in the bone marrow compartment. The purposes of manipulation of the HSC containing graft include (1) enrichment or purification of the desired stem cell population; (2) depletion of unwanted or detrimental associated cell populations; (3) enhancement of stem cell homing, engraftment, or function; (4) stem cell expansion; (5) immune manipulation of the stem cell graft; and (6) efficient gene transfer to the stem cell population. While stem cell manipulations may be required for successful application, there is a price to pay for each manipulation that is performed and in general, processing steps should be limited to those that are absolutely necessary. The most important sacrifice is cell yield. Even seemingly minimal manipulations such as gradient centrifugation can reduce the number of cells by 20–30%. Maintenance or "expansion" of the cells in culture media with or without specific growth factors may result in differentiation of the stem cells and reduction in their reconstituting capacity. Thus, great care must be taken in developing protocols involving manipulation of stem cells to preserve their number, capacity for self-renewal, and multipotentiality. The protocol detailed in this chapter describes one approach to isolation, enrichment, and gene transfer to the murine HSC population (4). This protocol represents a practical and proven method for high-efficiency gene transfer to murine HSC using lentiviral vectors.

1.1. Generation and Prenatal Application of Manipulated Stem Cells: An Overview

The theoretical advantages of in utero HSC transplantation (IUHCT) and in utero gene therapy (IUGT) have been outlined in numerous previous reviews (5, 6). They include the immature status of the fetal immune system prior to thymic processing of self-antigen, normal events in hematopoietic ontogeny that may provide opportunities for engraftment of donor cells without the need for myeloablation, and the very small size of the fetus, allowing very high relative doses of cells or vector particles to be delivered. In addition, for diseases that are manifest at birth, prenatal treatment offers the only hope of preempting clinical disease.

From the perspective of gene therapy, the same advantages apply, but in addition, there are opportunities to access stem cell populations during normal ontogeny that do not exist at other times in gestation (7). With respect to HSC for instance, in situ transduction can be achieved by systemic administration of lentiviral vectors (via E10 intracardiac injection in the mouse—Flake AW, Unpublished data). Similar in situ transduction has not been demonstrated in any postnatal model and it is probably related to fixation of HSC in the dorsal aorta at this early stage of hematopoietic ontogeny.

Unfortunately, there are also disadvantages to IUHCT and prenatal gene therapy. In reality, the fetal hematopoietic system presents a formidable competitive barrier to IUHCT. Fetal HSCs are present in excess in the circulation and rapidly populate available hematopoietic niches during the migration of hematopoiesis to the fetal liver and subsequently the bone marrow (8). The risk of toxicity of available myeloablative approaches is prohibitive in the fetus due to the rapid proliferation of all fetal tissue compartments and the high likelihood of developmental impact during the optimal time frame for IUHCT. This barrier has yet to be sufficiently overcome to provide high-level hematopoietic chimerism adequate for the treatment of most target diseases. An important consideration that has recently come to light is the potential for maternal immunization by IUHCT and possibly IUGT (9). This can in turn result in the induction by the maternal immune system of an adaptive immune response in the fetus/newborn resulting in loss of chimerism or gene expression. Obviously, for IUHCT, the use of maternal donor cells is a potential method to avoid the immune barrier because in the absence of a maternal immune response, no adaptive response occurs and tolerance based on deletional and regulatory mechanisms is induced. Concerns unique to gene therapy include the potential for germ line transduction, insertional mutagenesis, and developmental abnormality from ectopic or poorly regulated transgene expression ((10, 11), see also Chapters 15, 16, and 17). These concerns are if anything heightened in the fetus relative to the postnatal patient, particularly with systemic administration of viral vectors, and will need to be overcome before systemic IUGT becomes a clinical reality. However, manipulation of cells, ex vivo, by gene transfer with subsequent IUGT represents a somewhat safer option that may be clinically applicable in the foreseeable future.

In this chapter, we present a method for isolation and ex vivo transduction of murine hematopoietic cells using an abbreviated protocol with significantly fewer steps than traditional approaches. HSC enrichment has historically been accomplished by lineage depletion and positive selection for specific cell surface markers (1), or by taking advantage of dye efflux properties that are specific to HSC (12, 13), or a combination of both. The most commonly used strategy for cell surface marker selection has been the c-Kit$^+$,

Sca-1⁺, lineage marker negative (lin⁻/ˡᵒ) cell surface scheme to isolate the so-called KSL cells. While this scheme allows isolation of an enriched population of HSC, it requires a large number of antibodies and extensive manipulation of cells. More recently, the *signaling lymphocyte activation molecule* (SLAM) family of cell surface markers has been identified as useful for the isolation of highly enriched murine HSCs (14, 15). This HSC enrichment protocol is relatively simple, and results in an enriched HSC product that has been shown to have comparable repopulating capacity to KSL HSC. Protocols for lentiviral transduction of HSCs have evolved over the years with a recent trend toward abbreviated culture times and minimal cytokine exposure to avoid differentiation and loss of long-term repopulating capacity after transplantation. In addition to loss of repopulating capacity, transduction of HSCs has been shown to result in loss of homing capacity and may have other detrimental effects. Thus, the process of HSC enrichment and subsequent gene transfer may result in significant reduction of HSC repopulating capacity by a variety of mechanisms. We recently demonstrated the utility of a simple SLAM enrichment protocol and abbreviated transduction protocol for hematopoietic gene transfer studies (4). Specifically, we demonstrated that a SLAM-enriched HSC population was equal or better than KSL cells with respect to (1) transduction efficiency and (2) in vivo long-term repopulating capacity after lentiviral transduction. We describe this protocol below.

2. Materials

2.1. Isolation of Murine Hematopoietic Progenitors and Stem Cells

2.1.1. Bone Marrow Harvest

1. Adequate number of 4–8-week-old donor mice.
2. Instruments: Forceps, scissors.
3. 3- and 10-cc syringes.
4. 20- and 26-g needles.
5. 100×15-mm Petri dish.
6. 70-μm cell strainer.
7. 50-mL conical tubes.
8. Ice bucket.
9. 70% ethanol.
10. Cold phosphate-buffered saline (PBS).
11. 5-fluorouracil (5-FU).

2.1.2. Islolation of Bone Marrow Mononuclear Cells

1. Ficoll-Paque Plus (GE Healthcare, Uppsala, Sweden #17-1440-03).
2. Automatic pipettor.

3. 10- and 25-mL pipettes.

4. Micropipettor P20, P1000.

5. 20- and 1,000-μL pipette tips.

6. 50-mL conical tubes.

7. 1.7-mL microtube.

8. Cold PBS.

9. Trypan Blue (1:10 dilution in PBS).

10. Centrifuge.

11. Hemacytometer.

2.1.3. Flow Cytometric Sorting

1. Micropipettor P20, P200, P1000.

2. 20-, 200-, and 1,000-μL micropipette tips.

3. Automatic pipettor.

4. 10-mL pipettes.

5. 15-mL conical tubes.

6. Centrifuge.

7. Vortex.

8. Cold PBS.

9. Anti-mouse CD48 antibody, fluorochrome conjugated (usually FITC).

10. Anti-mouse CD150 antibody, fluorochrome conjugated (usually phycoerythrin, PE).

11. Anti-mouse CD45 antibody for each fluorochrome used.

12. Flow Cytometric Sorter (BD FACSAria, BD Biosciences, Inc., San Jose, CA).

2.1.4. Magnetic Sorting

1. 15-mL conical tubes.

2. Centrifuge.

3. Vortex.

4. Cold PBS.

5. Anti-mouse CD48 antibody, fluorochrome conjugated (usually FITC).

6. Anti-mouse CD150 antibody, fluorochrome conjugated (usually PE).

7. Magnetic beads for each fluorochrome used (Miltenyi Biotec, Auburn, CA, FITC—#130-048-701, PE—#130-048-801).

8. Magnetic Cell Separator (Miltenyi Biotec, Auburn, CA, QuadroMACS #130-090-976 w/multi-stand #130-042-303).

9. MACS buffer (PBS, 0.5% Bovine Serum Albumin, 2 mM EDTA, as per manufacturer's instructions).

10. LD depletion columns (Miltenyi Biotec, Auburn, CA #130-042-901).

11. LS selection columns (Miltenyi Biotec, Auburn, CA #130-042-401).

If performing MACS lineage depletion:

12. Lineage depletion kit (Biotin Lineage Antibody Cocktail, anti-Biotin Microbeads, Miltenyi Biotec, Auburn, CA #130-090-858).

2.2. Genetic Manipulation of Murine Hematopoietic Progenitors and Stem Cells

2.2.1. Lentiviral Transduction of HPS/HSC

1. 48-well non-tissue culture plate.

2. Incubator with 5% CO_2 at 37°C.

3. Retronectin (recombinant fibronectin) (Takara Bio, Inc., Shiga, Japan #T100 A/B).

4. StemPro-34 serum-free media (Invitrogen, Carlsbad, CA #10639-011).

5. Recombinant murine stem cell factor.

6. Recombinant murine thrombopoietin.

7. Lentiviral vector encoding for gene of interest and reporter gene.

2.2.2. In Utero Transplantation of Transduced HPS/HSC by Vitelline Vein Injections

1. Anesthesia system.

2. Oxygen.

3. Isoflurane.

4. Alcohol pads.

5. Cotton swabs.

6. Instruments: Forceps, scissor, needle driver.

7. 4–0 Vicryl Suture (Ethicon, Somerville, NJ).

8. Tape.

9. Micropipettor P200.

10. 200-µL micropipette tips.

11. Glass micropipettes—We recommend pulling your own pipettes to achieve an outer diameter between 70 and 90 µm using the following:

   (a) Micropipet puller (Sutter Instrument Co., Novato, CA; model P-30).

   (b) Micropipet beveller (Sutter Instrument Co., Novato, CA; model Bv-10) and micropipets (Fisher Scientific, Pittsburgh, PA).

   (c) Microscope reticule (0.5-mm range with 0.01 minor delineations and 0.05 major delineations is ideal for measuring the external diameter of the micropipets).

   (d) Stage micrometer (to ensure that the reticle delineations are correct).

12. An IM 300 programmable microinjector (MicroData Instrument, Inc., S. Plainfield, NJ) or tubing with attachment for micropipette.

13. Nitrogen gas for the microinjector.

14. Microinjector tubing.

15. Operating microscope (range 4–40×).

16. Flunixin Meglumine.

17. Clean cages.

18. Heating lamp.

19. Warm PBS.

20. Cells in a 0.5-mL clear microtube on ice.

3. Methods

3.1. Isolation of Murine Hematopoietic Progenitors and Stem Cells

3.1.1. Bone Marrow Harvesting

Mice between 4 and 8 weeks of age are the most commonly used source of HPS/HSC. Donor mice can be treated with 5-FU (150 mg/kg body weight) before harvest to eliminate the committed progenitors and force quiescent stem cells into the cell cycle since they must divide to replace the ablated hematopoietic system; however, it is well known that this manipulation can modify the surface marker profile of the cells (16).

1. Euthanize mice by cervical dislocation.

2. Position mice in prone position and spray their back with 70% ethanol.

3. Remove the skin from the lower extremities, separate the lower extremities from the body at the coxofemoral joint, and place them in cold PBS.

4. Remove the muscles around the femurs and tibias (using forceps and scissors), disarticulate the femorotibial joint, individualize each bone, and collect all bones in cold PBS. See Note 1.

5. Once all the bones have been cleaned and isolated, the bone marrow must be extracted by flushing the medullary cavity out of the diaphysis. For this purpose, a 26-gauge needle attached to a 3-ml syringe is inserted through the epiphysis of the bone into the marrow cavity, and ice-cold PBS is forcefully flushed through the marrow cavity. During this process, the bones must be submerged under cold PBS (in a Petri dish) so that the PBS and the casts of bone marrow that flush out of the bone do not create bubbles. In general, two to four flushes are required to extract all the marrow from each bone.

6. Once all the bones have been flushed, the bone marrow casts must be made into a single-cell suspension. Maintaining the Petri dish on ice, a 10-ml syringe is used to break apart the clumps by repeated aspiration of the bone marrow through a 20-gauge needle. Finally, the suspension is filtered through a 70-micron strainer in order to eliminate small debris. The cold PBS suspension is then placed in a 50-ml conical tube and diluted to achieve a 50-ml volume. See Note 2.

3.1.2. Isolation of Bone Marrow Mononuclear Cells

1. Using sterile technique, 10 ml of Ficoll is carefully layered under the cell suspension with a 10-ml pipette and an automatic dispenser.

2. The 50-ml tube is centrifuged for 30–40 min at room temperature, at $650 \times g$, with no brake at the end of the cycle. Following this, the tube is carefully retrieved from the centrifuge and the white layer of mononuclear cells can be easily seen on top of the Ficoll layer.

3. With a 10-ml pipette, the mononuclear cell layer is gently aspirated placing the tip of the pipette just on top of the layer. Aspirate around the circumference of the tube to retrieve the maximum number of cells (some Ficoll will unavoidably be aspirated with the cells).

4. Place the cells into a new 50-ml conical tube and fill it to a final volume of 50 ml in cold PBS.

5. Centrifuge at 4°C for 15 min/500 g to obtain a pellet. Pour out the PBS, dilute the pellet, and resuspend in 50 ml of cold PBS. Repeat the wash once. Pour out the PBS and resuspend the pellet in 3–5 ml of cold PBS.

6. Following this, a small aliquot of the suspension will be used to count the cells with a hemocytometer. On average, between 30 and 50 million bone marrow mononuclear cells (BMMNCs) should be obtained per adult mouse after Ficoll separation. See Note 3.

3.1.3. Isolation of Hematopoietic Progenitor/ Stem Cells (HPC/HSC)

There are two methods to isolate HSCs based on their surface markers: flow cytometric cell sorting and magnetic sorting. Magnetic sorting is generally more practical for a large number of cells and the purity is less than flow cytometric sorts. A common strategy is to use the magnetic sorter for lineage depletion followed by a flow cytometric positive selection of the vastly reduced number of cells. The method chosen depends upon the number and desired purity of the cells processed.

Flow Cytometry Cell Sorting

1. Once the BMMNCs have been turned into a single-cell suspension, concentrate them to 250 million cells/ml of cold PBS

in a 15-ml conical tube (ideally, the tube should have no more than 5 ml of suspension).

2. Add a fluorochrome-conjugated anti-mouse CD48 (usually FITC) and a different fluorochrome-conjugated anti-mouse CD150 (usually PE) at the concentration recommended by the manufacturer (generally, 0.1–0.5 µg of antibody per million cells).

3. Incubate the cells with antibodies for 30 min at 4°C. Dilute the suspension to a total of 15 ml in the 15-ml conical tube with cold PBS and centrifuge for 10 min at 500 g/4°C. Pour the PBS to eliminate the unbound antibodies, dilute the pellet, resuspend in cold PBS, and repeat the washing cycle. See Note 4.

4. Dilute the pellet to reach a concentration of 50–100 million cells/ml of PBS for FACS sorting.

5. In order to perform an adequate FACS sorting, a number of control samples must be prepared, which include unstained cells (negative control) and positive controls for all the used fluorochromes. See Notes 5 and 6.

Magnetic Cell Sorting

1. Once the BMMNCs have been turned into a single-cell suspension, concentrate them to 250 million cells/ml of cold PBS in a 15-ml conical tube (ideally, the tube should have no more than 5 ml of suspension).

2. Add a fluorochrome-conjugated anti-mouse CD48 (usually FITC) at the concentration recommended by the manufacturer. Incubate the cells for 30 min at 4°C. Dilute the suspension to a total of 15 ml in the 15-ml conical tube with cold PBS and centrifuge for 10 min at 500 g/4°C. Pour the PBS to eliminate the unbound antibodies, dilute the pellet, resuspend in cold PBS, and repeat the washing cycle. Finally, pour the PBS, and resuspend the pellet in 90 µL of PBS per 10^7 cells.

3. Add anti-FITC magnetic beads at the concentration recommended by the manufacturer (usually 10 µL per 10^7 cells). Mix well and incubate for 15 min at 4°C.

4. Resuspend cells at 1 ml per 10^7 cells and centrifuge at 300 g for 10 min. Eliminate the supernatant and resuspend the cells at 10^8 cells per 500 µL of PBS.

5. Proceed to perform the CD48+ depletion according to the manufacturer's directions. Briefly, prime the columns, place them in the magnetic field, apply the cell suspension, collect the cells that pass through the column, and rinse the columns with PBS. Each column has a maximum number of cells that can be passed and a maximum number of cells that can be held, so it is strongly recommended to work according to these limits. Additionally, it is recommended that a small aliquot of the

collected cells is run in a flow cytometer to assess the efficacy of the depletion. Given that the CD48 is expressed in a high percentage of the BMMNC, it is usually necessary to repeat the depletion process two or even three times.

6. After the last depletion cycle is performed, collect the CD48– cells and count them. Add a fluorochrome-conjugated anti-mouse CD150, usually phycoerythrin (PE), and incubate for 30 min at 4°C. Dilute the suspension to a total of 15 ml in the 15-ml conical tube with cold PBS and centrifuge for 10 min at 500 g/4°C. Pour the PBS to eliminate the unbound antibodies, dilute the pellet, resuspend in cold PBS, and repeat the washing cycle. Finally, pour the PBS and resuspend the pellet in 90 μL of PBS per 10^7 cells.

7. Add anti-PE magnetic beads at the concentration recommended by the manufacturer (usually 20 μL per 10^7 cells). Mix well and incubate for 15 min at 4°C.

8. Resuspend cells at 1 ml per 10^7 cells and centrifuge at $300 \times g$ for 10 min. Eliminate the supernatant and resuspend the cells at 10^8 cells per 500 μL of PBS.

9. Proceed to perform the CD150+ selection according to the manufacturer's directions. Briefly, prime the columns, place them in the magnetic field, apply the cell suspension, discard the cells that pass through the column, and rinse the columns with PBS. After that, remove the column from the magnetic field, place it on a suitable collection tube, and flush out the labeled CD150+ cells by firmly pushing the plunger into the column. Each column has a maximum number of cells that can be passed and a maximum number of cells that can be held, so it is strongly recommended to work according to these limits. It is recommended that a small aliquot of the collected cells is run in a flow cytometer to assess the efficacy of the selection. See Note 7.

10. There are different magnetic sorting companies and different systems within each company. Follow the directions strictly (incubation times, cell concentrations, reagents, buffers, etc.) to optimize the HSC isolation.

3.2. Genetic Manipulation of Murine Hematopoietic Progenitors and Stem Cells

3.2.1. Lentiviral Vector Transduction of HPS/HSC Using an Abbreviated Protocol

The HSCs can be manipulated in vitro to enhance the transduction efficiency; however, it is important to remember that their phenotype and behavior will be modified by all our interventions. Thus, an appropriate balance between transduction efficiency and manipulation must be found for each experiment's purpose. We describe here a brief transduction protocol that involves minimal cell manipulation.

1. Transduction will be carried on a 48-well, non-tissue culture plate. Prepare the wells by coating them with recombinant fibronectin (Retronectin®) according to the manufacturer's instructions.

2. Once the CD150+/CD48– cells are collected and counted, plate up to 5×10^5 cells in 500 μL of StemPro-34® per well. Supplement the media with 100 ng/mL of recombinant murine stem cell factor and 100 ng/mL of recombinant murine thrombopoietin.

3. Add the lentiviral vector at the desired multiplicity of infection (MOI), and incubate the cells for 14–16 h at 37°C in a 5% CO_2-humidified incubator. Ideally, an aliquot of cells should be plated without vector to serve as a negative control at the time of the transduction efficiency assessment.

4. After the incubation, retrieve the cells by gently pipetting the bottom of the wells so that the cells that have decanted become a homogeneous suspension. Place the suspension in a suitable collecting tube. Wash each well twice with cold PBS to remove the maximum number of cells. Centrifuge the cells and dilute them in the volume required for the transplantation. At this time, it is not possible to identify transduced and non-transduced cells because the time required for complete vector integration, transcription, and translation is longer than 14–16 h. See Note 8.

5. An aliquot of transduced cells should be washed to remove the unbound vector and replated in vector-free media in new uncoated wells to assess the transduction efficiency by flow cytometry after a final culture time of 72 h, using non-transduced cells as negative control. To assess the transduction efficiency by flow cytometry, there must be a way to identify at least one protein encoded in the vector backbone either by its own fluorescence (e.g., green fluorescent protein, red fluorescent protein, or others) or by means of a fluorochrome-conjugated antibody.

3.2.2. In Utero Transplantation of Transduced HPS/HSC by Vitelline Vein Injections

Once the cells have been transduced, they must be re-suspended in a volume that is suitable for the desired transplantation protocol. For adult tail-vein injections, a maximum volume of 300–400 μL can be used. For in utero transplantation, the final volume varies with the anatomical site of the transplantation. For vitelline vein injections, the maximum volume is 20 μL for E14 fetuses, but it may need to be reduced to 15 or even 10 for E13 fetuses. For in utero intra-hepatic injections at E14, the maximum volume to inject is 5 μL, similar to that for intraperitoneal injections. Other anatomical targets may require different volumes. The technique for vitelline vein injections as well as details of animal husbandry and anesthesia are described in Chapter 10.

4. Notes

1. Residual muscle and/or connective tissue can act as a sink for bone marrow cells during the flushing process. Careful cleaning of the bones increases cellular yield.

2. A typical BM harvest of the four long bones yields between 50 and 80×10^6 MNCs depending upon the age, sex, and strain of mouse. This represents only approximately 15% of BM cellularity. If higher numbers of cells are needed, particularly for stem cell enrichment, it is more economical to harvest the whole skeleton, although much more laborious to harvest the cells. We would recommend using this technique only if there is a team of people harvesting because dissection and crushing the iliac and vertebral bones are time consuming which can reduce the viability of the cells. A typical complete skeleton harvest yields $400–500 \times 10^6$ MNCs (17).

3. An alternative to Ficoll separation is red cell lysis with ACK lysis buffer. However, we have had no better yield of cells using this technique and it often increases the clumping of cells leading to difficult injections.

4. Samples should be protected from light during this process to avoid degradation of the fluorochromes, particularly if secondary conjugates are used. Aluminum foil can be used to prevent inadvertent light exposure.

5. Antibody to any strongly expressed marker can be used for positive controls. Because CD45 is highly expressed in hematopoietic cells, we recommend its use for this purpose.

6. In order to minimize cell loss during flow cytometric sorting, be sure to sort cells into a large enough volume to decrease the risk of cells sticking to the sides of the tube. In a 5-mL FACS tube, we recommend sorting into at least 2 mL.

7. To maximize cell recovery, flush the columns at least three times.

8. The transduction efficiency of this protocol in our hands is 60–70% of HSC as determined by GFP expression in aliquots of cells maintained in liquid suspension culture for 3 days to allow time for GFP expression. It is critically dependent upon the quality and titer of the vector utilized. Optimal transduction is with an MOI of around 100. We utilize an HIV lentiviral backbone with the MND promoter.

References

1. Spangrude GJ, Heimfeld S, Weissman IL (1988) Purification and characterization of mouse hematopoietic stem cells. Science 241:58–62

2. Flake AW (2004) The conceptual application of systems theory to stem cell biology: a matter of context. Blood Cells Mol Dis 32:58–64

3. Badillo AT, Flake AW (2007) The regulatory role of the stromal microenvironment in fetal hematopoietic ontogeny. Stem Cell Rev 2: 241–246

4. Laje P, Zoltick PW, Flake AW (2010) SLAM-enriched hematopoietic stem cells maintain long-term repopulating capacity after lentiviral transduction using an abbreviated protocol. Gene Ther 17(3):412–418

5. Flake AW, Zanjani ED (1999) In utero hematopoietic stem cell transplantation: ontogenic opportunities and biologic barriers. Blood 94:2179–2191

6. Roybal JL, Santore MT, Flake AW (2010) Stem cell and genetic therapies for the fetus. Semin Fetal Neonatal Med 15:46–51

7. Endo M, Henriques-Coelho T, Zoltick PW et al (2010) The developmental stage determines the distribution and duration of gene expression after early intra-amniotic gene transfer using lentiviral vectors. Gene Ther 17:61–71

8. Christensen JL, Wright DE, Wagers AJ, Weissman IL (2004) Circulation and chemotaxis of fetal hematopoietic stem cells. PLoS Biol 2:E75

9. Merianos DJ, Tiblad E, Santore MT et al (2009) Maternal alloantibodies induce a postnatal immune response that limits engraftment following in utero hematopoietic cell transplantation in mice. J Clin Invest 119:2590–2600

10. RAC (2000) Prenatal gene tranfer: scientific, medical, and ethical issues: a report of the Recombinant DNA Advisory Committee. Hum Gene Ther 11:1211–1229

11. Themis M, Waddington SN, Schmidt M et al (2005) Oncogenesis following delivery of a nonprimate lentiviral gene therapy vector to fetal and neonatal mice. Mol Ther 12: 763–771

12. Goodell MA, Brose K, Paradis G et al (1996) Isolation and functional properties of murine hematopoietic stem cells that are replicating in vivo. J Exp Med 183:1797–1806

13. Goodell MA, Rosenzweig M, Kim H et al (1997) Dye efflux studies suggest that hematopoietic stem cells expressing low or undetectable levels of CD34 antigen exist in multiple species. Nat Med 3:1337–1345

14. Kiel MJ, Yilmaz OH, Iwashita T et al (2005) SLAM family receptors distinguish hematopoietic stem and progenitor cells and reveal endothelial niches for stem cells. Cell 121: 1109–1121

15. Yilmaz OH, Kiel MJ, Morrison SJ (2006) SLAM family markers are conserved among hematopoietic stem cells from old and reconstituted mice and markedly increase their purity. Blood 107:924–930

16. Randall TD, Weissman IL (1997) Phenotypic and functional changes induced at the clonal level in hematopoietic stem cells after 5-fluorouracil treatment. Blood 89:3596–3606

17. Colvin GA, Lambert JF, Abedi M et al (2004) Murine marrow cellularity and the concept of stem cell competition: geographic and quantitative determinants in stem cell biology. Leukemia 18:575–583

Chapter 9

Animal Models for Prenatal Gene Therapy: Choosing the Right Model

Vedanta Mehta, Donald Peebles, and Anna L. David

Abstract

Testing in animal models is an essential requirement during development of prenatal gene therapy for clinical application. Some information can be derived from cell lines or cultured fetal cells, such as the efficiency of gene transfer and the vector dose that might be required. Fetal tissues can also be maintained in culture for short periods of time and transduced ex vivo. Ultimately, however, the use of animals is unavoidable since in vivo experiments allow the length and level of transgene expression to be measured, and provide an assessment of the effect of the delivery procedure and the gene therapy on fetal and neonatal development. The choice of animal model is determined by the nature of the disease and characteristics of the animal, such as its size, lifespan, and immunology, the number of fetuses and their development, parturition, and the length of gestation and the placentation. The availability of a disease model is also critical. In this chapter, we discuss the various animal models that can be used and consider how their characteristics can affect the results obtained. The projection to human application and the regulatory hurdles are also presented.

Key words: Disease models, Fetal size, Fetal growth, Fetal development, Gestation length, Lifespan, Fetal immunity, Immune responses, Gene therapy, Vectors, Gene transfer, Gene expression, Fetal number, Parturition, Preterm birth, Placenta, Placentation, Prenatal diagnosis

1. Introduction

The choice of whether to use animals and then selecting the appropriate animal model is important when developing prenatal gene therapy for clinical treatment. Cell lines can provide initial data as to the efficiency of gene transfer according to vector dose, vector constructs, or transduction enhancement agents. The information is limited, however, because the manipulations that need to be performed to produce cell lines alter their characteristics and growth

Charles Coutelle and Simon N. Waddington (eds.), *Prenatal Gene Therapy: Concepts, Methods, and Protocols*,
Methods in Molecular Biology, vol. 891, DOI 10.1007/978-1-61779-873-3_9, © Springer Science+Business Media, LLC 2012

so that they are likely to behave very differently to fetal cells. Culture of fetal cells is possible, and human cells can be sourced such as from termination of pregnancies. For example, Castillon et al. cultured 3D spheroid structures using human fetal airway epithelial cells and showed that after lentivirus transduction, the cells were able to reconstitute a well-differentiated human airway surface epithelium in SCID mice (1). Nevertheless, the characteristics of embryonic and fetal cells change with gestation, and therefore culture of cells from a fetus for even a short time can mean that their behaviour is not a true reflection of their status at the time of harvest.

Fetal tissues can be maintained in culture for short periods of time and transduced ex vivo. This can prove useful to assess methods of improving gene transfer after manipulation of the vector construct, or the use of transduction enhancement agents. Moreover, experiments can be multiplied to explore different experimental permutations using tissues taken from the same animal. This fulfils one of the principles of the 3Rs, to reduce the number of animal experiments, which is a widely accepted ethical framework for conducting scientific experiments using animals humanely (2). For example, to compare the effect of sodium caprate pretreatment or complexation with DEAE dextran on adenovirus-mediated gene transfer, sections of fetal sheep small bowel (3) and trachea (4) were maintained under culture conditions for 28 h after adenovirus infection.

Ex vivo gene transfer to human fetal tissues can also be performed. In a series of elegant experiments, Lim et al. studied gene transfer to sections of human fetal trachea from mid-gestation pregnancies, either in organ culture (5) or after xenotransplantation into SCID mice (6). Fetal trachea organ culture and xenografts were maintained for 4 weeks and 9 months, respectively, after gene transfer.

Ultimately, however, the use of animals is unavoidable. In vivo experiments allow the length and level of transgene expression to be measured. They also provide an assessment of the effect of the delivery procedure and the gene therapy on fetal and neonatal development. This cannot be achieved using in vitro or ex vivo experiments.

2. Factors to Consider When Choosing an Animal Model

When choosing an animal model in which to study prenatal gene therapy, it is important to consider a number of factors. These include the characteristics of the animal, such as its size, lifespan, and immunology, the number of fetuses and their development, and finally the length of gestation and the placentation. The availability of a disease model is also critical. Each of these factors is considered below. In many ways, each animal presents certain advantages and disadvantages and it is for the researchers to decide which animal best provides the characteristics that they wish to study. The advantages and disadvantages of the use of rodents, sheep, and non-human primates are addressed in more detail under the relevant chapter sections.

2.1. Animal Models of Disease

Over the past decade, a variety of small animal models of disease have been used to show the effectiveness of prenatal gene therapy for the correction of monogeneic diseases. These include Leber congenital amaurosis in mice (7) and chicken (8), Crigler–Najjar syndrome in rats (9), and haemophilia B (10) and mucopolysaccharidosis (11) in mice.

A few models of disease do exist in larger animals. There are a variety of canine single gene disorders that have proved extremely useful in exploring adult gene therapy. These include muscular dystrophy (12), haemophilia B (13), and dystrophic epidermolysis bullosa (14). Only one, canine α-l-iduronidase (α-ID) deficiency, a model of human mucopolysaccharidosis type 1, has been used to study prenatal gene therapy. The authors developed a retrovirus vector containing human α-ID cDNA, and injected it into the yolk sac and intraperitoneal cavity of seven pups in the same mother in midgestation using a combined laparotomy and ultrasound technique (15). Gene transfer to a variety of tissues was observed in the pups, but transgenic protein expression did not last into adulthood. The study illustrates some of the technical difficulties of working with dogs for prenatal gene therapy approaches, such as the large number of pups per litter, and the need for a combined laparotomy and ultrasound approach because of the relatively small size of the pups.

There are even fewer models of disease in larger animals. The β-mannosidosis goat (16) is one example, wherein deficiency of β-mannosidase results in the accumulation of tri- and disaccharides leading to prenatal neurodegeneration and death in the neonatal period. Recently, a haemophilia A sheep animal model has been re-established using reproductive technologies (17), over 20 years after a chance observation of excessive bleeding in lambs from one ewe within a flock. Transgenic livestock such as sheep have been generated using pronuclear injection or somatic cell nuclear transfer with genetically modified cells, although this process is extremely inefficient. More recently, transgenic sheep have been developed using microinjection of lentivirus vectors into in vivo-generated embryos (18). The work required to develop and maintain these animal colonies, however, is immense and costly. Hence, small animal models are essential to the initial assessment of a new vector delivery system for prenatal treatment of a disease. These models may not fully recapitulate human disease pathogenesis and long-term outcomes of prenatal gene therapy because of significant differences in fetal physiology and development, gestation length, placentation, and lifespan.

2.2. Characteristics of the Animal: Fetal Size and Lifespan

To study the effects of immune maturity on cellular transduction and transgene expression, it is necessary to access the fetus early in gestation, equivalent to the first trimester in humans. The small size of the mouse fetus at early gestation is the main challenge to parenteral vector delivery. Intracerebral (11), intracardiac (19), intrapertitoneal (20), intrahepatic (21), intramuscular (22), and intravenous (via yolk sac vessels) (23) routes of delivery are achievable from around mid-gestation onwards and most murine fetal injections have been performed at E14–18. However, the only practical route of vector delivery before this gestational age in mice is intra-amniotic (24) as the thoracic and peritoneal cavities are too underdeveloped to visualise at the equivalent of the first trimester (<E9). Since the development of the fetal mouse is so rapid and time dependent, small changes in the time of vector application can result in targeting very different organ systems.

This is not the case for larger animals such as the fetal sheep or non-human primate, where development proceeds at a slower pace. Minimally invasive techniques that are used in the clinical practise of fetal medicine have been successfully adapted for the sheep (see Subheading 11.1, Chap. 11) and non-human primate (see Subheading 12.1, Chap. 12). Another advantage of a larger animal is that it is possible to assess the effect of the gene therapy after its delivery while the animal is still in utero. Such serial sampling has been used to detect levels of human factor IX in the fetal sheep plasma after early and late gestation injection of AAV (25), and could feasibly be undertaken in clinical application to confirm transgenic protein expression in an affected fetus, for example.

The short lifespan of the mouse is advantageous when investigating transgenerational effects and the long-term effects of prenatal gene therapy. This became apparent in one study of prenatal gene therapy using a variety of lentivirus vectors, in which in utero injection of non-primate equine infectious anaemia virus vector resulted in a high frequency of liver tumours at least 1 year after birth, whereas injection of primate HIV-1 vector did not (26). Analysis of these tumours by Southern blotting showed each to be clonally derived. Provirus insertion sites and nearby genes were analysed to study their association with cancer development. The use of the fetal mouse as an in vivo model of genotoxicity is described further in Chap. 16.

2.3. Fetal Physiology and Development

2.3.1. Development of the Immune System

The temporal pattern of immune system development during pregnancy is an important consideration in choosing an animal model. One of the main aims in prenatal gene transfer is to achieve immune tolerance to the transgenic protein. The naive immune system in early gestation should provide for tolerance to the vector and transgene to be generated, but much will depend on the stage of the immune system development with respect to the timing of vector injection and the type of antigen introduced. In the mouse, a series of experiments demonstrated the different effects of age on the development of immune tolerance by comparing outcomes in fetal, neonatal, and adult mice injected with the same vector and transgene (27). There was an inverse relationship between anti-vector immunoglobulin (Ig) response and the level and longevity of transgenic protein expression, with mice injected in utero demonstrating sustained transgene activity in the absence of an Ig response to the capsid. On the other hand, injected neonatal and adult mice showed a progressively earlier onset of Ig production, which consequently abolished transgenic protein expression.

The classic experiments that tested the hypothesis of "actively acquired tolerance" were conducted in the fetal mouse (28, 29). Cells from the spleen, testis, or kidney of one strain of mouse were injected into the fetus of another strain at 16–17 days of gestation. The treated mice as adults did not reject skin grafts taken from the donor strain but could reject grafts from a third mouse strain. Injection of neonatal mice elicited neither tolerance nor immunity and a null period of immunological reactivity was proposed. Later experiments showed that either tolerance or immunity could be induced in the neonatal mouse depending on the dose of innoculum (30).

Attempts to induce tolerance in other animals such as the fetal sheep, for example, using the conditions adequate for tolerance induction in the fetal mouse failed to achieve comparable results. This is most likely because of differences in the maturation of the fetal immune response between the two species. In animals with shorter gestations such as the mouse which has a 21-day gestational period, the immune system completes maturation in the postnatal period. Human fetuses, and also sheep, are developmentally more mature than mouse fetuses at the corresponding gestational time points (31).

Much of the early work on the ontogeny of the fetal immune response was performed in the fetal sheep because of the easy surgical manipulation of the fetus, and the thick sheep placenta which does not permit passage of gammaglobulin from mother to fetus (32). This ensures that all antibodies in the fetal serum are derived from the fetus and not from the mother and provides an ideal opportunity to test how a deliberate experimental exposure of the fetus to antigen will affect the development of immune competence. Transplantation experiments in the fetal sheep show that skin allografts are uniformly rejected as early as 77 and 80 days of gestation, term being 145 days in the sheep (33, 34), and renal allografts are similarly rejected when placed at 70 days of gestation (35). Skin grafts at 65 days of gestation, however, were accepted for at least 21 days with no evidence of cellular infiltration, and permanent survival of skin grafts beyond term has been achieved in fetuses transplanted with adult skin allografts at 55 days of gestation (36).

In addition to the time of grafting, the nature of the antigen is also important in determining whether tolerance is induced. Fetal lambs at 55 days of gestation uniformly reject skin allografts from fetal sheep aged up to 85 days but retain skin allografts from fetal sheep at 95 days of gestation and adult sheep (37). Transplantation of other tissues, such as blood products or splenic allografts, into fetal sheep even at 50 and 60 days of gestation is unable to induce tolerance to donor cell challenges as adults (38, 39). The fetal sheep immune response to intrauterine infection also shows a different modality depending on the antigen. Exposure of a fetal sheep to Border disease results in a state of immulogical tolerance after birth and into adulthood, with persistent shedding of viral particles (40). In contrast, exposure of the fetus to Akabane virus infection in early gestation produces a viraemia that persists until 75 days of gestation when neutralizing antibodies appear and eliminate the virus (41). Silverstein and his colleagues determined that the fetal sheep responds to antigenic challenges according to the antigen presented (42, 43). There is a hierarchy of response among the antigens with the earliest antibody produced to bacteriophage φX from 60 days, to ferritin from 96 days and to albumin from 120 days of gestation, but responses to diphtheria toxoid, Salmonella typhosa organisms, and BCG could not be elicited until after birth. The appearance of a particular immune capability was not correlated with the stage of lymphopoiesis.

Experiments using serial fetal blood sampling via vascular catheters implanted into fetuses challenged with various antigens show that most fetuses acquired a capacity to respond to several antigens between 60 and 75 days of gestation. For all antigens tested, there was a higher antibody titre, a longer duration of response, and a wider range of antibody production in older fetuses, showing that the humoral immune response matures through gestation. A similar hierarchy was observed, but it was not particularly precise and there was also a wide variation among individual fetuses (44, 45).

In the human, mitogen-responsive lymphocytes are detectable in the peripheral blood by 0.3–0.4 G (46, 47), and functional natural killer (NK) cells develop at 0.7 G (47). In contrast, maturation of the immune system in the mouse fetus is relatively delayed. T and B lymphocytes appear at 0.85 G (48, 49), and they gain the ability to respond to antigens by a few weeks after birth (50). Morphological analysis of haemopoietic tissue in the liver of a variety of mouse strains shows that erythropoietic, granulopoietic, and lymphopoietic activity persists for approximately 2 weeks after birth (50). There are a number of other important differences between mouse and human immune development, such as the repertoire of expressed Ig isotypes, B- and T-cell development and regulation, and tolerance induction to grafts (51). These differences limit the parallels that can be drawn from experimental mouse data to predict possible human clinical outcomes, particularly in gene therapy.

2.3.2. Fetal and Maternal Immune Response to Vector and Transgene

Despite the notable evidence that prenatal gene transfer can produce long-term transgenic protein expression, the fetal immune response is still a relative barrier to long-term expression. One study in the fetal mouse evaluated multiple routes and viral dose combinations of adenovirus and adeno-associated virus vectors carrying the marker β-galactosidase gene, at 13–15 days post conception (52). In utero injection of either viral vector at any route and dose combination resulted in the generation of low titres of neutralising antibodies to the virus and transgenic protein. This primary immune response only partially blocked transgenic protein expression after the re-administration of viral vectors once post-natally. However, after a second post-natal re-administration, the immune response completely blocked transgenic protein expression. A further example of the potential problem of the fetal immune response can be seen in the rat Crigler–Najjar model that had been gene treated prenatally (53). These animals developed antibodies against bilirubin UDP-glucuronyltransferase, which may be related to the fact that the fetal injection was done late in fetal life.

The maternal immune response to the gene therapy has the potential to limit gene transfer and/or to reduce transgenic protein expression. Maternal allo-antibodies can cross over to the fetus, and have been shown to induce a post-natal immune response that limits engraftment following in utero haematopoietic cell transplantation in mice (54). In similar experiments using mice, maternal T cells were found to limit engraftment after in utero haematopoietic cell transplantation (55). A pre-existing maternal immune response might also be an important factor in preventing transgenic protein expression. Maternal IgG antibodies are able to cross the placental barrier in humans and other mammals. This is especially a consideration in AAV-mediated gene transfer, where the limiting factor seems to be pre-existing memory T-cell immunity to AAV2 in humans who are the only natural hosts to this virus (56). This can be circumvented by applying AAV serotypes that humans are not naturally exposed to, such as AAV8.

These fetal and potentially maternal immune responses to the vector and transgene protein are a reminder that prenatal gene transfer is still subject to immunologic barriers, which might relate to differences in bio-distribution, timing of transduction and expression, and levels of expression. Designing less immunogenic vectors and considering immune conditioning prior to gene delivery, a less desirable option, could be ways to partially overcome the problems. There is a need to elucidate the pathways of fetal immune development and regulation further.

2.4. Characteristics of the Pregnancy

2.4.1. Fetal Number

Many smaller animal species, such as rodents, guinea pigs, and rabbits, have a large number of offspring per litter. This can be a help but also a hindrance to experimental prenatal gene therapy. For treatment of animal models of human disease, it is advantageous to have a number of fetuses for correction per mother, since for most single-gene disorders, only 25% of the offspring will be homozygous and therefore manifest the phenotype. The large number of fetuses that are available, however, can make injection technically difficult due to the need to manipulate fetuses within the uterine horns, and can prolong the anaesthetic time required. Marking of injected fetuses has been performed by some operators using either colloidal carbon or fluorescent beads, which avoids the need to inject the whole litter. Generally, it is simpler however to genotype delivered animals while checking whether the phenotype has been corrected. The larger animals, such as fetal sheep, tend to have only one or two fetuses, which are often identifiable lying in the separate uterine horns. Sheep fetuses can also be marked in utero, for example, with colloidal carbon injection into the skin, although the development of the fleece in late gestation in sheep hinders their identification after birth.

2.4.2. Gestation Length

The gestation length is probably of more importance when considering the type of animal to use. Small animals tend to have short gestation lengths of 18–21 (mouse), 21–23 (rat), or 28–32 days (rabbit). Much of the development that would take place during fetal life in the human, for example that of the immune system or the lungs, occurs in the neonatal period in these animals. The exception to the rule is the guinea pig that has a 63- to 65-day gestation, but which has few if any known equivalents of human genetic disease. The short gestation length of small animals means that evaluation of the effect of prenatal application of gene therapy on fetal growth and development is not possible. Large animals such as sheep and non-human primates that have gestation lengths of 135 and 167 days, respectively, are necessary.

2.4.3. Parturition

Parturition or the onset of labour is an important consideration for any intervention during pregnancy. In all developed countries, preterm birth (<37 weeks of gestation) is associated with approximately 75% of newborn death or permanent disability (57). A preterm birth, particularly if it occurs at the extremes of prematurity, can result in a poor neonatal outcome, even if an affected fetus is cured of its disease. An example of this phenomenon was observed in a randomised control trial of the treatment of a structural fetal malformation, congenital diaphragmatic hernia (CDH) (58). Fetuses in the intervention arm of the trial received endoscopic placement of an inflatable balloon into the trachea to inflate the lungs and prevent pulmonary hypoplasia, while control fetuses received standard care. The rates of neonatal morbidity were no different in the groups. The underlying problem was a significantly higher rate of premature rupture of the membranes (PPROM) and preterm delivery in the group receiving the intervention than in the group receiving standard care (mean [±SD] gestational age at delivery, 30.8 ± 2.0 weeks vs. 37.0 ± 1.5 weeks; $P < 0.001$).

Thus, any intervention that significantly increases the risk of preterm birth, as a consequence of either the invasive procedures or inflammation, is not likely to improve the outcome for the neonate, and may make it worse.

The events leading to parturition in animals are very different from those in the human. Factors, such as infection, antepartum haemorrhage, and myometrial stretch, are significantly associated with preterm labour. In most cases, however, there is no overt, overriding pathological cause, and the mechanism of parturition appears to represent a premature triggering of the normal mechanisms occurring at term.

Much of our understanding of parturition is derived from rodent animal models. This suggests that there are two major mechanisms regulating the timing of parturition: the withdrawal of the steroid hormone progesterone and a proinflammatory response by the immune system. In most experimental animal models such as the sheep, rabbit, rat, and mouse, there is an abrupt withdrawal of progesterone from the maternal circulation before parturition. In the guinea pig and human, however, progesterone levels are maintained at a high and increasing concentration and do not fall at parturition (59). There is some evidence that a "functional" progesterone withdrawal may play a role in human parturition. This could occur, for example, at the level of progesterone receptor expression (60) or due to interaction between the progesterone receptor and transcription factors such as NF-κB (61).

Labour, whether at term or preterm, appears to be an inflammatory process. It is well known that significant intrauterine infection can cause preterm birth (62) and in this situation prompt delivery is indicated and is best for neonatal outcome. In the majority of spontaneous preterm births, however, clear evidence of infection is lacking. Research has suggested that it may be the magnitude of the inflammatory reaction rather than infection per se that is the cause. A number of studies have measured the levels of proinflammatory cytokines in the amniotic fluid of women, both at term and preterm, and also in the presence and absence of positive amniotic fluid bacterial cultures (63).

An important consideration in any prenatal gene therapy study, therefore, will be to determine whether the vector upregulates the inflammatory response leading to preterm birth. Animal models may not prove particularly useful in this regard. Introduction of bacteria via the cervix of pregnant rabbits or their breakdown products into the peritoneal cavity of pregnant mice generally results in preterm labour, but is associated with a dramatic fall in serum progesterone concentrations prior to the onset of parturition, and this may be the mechanism of action (64). In the spontaneous preterm labour syndrome where infection is not suspected to play a causative role, animal models are less useful. In rats, intravenous infusion of interleukin-1 or TNF-α to reach levels similar to those measured in normal human term delivery or preterm delivery in the absence of infection had no effect on uterine contractility or expression of parturition-associated genes (65). In primates, the evidence is more equivocal, with only 3 out of 11 animals delivering preterm after infusion of interleukin-1 into the amniotic cavity of rhesus monkeys preterm (66). The pregnant guinea pig may have some advantages over other models. Treatment with anti-progestin in mid-gestation, for example, increases the sensitivity of the uterus to oxytocin (67), which is a similar finding to human pregnancy. Nevertheless, direct extrapolation of the effect of prenatal gene therapy on preterm birth from animal models to the human situation will be difficult, and the assessment of the risk of preterm birth will be an important consideration in any phase I/II trials.

2.4.4. Placentation

Placentation is of critical importance to the movement of vector, antibodies, or, even theoretically, transduced cells between the fetus and the mother. Vector can cross from the fetus to the mother or vice versa by diffusion and/or active transport. Damage to the placental barrier by the vector or through the injection process could allow further movement. It must be remembered that inadvertent vector transfer to the mother can also occur as the vector-contaminated needle is being removed from the uterine cavity, when using closed uterine delivery techniques such as ultrasound-guided percutaneous injection, although judicious flushing of the needle is likely to prevent this. The differences in the placentation between species are shown in Table 1.

There are three factors to consider in placentation: the maternofetal contact area, maternofetal interdigitation, and maternofetal barrier (68). The extent of maternofetal and fetomaternal transfer depends somewhat on the size of the maternofetal contact area. Four main placental shapes have been identified. In *Placenta diffusa*, the maternal interdigitation extends over the entire surface of the chorionic sac; this is seen in some lower primate species. *Placenta cotyledonaria* has many spot-like regions of intense maternofetal interdigitations and is common to ruminants. In *Placenta zona*, the placenta forms a ring-like zone in the chorionic sac surrounding the fetus like a girdle; this is seen in carnivores. *Placenta discoidalis* represents the highest concentration of maternofetal interdigitation, which provides a single disc-like zone of close maternofetal contact and is seen in rodents, great apes, and humans.

The type of maternofetal interdigitation also differs between species. There are five types. The simplest form is combined with a large-surfaced diffuse placenta, called the *folded type*, and is seen in the pig. It is characterised by poor branching, ridge-like folds of the chorion that fit into corresponding grooves of the uterine mucosa. A more complex construction is seen in the *lamellar type* of placenta, typically found in carnivores. In such a case, the ridges multiply and branch into complicated systems of slender chorionic

Table 1
Placentation classification in various species (adapted from 87)

Species	Placental shape	Maternofetal indigitation	Placental barrier	Fetomaternal blood flow interrelations
Human (3rd trimester)	Discoidal	Villous	Haemomonochorial	Multivillous
Human (1st/2nd trimester)	Discoidal	Villous	Haemodichorial	Multivillous
Primates (higher): Great apes	Discoidal	Villous	Haemomonochorial	Multivillous
Primates (lower): Galago	Diffuse	Folded	Epitheliochorial	–
Rabbit	Discoidal	Labyrinthine	Haemodichorial	Counter-current
Guinea pig	Discoidal	Labyrinthine	Haemomonochorial	Counter-current
Rat	Discoidal	Labyrinthine	Haemotrichorial	Counter-current
Mouse	Discoidal	Labyrinthine	Haemotrichorial	Counter-current
Sheep	Cotyledonary	Villous	Epitheliochorial	Multivillous

lamellae, orientated in parallel to each other and separated by correspondingly branching endometrial folds. Increased exchange surface area is provided by a tree-like branching pattern of the chorion, resulting in the *placental villous tree*. The villi fit into corresponding endometrial crypts or are directly surrounded by maternal blood. This type of placenta exists in higher primates. The *trabecular type* of placenta represents an intermediate stage between the lamellar and the villous situation. It is characterised by branching folds from which leaf-like and finally finger-like villi branch off. This type of placenta has been described for some platyrrhine monkeys, such as callithrix. The most common and efficient type of interdigitation is found in the *labyrinthine type* of placenta and is typical for placentas of rodents. A tissue block of trophoblast is penetrated by web-like channels filled with maternal blood or fetal capillaries.

The maternofetal barrier refers to the tissue layers separating maternal and fetal circulations in the placenta (68). There is significant variation among different species in the completeness of this placental barrier depending on the extent of trophoblast invasion. In the *epitheliochorial placenta*, the fetal chorion directly faces the intact endometrial epithelium. The blastocyst does not invade the endometrium, and only remains attached to it. Six tissue layers separate maternal and fetal blood: maternal capillary endothelium, maternal endometrial connective tissue, and maternal endometrial epithelium and three layers of fetal origin, the trophoblast, chorionic connective tissue, and fetal endothelium. Syncytia are formed by fusion of fetal chorionic binucleate cells and maternal endometrial epithelial cells producing this type of placenta (69). This type of placenta is characteristic of ruminants and it provides a significant barrier. As a result, gammaglobulins do not pass from the mother to the fetus in the sheep, and the health and survival of the neonatal lamb are dependent on receiving antibodies from the mother via the colostrum in the first 24 h of life. In the *endotheliochorial placenta*, there is deeper invasion of trophoblasts and the corresponding removal of endometrial connective tissue. This placental barrier is characteristic for all carnivores. In the *haemochorial placenta*, the final step of trophoblast invasion results in erosion of the maternal vessels, which are finally completely destroyed and the trophoblastic surfaces now face the maternal blood directly (70). Depending on the number of trophoblastic epithelial layers, a more detailed subdivision has been proposed: *haemotrichorial* (rat (70)), *haemodichorial* (rabbit (70), human in the first trimester (71)), and *haemomonochorial* (guinea pig (72), non-human primate and human placenta at term (71)).

At term, the human placenta is more similar to the haemomonochorial placental barrier found in the guinea pig and certain types of non-human primates. This is because the human maternofetal barrier changes over the course of the last trimester of pregnancy as the cytotrophoblast degenerates leaving a single thin layer of syncytiotrophoblast. To study gene transfer from the mother to the fetus, as for example after maternal uterine artery transduction, the most appropriate animal model for human pregnancy during the second trimester would be similar to humans in the transport of substances across the placenta. These data suggest that the rabbit (haemodichorial) offers the desirable features of placentation, which are close to the human situation.

The movement of vector across the placental barrier may also depend on specific receptors being present. Recent research on lentiviruses, such as Feline Immunodeficiency Virus (FIV), suggests a role for a novel chemokine receptor CXCR4 in their transfer across the placenta (73). More is known about adenovirus infection of the trophoblast. Adenovirus virions normally infect susceptible species (e.g. humans, pigs, and cotton rats) by entering cells via the Coxsackie and Adenovirus Receptor (CAR) and $\alpha v \beta$-3 and -5 integrins (74). Koi et al. demonstrated that expression of CAR on human trophoblast cells varies with gestational age and trophoblast phenotype (75). CAR was found to be continuously expressed in invasive cytotrophoblast (fetal side) but not in syncytiotrophoblast, thus rendering the syncytiotrophoblast resistant to adenoviral infection and limiting transplacental transmission in the human. Two further studies in human placenta have showed that the ability of adenoviruses to infect trophoblasts was related to the state of trophoblast differentiation (76, 77). Recombinant adeno-viruses efficiently transduce the inner cytotrophoblast, but there is a significant reduction in the transduction efficiency of these vectors after the terminal differentiation of the mononucleated cytotrophoblast into the multinucleated syncytiotrophoblast. Thus, transfer of adenovirus vectors across the placenta is likely to be limited in the human.

2.5. Projection to Human Application (See also Chapter 17)

From the previous descriptions, it is clear that no one animal model is ideal for assessment of prenatal gene therapy application. A balance is needed, taking into consideration the gestational development of the organ to be targeted and how that relates to the human situation, the type of placentation, fetal size, and development of the immune system.

Toxicology studies will be needed using animals such as the pregnant rabbit, in which reproductive toxicology is commonly performed, with good historical datasets, and which is understood by the regulators. A variety of guidelines and regulations, such as those described by the Committee for Medicinal Products for Human Use (CHMP) of the European Medicines Agency, will need to be taken into consideration when planning preclinical study protocols. These could include, for example, the guidelines on the non-clinical testing for inadvertent germ line transmission of gene transfer vectors (78) or on the non-clinical studies required before the first clinical use of gene therapy medicinal products (79).

In addition to animal studies, the effect of gene therapy vectors on the human placenta can be assessed in vitro. Two models are available, cultured villous explants or perfused whole placental cotyledons. Villi isolated from different lobules of the placenta can be cultured in net-wells and submerged in growth medium. In this model, the syncytiotrophoblast routinely undergoes shedding in vitro after about 1 day of culture, but this barrier consistently regenerates through the differentiation of underlying cytotrophoblast cells 2 days later (80, 81). Cellular integrity and apoptosis can be assessed using specific markers, such as lactate dehydrogenase levels, released into the culture medium. The method of placental perfusion has been adapted so as to preserve cellular and tissue architecture while allowing a dual fetal- and maternal-side haemodynamic compartment to be maintained (82). Movement of substances applied to the maternal or fetal side of the placenta can be studied in the opposite side of the placenta using this model, over a 5- to 9-h time period after delivery of the placenta. This model has provided a wealth of data on the physiology of normal and pathological human placentae (83, 84) and may be useful in measuring spread of vector from the fetus to the mother or vice versa.

Phase I human trials are likely to face hurdles because of difficulties in testing pregnant women, where toxicological studies are usually contraindicated. Thus, when human application becomes possible, extensive unbiased parental counselling and informed consent are paramount because of the uncertainties about the efficacy and long-term safety of prenatal gene therapy which may not become evident until much later in the individual's life. This can be difficult when the decision to participate in a fetal gene therapy trial will occur close to the time of prenatal diagnosis of the condition. Because there may be risks to the health of the mother, the fetus, and possibly their future progeny, parents will also be required to consent themselves and their offspring to life-long follow-up.

Assuming that a safe and effective fetal gene therapy approach was to be possible, how might it work in practice for treatment of a fetal congenital condition? Figure 1 shows a possible scheme for a

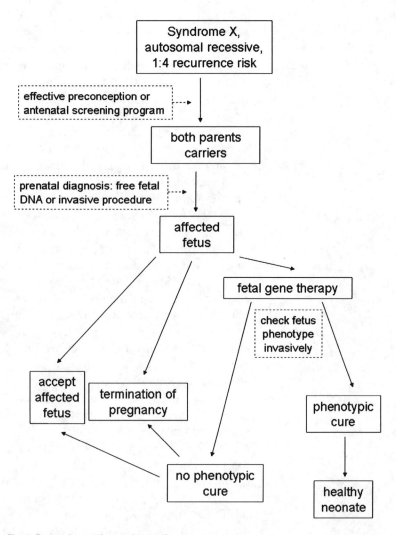

Fig. 1. Prenatal gene therapy in practice.

hypothetical syndrome X, an autosomal recessive condition which results in severe morbidity. Without an effective screening strategy with accurate prenatal diagnosis for syndrome X, many families would not know that they were at risk until an affected child was born. For the next pregnancy, the parents could choose to have prenatal diagnosis of syndrome X in the fetus prior to the gestational age for optimum gene therapy treatment by non-invasive prenatal diagnosis using cell-free fetal DNA if available or by chorionic villus sampling. The mother would undergo the invasive procedure to treat the fetus at the best time to target the affected organ. The option of further invasive testing to confirm expression of the curative gene product later in the pregnancy could be available.

An alternative strategy is preimplantation genetic diagnosis (PGD), which is often proposed as the most sensible option for parents at risk of having an affected fetus. The main limitations of IVF and PGD are the ovulation induction and invasive procedures that the woman is required to have. Only 20–30% of couples achieve a pregnancy per cycle (85) and some embryos will be disposed of, which for some individuals is of concern (86). In addition, it provides no help for a couple already pregnant with a wanted but affected fetus, who cannot countenance termination of pregnancy.

References

1. Castillon N, Avril-Delplanque A, Coraux C et al (2004) Regeneration of a well-differentiated human airway surface epithelium by spheroid and lentivirus vector-transduced airway cells. J Gene Med 6:846–856

2. Robinson V (2005) Finding alternatives: an overview of the 3Rs and the use of animals in research. Sch Sci Rev 87(319):1–4

3. David AL, Peebles DM, Gregory L et al (2006) Clinically applicable procedure for gene delivery to fetal gut by ultrasound-guided gastric injection: toward prenatal prevention of early-onset intestinal diseases. Hum Gene Ther 17:767–779

4. Peebles D, Gregory LG, David A et al (2004) Widespread and efficient marker gene expression in the airway epithelia of fetal sheep after minimally invasive tracheal application of recombinant adenovirus in utero. Gene Ther 11:70–78

5. Lim F-Y, Martin BG, Sena-Esteves M et al (2002) Adeno-associated virus (AAV)-mediated fetal gene transfer in respiratory epithelium and submucosal gland cells in human fetal tracheal organ culture. J Pediatr Surg 37:1051–1057

6. Lim F-Y, Kobinger GP, Weiner DJ et al (2003) Human fetal trachea-SCID mouse xenografts: efficacy of vesicular stomatitis virus-G pseudotyped lentiviral-mediated gene transfer. J Pediatr Surg 38:834–839

7. Dejneka NS, Surace EM, Aleman TS et al (2004) In utero gene therapy rescues vision in a murine model of congenital blindness. Mol Ther 9:182–188

8. Williams ML, Coleman JE, Haire SE et al (2006) Lentiviral expression of retinal guanylate cyclase-1 (RetGC1) restores vision in an avian model of childhood blindness. PLoS Med 3:e201

9. Seppen J, van der Rijt R, Looije N et al (2003) Long-term correction of bilirubin UDPglucuronyltransferase deficiency in rats by in utero lentiviral gene transfer. Mol Ther 8:593–599

10. Waddington SN, Nivsarkar MS, Mistry AR et al (2004) Permanent phenotypic correction of hemophilia B in immunocompetent mice by prenatal gene therapy. Blood 104: 2714–2721

11. Karolewski BA, Wolfe JH (2006) Genetic correction of the fetal brain increases the lifespan of mice with the severe multisystemic disease mucopolysaccharidosis type VII. Mol Ther 14:14–24

12. Wells DJ, Wells KE (2005) What do animal models have to tell us regarding Duchenne muscular dystrophy? Acta Myol 24:172–180

13. Arruda VR, Stedman HH, Nichols TC et al (2005) Regional intravascular delivery of AAV-2-FIX to skeletal muscle achieves long-term correction of hemophilia B in a large animal model. Blood 105:3458–3464

14. Baldeschi C, Gache Y, Rattenholl A et al (2003) Genetic correction of canine dystrophic epidermolysis bullosa mediated by retroviral vectors. Hum Mol Genet 12:1897–1905

15. Meertens L, Zhao Y, Rosic-Kablar S et al (2002) In utero injection of alpha-L-iduronidase-carrying retrovirus in canine mucopolysaccharidosis type I: infection of multiple tissues and neonatal gene expression. Hum Gene Ther 13:1809–1820

16. Leipprandt JR, Kraemer SA, Haithcock BE et al (1996) Caprine β-mannosidase: sequencing and characterization of the cDNA and identification of the molecular defect of caprine β-mannosidosis. Genomics 37:51–56

17. Porada CD, Sanada C, Long CR et al (2010) Clinical and molecular characterization of a re-established line of sheep exhibiting hemophilia A. J Thromb Haemost 8:276–285

18. Tessanne K, Long C, Spencer T et al (2011) 337 production of transgenic sheep using recombinant lentivirus microinjection of in vivo produced embryos. Reprod Fertil Dev 23:264

19. Christensen G, Minamisawa S, Gruber PJ et al (2000) High-efficiency, long-term cardiac expression of foreign genes in living mouse embryos and neonates. Circulation 101:178–184

20. Lipshutz GS, Gruber CA, Cao Y et al (2001) In utero delivery of adeno-associated viral vectors: intraperitoneal gene transfer produces long-term expression. Mol Ther 3:284–292

21. Lipshutz GS, Flebbe-Rehwaldt L, Gaensler KML (1999) Adenovirus-mediated gene transfer in the midgestation fetal mouse. J Surg Res 84:150–156

22. Gregory LG, Waddington SN, Holder MV et al (2004) Highly efficient EIAV-mediated in utero gene transfer and expression in the major muscle groups affected by Duchenne muscular dystrophy. Gene Ther 11:1117–1125

23. Waddington SN, Buckley SMK, Berloehr C et al (2004) Reduced toxicity of F-deficient Sendai virus vector in the mouse fetus. Gene Ther 11:599–608

24. Endo M, Henriques-Coelho T, Zoltick PW et al (2010) The developmental stage determines the distribution and duration of gene expression after early intra-amniotic gene transfer using lentiviral vectors. Gene Ther 17:61–71

25. David AL, McIntosh J, Peebles DM et al (2011) Recombinant adeno-associated virus-mediated in utero gene transfer gives therapeutic transgene expression in the sheep. Hum Gene Ther 22:419–426

26. Themis M, Waddington SN, Schmidt M et al (2005) Oncogenesis following delivery of a nonprimate lentiviral gene therapy vector to fetal and neonatal mice. Mol Ther 12:763–771

27. Sabatino DE, MacKenzie TC, Peranteau WH et al (2007) Persistent expression of hFIX after tolerance induction by in utero or neonatal administration of AAV-1-F.IX in hemophilia B mice. Mol Ther 15:1677–1685

28. Billingham RE, Brent L, Medawar PB (1953) Actively acquired tolerance of foreign cells. Nature 172:603–606

29. Billingham RE, Brent L, Medawar PB (1956) Quantitative studies on tissue transplantation immunity III Actively acquired tolerance. Phil Trans R Soc Lond B B239:357–369

30. Howard JG, Michie D (1962) Induction of transplantation immunity in the newborn mouse. Transplant Bull 29:1–6

31. Holladay SD, Smialowicz RJ (2000) Development of the murine and human immune system: differential effects of immunotoxicants depend on time of exposure. Environ Health Perspect 108:463–473

32. Morris B (1986) The ontogeny and comportment of lymphoid cells in fetal and neonatal sheep. Immunol Rev 91:219–233

33. Schinkel PG, Ferguson KA (1953) Skin transplantation in the foetal lamb. Aust J Biol Sci 6:533

34. Silverstein AM, Prendergast RA, Kraner KL (1964) Fetal response to antigenic stimulus IV. Rejection of skin homografts by the fetal lamb. J Exp Med 119:955–964

35. Neiderhuber JE, Shermeta D, Turcotte JG, Pikas PW (1971) Kidney transplantation in the foetal lamb. Transplantation 12:161–166

36. McCullagh P (1988) Immunological tolerance of sheep to skin allografts. Transplantation 46:280–285

37. McCullagh P (1989) Inability of fetal skin to induce allograft tolerance in fetal lambs. Immunology 67:489–495

38. Moore NW, Rowson LEA (1961) Attempts to induce tissue tolerance in sheep. Res Vet Sci 2:1

39. Mitchell RM (1959) Attempts to induce tolerance of renal homografts in sheep by intra-embryonic injection of spleen cells. Transplant Bulletin 6:424–426

40. Nettleton PF (2000) Border disease. In: Martin WB, Aitken ID (eds) Diseases of sheep. Blackwell Science, Oxford, pp 95–102

41. McClure S, McCullagh P, Parsonson IM et al (1988) Maturation of immunological reactivity in the fetal lamb infected with Adabane virus. J Comp Pathol 99:133–143

42. Silverstein AM, Uhr JW, Lukes RJ (1963) Fetal response to antigenic stimulus II. Antibody production by the fetal lamb. J Exp Med 117:799–812

43. Silverstein AM, Thorbecke GJ, Kraner KL, Lukes RJ (1963) Fetal response to antigenic stimulus. III. Gamma-globulin production in normal and stimulated fetal lambs. J Immunol 91:384–395

44. Fahey KJ, Morris B (1978) Humoral immune responses in foetal sheep. Immunology 35:651–661

45. Fahey KJ, Morris B (1974) Lymphopoiesis and immune reactivity in the fetal lamb. Ser Haematol 7:548–567

46. Stites DP, Carr MC, Fudenberg HH (1974) Ontogeny of cellular immunity in the human fetus: development of responses to phytohemagglutinin and to allogeneic cells. Cell Immunol 11:257–271

47. Toivanen P, Uksila J, Leino A et al (1981) Development of mitogen responding T cells and natural killer cells in the human fetus. Immunol Rev 57:89–105

48. Velardi A, Cooper MD (1984) An immunofluorescence analysis of the ontogeny of myeloid, T, and B lineage cells in mouse hemopoietic tissues. J Immunol 133:672–677

49. Tyan ML, Herzenberg LA (1968) Studies on the ontogeny of the mouse immune system. II. Immunoglobulin-producing cells. J Immunol 101:446–450

50. Grossi CE, Velardi A, Cooper MD (1985) Postnatal liver hemopoiesis in mice: generation of pre-B cells, granulocytes, and erythrocytes in discrete colonies. J Immunol 135:2303–2311

51. Mestas J, Hughes CC (2004) Of mice and not men: differences between mouse and human immunology. J Immunol 172:2731–2738

52. Jerebtsova M, Batshaw ML, Ye X (2002) Humoral immune response to recombinant adenovirus and adeno-associated virus after *in utero* administration of viral vectors in mice. Pediatr Res 52:95–104

53. Seppen J, van Til NP, van der Rijt R et al (2006) Immune response to lentiviral bilirubin UDP-glucuronosyltransferase gene transfer in fetal and neonatal rats. Gene Ther 13:672–677

54. Merianos DJ, Tiblad E, Santore MT et al (2009) Maternal alloantibodies induce a postnatal immune response that limits engraftment following in utero hematopoietic cell transplantation in mice. J Clin Invest 119:2590–2600

55. Nijagal A, Wegorzewska M, Jarvis E et al (2011) Maternal T cells limit engraftment after in utero hematopoietic cell transplantation in mice. J Clin Invest. doi:10.1172/JCI44907

56. Manno CS, Pierce GF, Arruda VR et al (2006) Successful transduction of liver in hemophilia by AAV-factor IX and limitations imposed by the host immune response. Nat Med 12:342–347

57. Goldenberg RL, Culhane JF, Iams JD, Romero R (2008) Epidemiology and causes of preterm birth. Lancet 371:75–84

58. Harrison MR, Keller RL, Hawgood SB et al (2003) A randomised trial of fetal endoscopic tracheal occlusion for severe fetal congenital diaphragmatic hernia. N Eng J Med 349:1916–1924

59. Mitchell BF, Taggart MJ (2009) Are animal models relevant to key aspects of human parturition? Am J Physiol Regul Integr Comp Physiol 297:R525–R545

60. Merlino AA, Welsh TN, Tan H et al (2007) Nuclear progesterone receptors in the human pregnancy myometrium: evidence that parturition involves functional progesterone withdrawal mediated by increased expression of progesterone receptor-A. J Clin Endocrinol Metab 92:1927–1933

61. Kalkhoven E, Wissink S, van der Saag PT, van der Burg B (1996) Negative interaction between the RelA(p65) subunit of NF-kappaB and the progesterone receptor. J Biol Chem 271:6217–6224

62. Goldenberg RL, Hauth JC, Andrews WW (2000) Intrauterine infection and preterm delivery. N Engl J Med 342:1500–1507

63. Romero R (2006) The preterm parturition syndrome. BJOG 113:17–42

64. Fidel PI Jr, Romero R, Maymon E, Hertelendy F (1998) Bacteria-induced or bacterial product-induced preterm parturition in mice and rabbits is preceded by a significant fall in serum progesterone concentrations. J Matern Fetal Med 7:222–226

65. Mitchell BF, Zielnik B, Wong S, Roberts CD, Mitchell JM (2005) Intraperitoneal infusion of proinflammatory cytokines does not cause activation of the rat uterus during late gestation. Am J Physiol Endocrinol Metab 289:E658–E664

66. Baggia S, Gravett MG, Witkin SS, Haluska GJ, Novy MJ (1996) Interleukin-1 beta intra-amniotic infusion induces tumor necrosis factor-alpha, prostaglandin production, and preterm contractions in pregnant rhesus monkeys. J Soc Gynecol Investig 3:121–126

67. Chwalisz K, Fahrenholz F, Hackenberg M, Garfield R, Elger W (1991) The progesterone

antagonist onapristone increases the effectiveness of oxytocin to produce delivery without changing the myometrial oxytocin receptor concentrations. Am J Obstet Gynecol 165: 1760–1770

68. Benirschke K, Kaufmann P (1990) Placental types in pathology of the human placenta. Springer, New York

69. Wooding FB (1992) Current topic: the synepitheliochorial placenta of ruminants: binucleate cell fusions and hormone production. Placenta 13:101–113

70. Enders AC (1965) A comparative study of the fine structure of the trophoblast in several hemochorial placentas. Am J Anat 116:29–67

71. Hamilton WJ, Boyd JD (1970) The human placenta. Heffer and Sons, Cambridge

72. Kaufmann P, Davidoff M (1977) The guinea pig placenta. Adv Anat Embryol Cell Biol 53:1–90

73. Scott VL, Burgess SC, Shack LA, Lockett NN, Coats KS (2008) Expression of CD134 and CXCR4 mRNA in term placentas from FIV-infected and control cats. Vet Immunol Immunopathol 123:90–96

74. Bergelson JM (1997) Isolation of a common receptor for Coxsackie B viruses and adenoviruses 2 and 5. Science 275:1320–1323

75. Koi H, Zhang J, Makrigiannakis A et al (2001) Differential expression of the coxsackievirus and adenovirus receptor regulates adenovirus infection of the placenta. Biol Reprod 64:1001–1009

76. MacCalman CD, Furth EE, Omigbodun A et al (1996) Transduction of human trophoblast cells by recombinant adenoviruses is differentiation dependent. Biol Reprod 54: 682–691

77. Parry S, Holder J, Strauss JR (1997) Mechanisms of trophoblast-virus interaction. J Reprod Immunol 37:25–34

78. Committee for Medicinal Products for Human Use (2006) Guideline on non-clinical testing for inadvertent germline transmission of gene transfer vectors (273974). European Medicines Agency, London

79. Committee for Medicinal Products for Human Use (2008) Guideline on the non-clinical studies required before first clinical use of gene therapy medicinal products (125459). European Medicines Agency, London

80. Siman CM, Sibley CP, Jones CJ et al (2001) The functional regeneration of syncytiotrophoblast in cultured explants of term placenta. Am J Physiol Regul Integr Comp Physiol 280: R1116–R1122

81. Crocker IP, Tansinda DM, Baker PN (2004) Altered cell kinetics in cultured placental villous explants in pregnancies complicated by preeclampsia and intrauterine growth restriction. J Pathol 204:11–18

82. Brownbill P, Edwards D, Jones C et al (1995) Mechanisms of alphafetoprotein transfer in the perfused human placental cotyledon from uncomplicated pregnancy. J Clin Invest 96: 2220–2226

83. Brownbill P, Mills TA, Soydemir DF, Sibley CP (2008) Vasoactivity to and endogenous release of vascular endothelial growth factor in the in vitro perfused human placental lobule from pregnancies complicated by preeclampsia. Placenta 29:950–955

84. Sibley CP, Birdsey TJ, Brownbill P et al (1998) Mechanisms of maternofetal exchange across the human placenta. Biochem Soc Trans 26:86–91

85. Wells D, Delhanty JD (2001) Preimplantation genetic diagnosis: applications for molecular medicine. Trends Mol Med 7:23–30

86. Snowdon C, Green JM (1997) Preimplantation diagnosis and other reproductive options: attitudes of male and female carriers of recessive disorders. Hum Reprod 12:341–350

87. Benirschke K, Kaufmann P, Baergen RN (2006) Pathology of the human placenta. Springer, New York

Chapter 10

Animal Models for Prenatal Gene Therapy: Rodent Models for Prenatal Gene Therapy

Jessica L. Roybal, Masayuki Endo, Suzanne M.K. Buckley, Bronwen R. Herbert, Simon N. Waddington, and Alan W. Flake

Abstract

Fetal gene transfer has been studied in various animal models, including rabbits, guinea pigs, cats, dogs, and nonhuman primate; however, the most common model is the rodent, particularly the mouse. There are numerous advantages to mouse models, including a short gestation time of around 20 days, large litter size usually of more than six pups, ease of colony maintenance due to the small physical size, and the relatively low expense of doing so. Moreover, the mouse genome is well defined, there are many transgenic models particularly of human monogenetic disorders, and mouse-specific biological reagents are readily available. One criticism has been that it is difficult to perform procedures on the fetal mouse with suitable accuracy. Over the past decade, accumulation of technical expertise and development of technology such as high-frequency ultrasound have permitted accurate vector delivery to organs and tissues. Here, we describe our experiences of gene transfer to the fetal mouse with and without ultrasound guidance from mid to late gestation. Depending upon the vector type, the route of delivery and the age of the fetus, specific or widespread gene transfer can be achieved, making fetal mice excellent models for exploratory biodistribution studies.

Key words: Rodents, In utero gene delivery, Fetal gene therapy, Mating, Injection procedures, Tissue and biological fluid sampling, Biodistribution

1. Introduction

1.1. Overview

Rodent models, mainly mice and rats, have been used extensively in preclinical studies on prenatal gene therapy (1). The obvious advantages of the rodent model include the relatively low cost, short gestation, large litter size, defined genome, availability of reagents (such as anti-rodent antibodies, cytokines, etc.), and availability of rodent models of human disease. The disadvantages include small size and phylogenetic disparity from the human. From the perspective of development of most organ systems and

Charles Coutelle and Simon N. Waddington (eds.), *Prenatal Gene Therapy: Concepts, Methods, and Protocols*,
Methods in Molecular Biology, vol. 891, DOI 10.1007/978-1-61779-873-3_10, © Springer Science+Business Media, LLC 2012

their corresponding function, the murine model is stage by stage very similar to the human, although the gestational timing is disparate. From the perspective of thymic and immunological development, the embryonic day 15 (E15) mouse fetus is highly analogous to the 13- to 14-week human fetus (2, 3); these time points correspond to 85 and 35% of gestation, respectively. However, in terms of central nervous system development, the period of human gestation from 23 to 36 weeks is equivalent to postnatal days 3–7 in mice and rats (4). These differences must be considered in the design and interpretation of prenatal gene transfer experiments in rodent models. As in all animal model species, specific tropisms and differences in receptor expression for various gene vectors limit the direct applicability of rodent findings to clinical studies. Thus, therapeutic studies in rodent models of human disease can only be considered "proof of principle" and need validation in phylogenetically closer species prior to consideration of human application. There is also the issue of relative scale in the rodent model as it relates to doses of vector particles that can be delivered, growth of the animal into adulthood, and life expectancy. A lifetime of gene expression in the rodent model would represent only relatively short-term expression in the human. These issues of scale may be one of the primary reasons that many of the findings of gene therapy studies in the rodent have not translated well into larger animal models. Finally, there are specific organs that may not represent a valid model system for human applications. For instance, the placenta of rodents is considerably different from that of primates, and may not be useful for placenta-targeted gene therapy studies or studies of transplacental transfer of vector (See also Chapter 9).

Despite the disadvantages, the rodent model has contributed greatly to our current understanding of prenatal gene therapy and will continue to do so. Arguably, the greatest contribution has been in testing new gene transfer constructs in rodent models of human disease. Over the past decade, there have been several studies demonstrating partial or complete correction of disease phenotype by in vivo gene therapy (5–12).

1.2. Techniques and Technology

Advances in surgical technique and ultrasound technology have allowed major progress in our capability to manipulate the rodent model at developmental stages that are relevant to human application. Whereas initial attempts at gene transfer in the rodent model were performed relatively late in gestation targeting easily accessible compartments (intraperitoneal, intramuscular, intraventricular), in the past 10 years injection techniques have become much more precise.

The ability to perform intravascular injections into the vitelline vein at E14–15 (Fig. 1), first described by Schachtner and colleagues (13), was a major advance allowing systemic administration of vector prior to maturation of the immune system and during

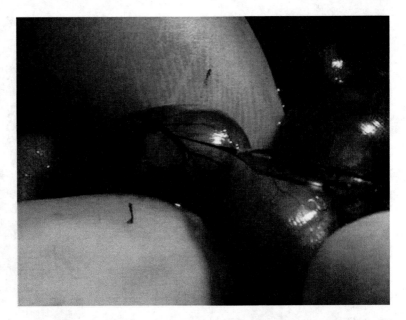

Fig. 1. Image of an E14 vitelline vein injection illustrated by injection of India ink. The uterus is stabilized between the thumb and forefinger with the vein aligned appropriately for injection with the other hand.

a period of critical development for many organ systems (6). The technique of Schachtner and colleagues describes laparotomy and uterotomy for access to the vitelline vessel with a pulled glass micropipette. All authors here omit the uterotomy as an unnecessary step, which adds unacceptable time to the operation. Some (Waddington SN, Buckley SMK) have tended to use steel 33- or 34-gauge Hamilton needles since they are adequate for intravenous and intraparenchymal injection yet are sufficiently durable to be used for late gestation intracranial, intramuscular, or intrathoracic injection with no fear of breakage on penetrating the skin or ossifying tissue. Others (Roybal JL, Endo M, Flake AW) have preferred the use of pulled glass micropipettes, which ensure greater accuracy of injection. Both methods are detailed. The advent of high-frequency ultrasound systems (ultrasound biomicroscopy) with small probes and micropipettes has allowed the development of early gestational models for gene transfer (Fig. 2).

These include injections into the amniotic fluid (E8) (14–17) and early gestational systemic administration via the intracardiac route (E9–E10) (18) as well as other cavities such as the extracoelomic cavity, embryonic foregut (Flake AW, unpublished observations), and intrathoracic cavity (19, 20) to be accurately performed. The variety of techniques currently available make targeting of stem cells for gene transfer possible in almost any organ when it is most accessible for manipulation. In addition to the obvious therapeutic implications, the ability to transfer genes to stem cell populations early in development should offer new

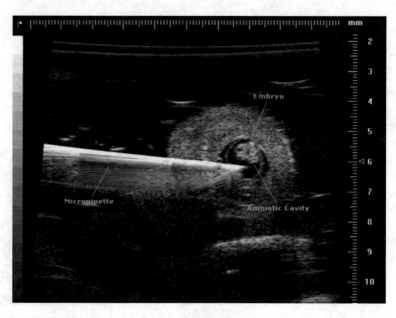

Fig. 2. Image of an ultrasound-directed intra-amniotic injection of an embryonic mouse at E8. The amniotic cavity, micropipette, and fetus are labeled. The large tics on the scale on the *right* represent 1-mm-size increments.

experimental insights into normal and abnormal development. In addition, because of the relatively high efficiency of gene transfer to stem cell populations using these approaches, the rodent model may offer a relatively rigorous assay for biodistribution and safety studies of vector-based gene transfer.

2. Materials

2.1. Surgical Supplies for IUGT Under Ultrasound Guidance

Ultrasound guidance is necessary for early gestational time points and later in gestation when accurate targeting of a specific organ is required. Specifically, we use ultrasound guidance for E8–9 intra-amniotic or extracoelomic injections, E9–10 intracardiac injections, E14 intrathoracic lung bud injections, and intrahepatic or intramuscular injections after E12. While it is unnecessary to have all supplies sterilized, clean surgical technique is important. All reusable instruments are cleaned with alcohol and MB-10 disinfectant prior to use. Clean gloves are used.

2.2. Materials for Ultrasound-Guided Procedures

1. Ultrasound with high frequency transducer, MS 550 or MS700 (Visualsonics, Toronto, Canada).
2. Vevo Integrated Rail System with heated table and needle-guided injection system (Visualsonics).

3. Glass micropipettes (outer diameter, 1.14 mm, inner diameter, 0.53 mm; Humagen, Charlottesville, VA).

4. Clear ultrasound gel (Aquasonic®, Parker Laboratories, Fairfield, NJ).

5. Ultrasound gel warmer.

6. Anesthesia system.

7. Isoflurane.

8. Oxygen.

9. Molding clay (to hold the injection needle during setup).

10. Instruments: Forceps, scissors, and needle holder.

11. Cotton gauze (4×4).

12. Clear plastic tape.

13. Cotton swabs.

14. 4-0 Vicryl suture.

15. Warm phosphate-buffered saline (PBS).

16. Mineral oil.

17. Parafilm.

18. 20–200-μl adjustable pipette with tips.

19. Viral vector on ice.

20. Clean cages.

21. Flunixin Meglumine.

22. Heating lamp.

2.3. Additional Materials for E14–15 Vitelline Vein Injections

1. An IM 300 programmable micro-injector (MicroData Instruments, Inc., South Plainfield, NJ, USA).

2. Pressurized N_2 gas for the micro-injector.

3. Pipettes as above for gene transfer experiments. For injection of manipulated stem cells, larger pipettes are required. We recommend pulling your own pipettes to achieve an outer diameter between 70 and 90 m using the following:

   (a) Micropipette puller (Sutter Instrument Co.; model P-30).

   (b) Micropipette beveller (Sutter Instrument Co.; model Bv-10) and micropipettes (Fisher Scientific, Pittsburgh, PA).

   (c) Micro-injector or tubing with an attachment for the micropipette.

   (d) Microscope reticule (0.5-mm range with 0.01 minor delineations and 0.05 major delineations is ideal for measuring the external diameter of the micropipettes).

   (e) Stage micrometer (to ensure that the reticle delineations are correct).

4. Operating microscope (range 4–40×).

2.4. Additional Materials for IUGT Without Ultrasound Guidance

1. Rx Honing Machine Corp. Model 2.3 Serial No RF-1715 with diamond sharpening wheel for sharpening of Hamilton needles.
2. 100-μl Hamilton syringe (Model 1710RN).
3. 27-G Removable (RN) Hamilton needle.
4. 33-G Removable (RN) Hamilton needle (see Note 1).
5. INTRAMEDIC* Polyethylene Tubing, Clay Adams PE10 (0.011″ internal diameter, 0.024″ external diameter).
6. Epoxy resin.

2.5. Specimen Sampling for Analysis of Gene Transfer

1. Petri dish, glass or plastic.
2. Thermostatically controlled heating chamber.
3. Inhalation anesthetic vaporizer, nose cone, and induction chamber.
4. Heating mat.
5. 20-gauge needle or small scalpel blade.
6. 25-gauge needle.
7. 1-ml syringe.
8. Small pointed dissection scissors.
9. Cryotubes.
10. Liquid nitrogen or dry ice plus ethanol.
11. Paraformaldehyde or formalin.
12. Sucrose.
13. Optimum cutting temperature (OCT) compound.

3. Methods

3.1. Animal Husbandry/Breeding

Careful time-dating of mice is an important aspect of IUGT. Successful breeding and timed mating are essential in most in vivo work, but can be expensive and time consuming. In general, in-house breeding programs should be established, as the time-dating practices of commercial sources are inconsistent and often inaccurate. The following techniques have been used to optimize this process, giving a 30–50% plug positive rate per mating, with almost the same success in actual pregnancies.

1. While many animal facilities are introducing individually ventilated cages (IVCs) to reduce allergens, this also reduces the scents and pheromones in the animal's larger environment and thus some of the mating cues. Housing grouped females on the same rack as individually housed stud males aids in the

transfer of these scent cues. Males can also be "primed" for mating using feces from females.

2. Young females of 6–8 weeks in age are optimal for timed mating (see Note 2). While older females can be used, it is important that the females are not larger than the male as this can result in bullying or injury (such as testicular removal) to the male. Females should be added to the male's cage for the allotted mating time (normally overnight as animals tend to mate at dusk or around midnight) and checked for signs of a sperm plug when removed. Animals are normally mated in a 1:1 or 1:2 (male:females); larger ratios are unlikely to increase plugging rate but rather result in bullying of the male.

3. To improve the efficiency of the timed mating procedure both in animal numbers and costs, successful matings should be noted on the male's cage. Thus, impotent males can be replaced quicker, while good breeders can be maintained. Males are normally used from 7 weeks of age until 8 months but a good breeding male can go over a year.

4. While males in general have a natural instinct to mate, it can take them some time to perfect their technique. Thus, the use of "fluffers" or spare females to allow the male to practice should increase the plugging rate for actual timed matings. Furthermore, the male cage should not be cleaned out for at least 2 days prior to mating to ensure that the male is comfortable in his scented territory. With this in mind, a "jealousy mating" technique can also be employed. Here, a female is placed in one male's cage for 10–15 min, allowing her to get coated in his scent. The female is then removed and placed in an alternative male's cage, with her posterior held to his nose to focus him. Mounting behavior is seen soon after the female is introduced to this second male. Since scent is so important for successful mating, it is important that strong scents and perfumes are not worn by personnel setting up the mating (see Note 3).

5. The day of separation is considered gestational day E0.

3.2. Preoperative Care

1. Identification of pregnant females is done preoperatively by means of palpation. Fetuses can be reliably palpated beginning at gestational age E8, but the pregnant dam should be anesthetized to facilitate palpation of the early gestational fetuses. Palpation is performed using one's thumb and index finger across the lower abdomen of the mouse. The feeling of beads on a string identifies a pregnant female.

2. Pregnant females are set aside for injection and nonpregnant females are returned to the breeding colony the following week.

3. The timing of injection depends on the organ targeted. For example, the amniotic cavity is accessible beginning at E8, the heart is accessible at E9 onward, and the lung and liver are accessible at E12.

3.3. Preoperative Preparation of Supplies

Prior to the induction of anesthesia:

1. The ultrasound gel is warmed in the gel warmer.
2. The operating stand is set to 37°C (see Note 4).
3. The injection needle is prepared. The needle is backfilled with mineral oil, and then connected to the micropipette holder which is attached to a three-axis micro-injector unit (see Note 5).
4. The viral vector is then prepared. The frozen vector is thawed on ice and transferred to a piece of Parafilm®, and the micropipette is then filled with virus vector (see Note 6).

3.4. Anesthesia

There are multiple types of anesthesia available for mice. For in utero gene or stem cell therapy, we prefer an inhalational anesthetic. A vaporizer releasing isoflurane results in rapid sedation and recovery. Without a vaporizer, it is difficult to control the dose of isoflurane and overdose is a hazard. Methoxyflurane is expensive and difficult to obtain, but a vaporizer is not required and both sedation and recovery are rapid. Once mastered, the procedures described are short taking 20 min or less and anesthesia should be administered accordingly.

1. Oxygen is set to 2.5% throughout the entire procedure.
2. Isoflurane is set at 4% for induction and 2% for maintenance.

3.5. Operation for Procedures Under Ultrasound Guidance

The procedures described in this section use glass micropipettes and an automated injector combined with ultrasound guidance.

1. Once the mouse is asleep, it is placed supine on the operating stand. Arms and feet are secured with tape. Tape is placed over the groin to prevent urine or feces from contaminating the field.
2. The abdomen is cleaned with an alcohol pad.
3. A 1-cm lower-midline laparotomy is made.
4. The peritoneum is grasped, elevated, and incised.
5. The uterus is externalized and the number of viable fetuses in each horn recorded.
6. One horn is then returned to the peritoneal cavity while the other horn is injected. Pre-warmed ultrasound gel is generously applied to the exposed fetuses.
7. Each fetus is positioned to allow a clear view of the area of interest (i.e., the heart) in B-mode.

8. Under two-dimensional visualization, the micropipette tip is advanced through the uterine wall into the target organ (see Note 7).

9. A set volume of viral vector is injected using the automated injection system. The volume of virus injected depends on the gestational age of the fetus. Typically, 350 nl of vector is injected in E8 mice, 700 nl in E9–10 mice, and 1,050 nl in E11–12 mice (see Note 8).

10. Under ultrasound visualization, the micropipette tip is retracted from the fetus and a new fetus is positioned beneath the probe.

11. Once all fetuses in one uterine horn have been injected, the ultrasound gel is cleaned off, and they are returned to the peritoneal cavity. The second horn is then externalized and each fetus injected. There are usually 6–11 fetuses per dam.

12. The incision is closed in two layers with absorbable suture.

13. The entire procedure takes about 20 min from incision until closure.

3.6. Operation for E14–15 Vitelline Vein injection

1. The anesthetic techniques and exposure of the uterine horns are identical to that described above.

2. This procedure targets the vitelline veins for systemic administration of cells or vector. The vitelline veins are relatively fixed within the membrane of the visceral yolk sac proximal to their convergence and subsequent connection to the portal circulation of the fetus. The veins can be differentiated from other vessels within the membranes by their size and course into the umbilical stalk, as well as their minimal movement when the uterus is manipulated. The tip of the micropipette is inserted bevel down into the uterus and through the vessel wall using either the dissecting microscope or loupe magnification (3.5×). Once a back flash of blood is seen entering the micropipette, the cells or vector is injected using the programmable injector (see Notes 9–11).

3.7. IUGT Without Ultrasound Guidance

3.7.1. Preparation of Needles

1. Hamilton needles come in two styles. The first is a steel hub (Fig. 3b) with a PTFE grommet. The second is cemented in a plastic hub, which replaces both the steel and PTFE parts. The former is the one we prefer and is depicted in Fig. 3.

2. Using the diamond wheel of the honing machine, the 33-gauge small hub needle is sharpened to a single bevel. For extra sharpness, it is then double beveled: Once the first bevel is made, the needle is rotated 90° and a second, smaller bevel is made. The needle is then rotated 180° and a second, small bevel is made (Fig. 3c) (see Note 1).

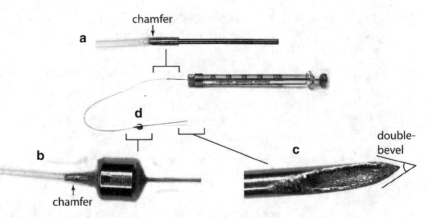

Fig. 3. Image of the arrangement of the Hamilton syringe barrel and needle when prepared for two-operator injections showing (**a**) the blunted 27-gauge with a chamfer to permit sheathing of the polyethylene tubing (**b**) the stub end of the 33-gauge needle, chamfered to permit sheathing with the other end of the polyethylene tubing (**c**) the sharpened end of the 33-gauge needle, illustrating the double bevel and (**d**) the entire barrel and needle assembly.

3. A chamfer is added to the stub protruding from the other side of the needle hub to facilitate sheathing with the polyethylene tubing (Fig. 3b).

4. The polyethylene tubing is pushed onto the chamfered stub end (Fig. 3b). It helps to have the tubing widened by prior insertion of a 27-gauge disposable needle. A small drop of cyanoacrylate adhesive (e.g., "Superglue") should be placed on the stub end once the tubing has been placed upon it sufficiently to occlude the orifice (so that the adhesive does not enter the orifice). The adhesive temporarily acts as a lubricant to allow the tubing to slide along the shaft of the stub end as far as possible.

5. The point of the 27-gauge needle is flattened (i.e., the point is ground perpendicular to the needle shaft) and a chamfer is added to the now flattened needle end to facilitate sheathing with the polyethylene tubing (Fig. 3a).

6. The knurled collar is inserted onto the tubing.

7. The polyethylene tubing is pushed onto the 27-gauge needle for approximately 5 mm; there is no need for adhesive as there is tight fit between the tubing and the needle.

8. The knurled knob is fastened down onto the syringe barrel, thus securing the 27-gauge needle and providing the finished product (Fig. 3d).

3.7.2. Intra-vitelline Injection

1. The anesthetic techniques and exposure of the uterine horns are identical to those described in Subheading 3.5.

2. The vitelline vessels are located as described in Subheading 3.6. Using a dissecting microscope, the tip of the needle is inserted bevel up through the myometrium at an angle of around 20°.

The needle can be slid in a reciprocating fashion over a vitelline vessel, beginning with a low angle of entry and increasing the angle, while gently reciprocating the needle, until the vessel is penetrated. In a vessel with a diameter larger than the needle, blood can be seen swirling around the needle orifice, particularly if the needle is gently canted up and down. In vessels with a diameter smaller than the needle, vasoconstriction occurs and the vessel will glove the needle snugly.

3. While one operator remains holding the needle in place, the other operator can press the plunger into the syringe barrel. Barrels are graduated up to 100 or 250 µl to aid accurate dispensation of the injectate.

3.7.3. Intra-amniotic Injection

1. The anesthetic techniques and exposure of the uterine horns are identical to those described in Subheading 3.5.

2. Either a conventional needle-in-barrel assembly or a bespoke needle/barrel assembly (described in Subheading 3.7.1) can be used for this procedure. At around day 16, it is possible to see the fetus clearly through the uterine wall and it is possible to discern the fore and hindlimbs and snout. The needle is inserted perpendicular to the uterine wall, ideally passing between the forelimbs and head of the fetus and out through the opposite wall of the uterus. The needle is then drawn back in so that the tip rests near the mouth of the fetus. Penetration of the fetus must be avoided. This procedure is necessary to ensure penetration of the amniotic sac.

3.8. Post-procedure

1. The mouse receives preemergent analgesic (Flunixin Meglumine, 2.5 mg/kg) via subcutaneous injection.

2. After the procedure, the mouse is placed under a heat lamp until it is awake and moving.

3.9. Group Sizes

1. For each injection day, 20 female mice are time-dated. The number of females that become pregnant can range from 1 to 7 out of 20.

2. The number of females injected depends on a number of factors. First, approximately 40–46% of fetuses are born after intra-amniotic injection and 30–35% are born after intracardiac injection ((9), unpublished data). With the early gestational injections, most of the fetuses that abort do so early, limiting the utility of a cesarean section for late gestational analysis. Second, vectors can differ widely in toxicity; adenovirus vectors are more toxic than VSVg-pseudotyped lentivirus vectors, whereas adeno-associated virus vectors appear to be the least toxic of all. Third, if a genetically manipulated model is used and the disease state is inherited in a Mendelian fashion, only 25% of fetuses will have the knockout disease phenotype. Although use of a high-resolution ultrasound increases the

likelihood of accurate injection, occasionally poor injections occur. The use of a reporter vector, such as GFP, is useful to identify accurate versus missed injections. Generally, we aim for at least 15 surviving injected fetuses per experimental group, which allows enough for postnatal evaluation at five different time points at least.

3. Control mice include (a) non-injected mice and (b) mice injected with a reporter vector that has the same backbone as the experimental vector but no transgene. If a transgenic strain is used, a third necessary control is a group of wild-type mice injected with the experimental vector.

3.10. Specimen Sampling for Analysis of Gene Transfer

Depending upon the nature of the gene transfer, it may be possible to detect gene products which are metabolic consequences of gene transfer in urine or blood or in isolated tissues of the treated animals. The advantage of collecting urine or blood is that it can be done serially and a dynamic profile can be obtained for each mouse. Notwithstanding the continual analysis, endpoint analysis can provide information about gene expression and vector genome presence in each tissue/organ. This is discussed in more detail in Chap. 13.

3.10.1. Urine Collection

Continual sampling of some body fluids is possible. Urine can be collected by holding the mouse above a Petri dish, ideally by scuffing, until it micturates of its own accord.

3.10.2. Continual Blood Sampling

Blood collection, with recovery, can be performed on mice by several different routes, including the retrobulbar plexus puncture, tail tip amputation, saphenous or lateral tarsal vein puncture, and lateral tail vein puncture (see ref. 21).

1. Mice are placed in a thermostatically controlled heating chamber at 42°C. They should remain in the chamber for around 10 min.

2. The mouse is anesthetized using inhalation anesthesia. Ideally, the induction chamber should be kept in the heating chamber so that the mouse does not cool down during induction of anesthesia.

3. The mouse is transferred to an anesthetic nose cone on a heating mat.

4. One of the two lateral tail veins (which run down the left and right sides of the tail) is punctured using a large gauge (e.g., 20 G) needle or the point of a small scalpel (see Notes 12 and 13).

5. 100 μl of blood can easily be obtained by this method; it is recommended that blood samples be taken no more frequently than once per week as recommended by Diehl and colleagues (21). Higher volumes may require longer recovery times.

6. The mouse is returned to a cool cage to recover. Gentle pressure can be applied to the puncture wound.

3.10.3. Terminal Blood Collection

We have found that cardiac puncture with a closed chest wall yields approximately 1 ml of whole blood.

1. The mouse is anesthetized using inhalation or injectable anesthesia.

2. A 25-gauge needle is inserted between the ribs on the right side of the mouse, caudal and ventral to the axilla.

3. A pulsatile back flash can often be seen through the translucent plastic of the needle hub, indicating that the needle tip has entered the heart.

4. Without moving the needle barrel, the syringe plunger can be gently pulled to withdraw up to 1 ml of whole blood (see Note 14).

5. Since cardiac puncture is a non-recovery procedure, the mouse must be euthanized; cervical dislocation is sufficient unless cervical tissues are required for analysis in which case lethal overdose of anesthesia may be used.

3.10.4. Tissue Sampling and Processing

Tissues may be collected for a wide range of analyses, including mRNA detection and quantitation, DNA detection and quantitation, ELISA, Western blot, electrophoresis mobile shift assays (EMSA), and histochemistry and immunohistochemistry (see Chap. 13). Each requires tissue to be collected according to specific guidelines, and some tissues are subject to unique collection requirements.

For biodistribution analysis of vector and/or transgene expression, it is important to consider the vector target organs when collecting tissues as it is most sensible to first collect from the tissues, which are expected to contain the least amount of vector or exhibit the lowest level of expression. Concentration of vector genomes is majorly determined by vector type; expression levels are a function of both vector type and the nature of the expression cassette. The gestational age of intervention and the route of injection are also key determinants. Although these factors have been considered in recent reviews (1, 22, 23), some salient points can be noted. First, the earlier in gestation that the gene transfer takes place, the more widespread the gene expression can be expected (14). Second, intra-parenchymal injections tend to result in localized expression (24), whereas systemic administration results in widespread expression, although the liver tends to be a major vector sink. Third, AAV tends to spread most extensively through fetal tissues (particularly AAV8 and AAV9), whereas lentivirus and adenovirus vectors tend to spread less well.

1. Once the animal has been euthanized, it is generally best to place the carcass on ice to limit RNA degradation and progression of other obfuscating catabolic processes (see Note 15).

2. Tissues to be collected for analysis of whole body vector distribution include, but are not restricted to, brain, eyes, upper airways (specifically trachea), lungs, heart, thymus, liver, pancreas, spleen, kidneys, small and large intestine, gonads, skeletal muscle, and bone marrow (see Notes 16 and 17).

3. The brain is removed by inserting pointed scissors into the foramen magnum and cutting in the horizontal (as opposed to the sagittal or coronal planes). However, for fine histology, it may be preferable to nibble at the cranium in the horizontal plane to avoid damaging the brain.

4. The abdominal viscera are accessed by performing a laparotomy from the sternum to the genitalia.

5. The thoracic organs are accessed by cutting the muscle transversely caudal to the diaphragm. Two longitudinal cuts are made through the rib cage so that the ventral wall can be lifted away (see Note 18).

6. Pieces of tissue are placed in tin foil or cryotubes for flash freezing in either liquid nitrogen or dry ice/ethanol slurry. This form of storage is appropriate for subsequent analysis of mRNA, DNA, protein ELISA, EMSA, Western blot, etc.

7. Separate pieces of tissue are collected and placed directly in 4% paraformaldehyde solution or 10% buffered formalin solution (see Note 19). For long-term storage, the tissue can be transferred to 70% ethanol for subsequent embedding in paraffin wax when immunohistochemical or histochemical analyses are to be performed. Wax embedding and tissue sectioning are performed routinely by a histology laboratory. Alternatively, instead of transferring to ethanol, the tissue is transferred to buffered 30% sucrose solution for cryosectioning and immunohistochemical or histochemical analyses. This is particularly appropriate for preparation of brain for cryosectioning on a sledge microtome.

8. Separate pieces of tissue can be placed in OCT and frozen in isopentane cooled with liquid nitrogen. This is appropriate for most tissues, where cryosectioning on a cryostat will be performed and where immunohistochemistry will be used to detect sensitive antigens, which are disrupted by fixatives (see Note 19).

9. Bone marrow is collected by removing both femurs. The femurs are cleaned of adherent tissue and the upper extremities and distal extremities are excised. The bone marrow can then be flushed out of the shaft of the femur into a centrifuge tube by inserting a 25-gauge needle into the one end of the marrow space and forcing PBS through the shaft.

4. Notes

1. It is possible to purchase the plain stainless steel tubing, which forms the needle barrel; this is likely to be more economical. Injection needles are ideally 2-cm long.

2. Experienced but young mothers should be utilized (second to third litter). Strains of mice should be carefully selected for the experimental needs, as some strains are much better breeders than others. Likewise, for postnatal survival studies, some strains provide much better maternal care than others. Particularly when transgenic models of disease are utilized, breeding of homozygous animals may be impossible or the survival of pups after in utero manipulation may be limiting. Practices, such as heterozygous breeding with postnatal establishment of genotype and cesarean delivery of pups with placement on surrogate mothers, are commonly needed to enable success.

3. Finally, environmental influences can profoundly affect the success of breeding programs. Changes in cage cleaning and bedding practices, diet, or neighboring animal species can completely derail a previously successful program.

4. Pregnant anesthetized mice are prone to hypothermia, which will result in loss of pregnancy or maternal death. All efforts must be made to keep the mother warm during and after the anesthetic and to limit operative time to 20 min or less.

5. The correct size and a sharp bevel of the pipette are essential. A microscope fitted with a reticle is used to examine the smoothness, sharpness, and the external diameter of the micropipettes. After pulling and grinding, 2-mm micropipettes have an external diameter of between 80 and 100 μm and 1.0-mm micropipettes have an external diameter between 40 and 60 μm. By changing the length of the tip, the final external diameter of the tip can be altered to accommodate different stem cell concentrations. All pipette tips should be examined under the dissecting microscope prior to use. Jagged or broken tips should be discarded.

6. AAV vector tends to be stable, in appropriate buffer, for days or even weeks when refrigerated at 4°C. Since VSV-G pseudo-typed lentivirus vectors tend to lose more than half their biological activity with each freeze–thaw cycle, and other pseudotypes even more so, we find that using fresh vector results in higher levels of expression. However, this requires careful timing to match vector harvesting with the age of gestation of intervention.

7. The Vevo Biomicroscope System comes with a micromanipulator system for injection. However, we prefer to position the ultrasound probe mechanically but to use a freehand technique for the injections. This provides additional degrees of freedom and saves time so that multiple injections can be performed.

8. Establishment of the optimal volume for a given experiment requires pilot experimentation. The desire to maximize volume should be resisted. Excessive volume results in high rates of fetal loss and frequently inaccurate localization of the vector.

9. The success of these injections depends upon the skill of the injector and the integrity of the vitelline veins. Prior to 14 days and after 15 days, the vitelline veins are either too small or have begun to involute, respectively, so the quality of the injections is usually best between E13.5 and E14.5. There is a learning curve of at least ten pregnant mice to become accomplished with these injections. We assess competency using GFP-positive cells and harvesting the litters after 1 h to assess the percentage of cells in the fetal liver by flow cytometry. A variance of up to 20% with one or two missed fetuses out of each ten injected is deemed acceptable.

10. A short length of tubing with a small attachment at the end for the micropipette is suitable for IUHCT. If this model is used extensively, we strongly recommend the use of a micro-injector and two people to perform the procedure. Two people allows one individual to focus entirely on positioning and maintaining the tip of the micropipette in the vein, while the other person can control the automatic injector, and to visually confirm injection of the preset volume. The use of the automated injector standardizes the procedure so that each injection maintains the same volume, pressure, and duration of injection.

11. To standardize the injections, the glass micropipettes are labeled with accurate volume delineations. The equation to determine the length on a micropipette that is equal to a given volume is $h = v/\pi r^2$, where h = height, v = volume, and r = radius. Once the internal diameter of the micropipette is known, the height for a given volume is calculated. A graph consisting of a horizontal line with demarcations corresponding to the calculated height is created, printed, and then used as a template to mark the volume demarcations onto the micropipettes.

12. Blood can be collected from conscious mice; however, to minimize stress to both the animal and the operator, it is preferable to use anesthesia.

13. Care must be taken to only puncture the vein and not to cut through the vein and damage nerves, ligament, or even bone. When bleeding many mice, it is advisable to change the needle/scalpel often to ensure a clean cut.

14. It may be necessary to rotate the needle or push or pull the needle slightly since, during the blood withdrawal, the mouth of the needle may become blocked by the collapsing cardiac wall.

15. If a large group of animals is to be culled for full body analysis, it may be best to euthanize and harvest sequentially rather than euthanize all at once, to limit degradative processes that occur postmortem.

16. For large organs with a relatively homogenous macrostructure, it is possible to collect several pieces, which may be stored in different ways. However, for some organs, e.g., brain and eye, which have regional specialization, it may not be possible to collect the tissue for more than one purpose.

17. It may be preferable, to prevent contamination with fur, to skin the animal before dissection. This can be achieved by making a lateral cut in the skin overlying the abdominal spine and separating the skin in a degloving manner, pulling anteriorly and caudally. Care must be taken with the caudal section of skin not to draw the intestines out through the pelvic outlet.

18. It is not a good idea to perform a single longitudinal incision through the sternum as it is difficult to spread the ribs in order to access the thoracic organs.

19. For preparation of lungs for cryosectioning on a cryostat, it is necessary to first inflate the lungs (or lung lobe) with 50% OCT. This eliminates large airspaces, which prevent effective cryosectioning.

References

1. Waddington SN, Kramer MG, Hernandez-Alcoceba R et al (2005) In utero gene therapy: current challenges and perspectives. Mol Ther 11:661–676

2. Haynes BF, Martin ME, Kay HH et al (1988) Early events in human T cell ontogeny. Phenotypic characterization and immunohistologic localization of T cell precursors in early human fetal tissues. J Exp Med 168:1061–1080

3. Takahama Y (2006) Journey through the thymus: stromal guides for T-cell development and selection. Nat Rev Immunol 6:127–135

4. Hagberg H, Peebles D, Mallard C (2002) Models of white matter injury: comparison of infectious, hypoxic-ischemic, and excitotoxic insults. Ment Retard Dev Disabil Res Rev 8:30–38

5. Seppen J, van der Rijt R, Looije N et al (2003) Long-term correction of bilirubin UDP glucuronyltransferase deficiency in rats by in utero lentiviral gene transfer. Mol Ther 8:593–599

6. Waddington SN, Mitrophanous KA, Ellard F et al (2003) Long-term transgene expression by administration of a lentivirus-based vector to the fetal circulation of immuno-competent mice. Gene Ther 10:1234–1240

7. Dejneka NS, Surace EM, Aleman TS et al (2004) In utero gene therapy rescues vision in a murine model of congenital blindness. Mol Ther 9:182–188

8. Rucker M, Fraites TJ Jr, Porvasnik SL et al (2004) Rescue of enzyme deficiency in embryonic diaphragm in a mouse model of metabolic myopathy: Pompe disease. Development 131:3007–3019

9. Karolewski BA, Wolfe JH (2006) Genetic correction of the fetal brain increases the lifespan of mice with the severe multisystemic disease mucopolysaccharidosis type VII. Mol Ther 14:14–24

10. Sabatino DE, Mackenzie TC, Peranteau W et al (2007) Persistent expression of hF.IX after tolerance induction by in utero or neonatal administration of AAV-1-F.IX in hemophilia B mice. Mol Ther 15:1677–1685

11. Niiya M, Endo M, Shang D et al (2008) Correction of ADAMTS13 deficiency by in utero gene transfer of lentiviral vector encoding ADAMTS13 genes. Mol Ther 17:34–41

12. Koppanati BM, Li J, Reay DP et al (2010) Improvement of the mdx mouse dystrophic phenotype by systemic in utero AAV8 delivery of a mini dystrophin gene. Gene Ther 17(11):1355–1362

13. Schachtner S, Buck C, Bergelson J et al (1999) Temporally regulated expression patterns following in utero adenovirus-mediated gene transfer. Gene Ther 6:1249–1257

14. Endo M, Henriques-Coelho T, Zoltick PW et al (2010) The developmental stage determines the distribution and duration of gene expression after early intra-amniotic gene transfer using lentiviral vectors. Gene Ther 17:61–71

15. Endo M, Zoltick PW, Chung DC et al (2007) Gene transfer to ocular stem cells by early gestational intraamniotic injection of lentiviral vector. Mol Ther 15:579–587

16. Endo M, Zoltick PW, Peranteau WH et al (2007) Efficient in vivo targeting of epidermal stem cells by early gestational intraamniotic injection of lentiviral vector driven by the keratin 5 promoter. Mol Ther 16:131–137

17. Stitelman DH, Endo M, Bora A et al (2010) Robust in vivo transduction of nervous system and neural stem cells by early gestational intra amniotic gene transfer using lentiviral vector. Mol Ther 18:1615–1623

18. Roybal JL, Endo M, Radu A et al (2011) Early gestational gene transfer of IL-10 by systemic administration of lentiviral vector can prevent arthritis in a murine model. Gene Ther 18:719–726

19. Henriques-Coelho T, Gonzaga S, Endo M et al (2007) Targeted gene transfer to fetal rat lung interstitium by ultrasound-guided intrapulmonary injection. Mol Ther 15: 340–347

20. Gonzaga S, Henriques-Coelho T, Davey M et al (2008) Cystic adenomatoid malformations are induced by localized FGF10 overexpression in fetal rat lung. Am J Respir Cell Mol Biol 39:346–355

21. Diehl KH, Hull R, Morton D et al (2001) A good practice guide to the administration of substances and removal of blood, including routes and volumes. J Appl Toxicol 21: 15–23

22. Buckley SMK, Rahim AA, Chan JKY et al (2011) Recent advances in fetal gene therapy. Ther Deliv 2:461–469

23. Roybal JL, Santore MT, Flake AW (2010) Stem cell and genetic therapies for the fetus. Semin Fetal Neonatal Med 15:46–51

24. Rahim AA, Wong AM, Ahmadi S, et al (2011) In utero administration of Ad5 and AAV pseudotypes to the fetal brain leads to efficient, widespread and long-term gene expression. Gene Ther (in press)

Chapter 11

Animal Models for Prenatal Gene Therapy: The Sheep Model

Khalil N. Abi-Nader, Michael Boyd, Alan W. Flake, Vedanta Mehta, Donald Peebles, and Anna L. David

Abstract

Large animal experiments are vital in the field of prenatal gene therapy, to allow translation from small animals into man. Sheep provide many advantages for such experiments. They have been widely used in research into fetal physiology and pregnancy and the sheep fetus is a similar size to that in the human. Sheep are tolerant to *in utero* manipulations such as fetoscopy or even hysterotomy, and they are cheaper and easier to maintain than non-human primates. In this chapter, we describe the animal husbandry involved in generating time-mated sheep pregnancies, the large number of injection routes in the fetus that can be achieved using ultrasound or fetoscopic-guided injection, and laparotomy when these more minimally invasive routes of injection are not feasible.

Key words: Sheep, Ultrasound, Fetoscopy, Amniocentesis, Fetal blood sampling

1. Introduction

If fetal gene therapy is to become clinically applicable, developments in vector technology must be accompanied by improvements in minimally invasive methods of delivering vectors to the fetus. Traditionally, invasive surgical techniques such as maternal laparotomy or hysterotomy have been performed to access the fetus in small- and even large-animal models. However, in clinical practice, minimally invasive techniques such as ultrasound-guided injection, or even fetoscopy, could be used to deliver gene therapy to the fetus with less morbidity and mortality. It is likely that non-human primates will be the ultimate animal model that will be used for safety studies in the immediate preparation for a clinical trial of fetal gene therapy. However, the high maintenance costs and breeding conditions prohibit their use in the routine development of novel injection techniques.

Charles Coutelle and Simon N. Waddington (eds.), *Prenatal Gene Therapy: Concepts, Methods, and Protocols*,
Methods in Molecular Biology, vol. 891, DOI 10.1007/978-1-61779-873-3_11, © Springer Science+Business Media, LLC 2012

Sheep are much easier to breed and maintain and are a well-established animal model of human fetal physiology. Sheep have a consistent gestation period of 145 days; the development of the fetus and of the immune system is very similar to humans. Using the pregnant sheep, we have adapted ultrasound-guided injection techniques from fetal medicine practice and developed new methods to deliver gene therapy to the fetal sheep including ultrasound-guided intratracheal injection to target the distal respiratory epithelium (1, 2) and ultrasound-guided intragastric injection to target the intestinal mucosa (3). Fetoscopic techniques can be utilized when improved visualization is needed beyond the resolution of ultrasound guidance or when surgical interventions are required to facilitate optimal delivery of vector. Examples of fetoscopic techniques include performing intravascular injections prior to 70 days gestation in the sheep, targeting the umbilical veins at their convergence at the cord (Flake AW, unpublished observation), or placement of an intratracheal balloon at the time of vector installation to enhance pulmonary epithelial transduction (4). The combination of ultrasound guidance and fetoscopy is often utilized when small fetoscopes are employed. Ultrasound provides the big picture allowing guidance of the scope with its relatively limited field of view to the point of intervention.

These techniques are described in detail in this chapter and the fetal organs that can be targeted via different routes of injection are listed in Table 1. It is likely that many if not all of these methods could be adapted for use in other large animals such as the macaque, baboon, goat, or dog, and are likely to target similar organs to those in the fetal sheep.

Maternal mortality in the pregnant sheep is negligible and fetal mortality was between 3 and 15%, depending on the route of injection. Over 90% of the fetal mortality is due to iatrogenic infection, usually with known fleece commensals (5). Thus, it is vital that any invasive procedure is performed with stringent aseptic technique. In clinical practice, cleaning by wiping the maternal abdominal skin with an antiseptic solution is sufficient. In the sheep, the fleece needs to be scrubbed as an additional step, as described below.

Ultrasound-guided invasive procedures in sheep such as tracheal injection had a complication rate of 6%, which was related to blood vessel damage within the thorax (6). Intracardiac and umbilical vein injection had an unacceptably high procedure-related fetal mortality in first-trimester fetal sheep (5) and umbilical vein injection was only reliably achieved from 70 days of gestation, equivalent to 20 weeks of gestation in humans. The relevant time windows for the different application routes in humans still need to be established with respect to technical feasibility, fetal physiology, and the development of the fetal immune system.

Table 1
Organ targeted prenatal gene transfer

Organ to be targeted	Injection routes	Limiting factors
Lung	Intra-amniotic	Large dilution effect of AF; relies on fetal BMs
	Intrapulmonary	Expression remains local to injection site
	Intratracheal	Transthoracic injection and fetoscopic tracheal delivery from mid-gestation onwards
Blood	Umbilical vein	
	Intraperitoneal	Increased risk of germline gene transfer when performed early in gestation
Skeletal Muscle	Intramuscular	May need multiple injection sites
	Intraperitoneal	Also targets diaphragm
	Hydrothoracic	Technically difficult in man
Liver	Intraperitoneal	Rapid division of fetal liver may limit level of transduction by vectors that only minimally integrate
	Intrahepatic	Transduction of liver injection sites only
Brain	Intraventricular	Technically difficult in man
	Umbilical vein or intraperitoneal	Relies on vector crossing the blood brain barrier; may be gestational age dependent
Skin	Intra-amniotic	Delivery early in gestation required to target deeper epidermal layers
Sensory organs	Intra-amniotic	Early stage of gestation critical to reach developing ear or eye
	Otocyst or retinal injection	Technically difficult in man
Gut	Intra-amniotic	Large dilution effect of AF; relies on fetal BMs
	Intragastric	Potential limiting effect of gastric fluid on vector
	Oropharyngeal	Fetoscopically guided
Heart	Intraperitoneal	
	Intramyocardial	Technically challenging in the fetus
Renal	Intra-amniotic	Transgenic protein expressed in kidney, but produced elsewhere
	Urinary tract	Never evaluated
Uteroplacental circulation	Uterine artery	

AF amniotic fluid, *BMs* breathing movements

2. Materials

2.1. Generation of Time-Mated Sheep

1. Ewes and rams in a ratio of 10:1.
2. Vaccination to *Chlamydia abortus* (Enzovax®, Intervet UK) and toxoplasmosis (Toxovax®, Intervet UK).
3. Chronogest® CR sponges (Intervet UK Limited, Milton Keynes, UK) containing 20 mg Flugestone acetate.
4. Sponge applicator.
5. Ram Mating Raddle (Agrimark Limited, Isle of Man, UK).
6. Super Ewe & Lamb UFAS compound feed (J & W Atlee Ltd, Dorking, Surrey, U K).

2.2. Confirmation of Pregnancy and Gestational Age by Ultrasound

1. Ultrasound coupling gel (Electro Medical Supplies Ltd, Oxon, UK).
2. Clippers for fleece (Masterclip Duo, Outlandish Items Ltd, Leicestershire, UK).
3. Ultrasound scanner with a curved-array 3.0–5.0 MHz abdominal probe with real-time image display, cine loop facility, B-mode, M-mode pulse wave Doppler flow mode and color flow mode, e.g., Logiq 400 (GE Medical Systems, Milwaukee, USA), or Acuson 128 XP10 ultrasound scanner (Siemens, Bracknell, UK).
4. Ultrasound charts in use for sheep (7, 8).

2.3. Sheep Anesthesia

1. Wood chips for bedding (GPM Shavings, East Angus, Canada).
2. A large sling with poles.
3. A balance to weigh the sheep (needs to measure up to 150 kg).
4. Clippers for fleece (Masterclip Duo, Outlandish Items Ltd, Leicestershire, UK).
5. 19-Gauge butterfly winged perfusion set (Terumo Europe NV, Leuven, Belgium).
6. BD Vacutainer EDTA, 9NC, and SST tubes (BD Vacutainer systems, Plymouth, UK).
7. 15 ml of 10% w/v thiopental sodium i.v. (Thiovet, Novartis Animal Health UK Ltd, Hertfordshire, UK).
8. 2–2.5% Isoflurane (Isoflurane-Vet, Merial Animal Health Ltd, Essex, UK).
9. 100% Oxygen.
10. An oral gastric tube (Arnold Veterinary Products, Harlescott, Shrewsbury, UK).
11. A laryngoscope with a 30-cm blade (Penlon, UK).
12. A 9.0-mm cuffed endotracheal tube (Portex, UK).

13. A semiclosed anesthetic system that includes a capnograph (Ohmeda Anethesia system, BOC Healthcare).

14. A stethoscope (Littman, 3 M, Bracknell, UK).

15. A pulse oximeter (5250 RGM, Ohmeda).

16. A surgical theater table that can be raised.

17. A ventilator (Manley MP5, Blease Medical Equipment Ltd, Chesham, Bucks, UK).

2.4. Ultrasound-Guided Procedures

1. Povidone iodine antiseptic cleaning solution (Grampian Pharmaceuticals Ltd, UK).

2. Chlorhexidine gluconate cleaning scrub (Hibiscrub, AstraZeneca Ltd, UK).

3. Povidone iodine antiseptic solution (Grampian Pharmaceuticals Ltd, UK).

4. Sterile drapes.

5. A 3.0–5.0-MHz curved-array abdominal probe on an ultrasound scanner with real-time image display, cine loop facility, B-mode, M-mode pulse wave Doppler flow mode, and color flow mode, e.g., Logiq 400 (GE Medical Systems, Milwaukee USA), or Acuson 128 XP10 ultrasound scanner (Siemens, Bracknell, UK).

6. Sterile ultrasound coupling gel (Electro Medical Supplies Ltd, Oxon, UK).

7. A no. 11 blade (Swann-Morton, Sheffield, UK).

8. 18-G, 20-G, 15-cm long and 22-G, 9-cm long echotip needles with a lancet tip (Cook Medical, UK).

9. Sterile physiological saline (0.9% NaCl).

10. 1-, 2-, and 5-ml syringes.

11. Viricide agent, e.g., Virkon (Antec International Limited).

2.5. Fetoscopic Procedures

1. Sterile preparation as described for ultrasound-guided and laparotomy procedures.

2. If ultrasound guidance is required, ultrasound equipment as described for ultrasound-guided procedures above.

3. If laparotomy is required, equipment as described for laparotomy procedures below.

4. Storz (Storz Endoskope, Tuttlingen, Germany) compatible videocart with minimum requirements of: (1) high definition monitor, ideally configured for picture in picture with the ultrasound image displayed on the same screen, (2) camera control unit, (3) camera head preferably 3-chip, (4) suction (optional) and irrigation system, (5) cold light source, and (6) video recording unit.

5. Storz fetoscopy set appropriate for planned procedure. We recommend the use of a remote eyepiece for maximum range of motion. Each set consists of a telescope and an operating sheath with various working channels, and accessories for specific procedures. Telescopes range from the straight forward 0° 1 mm, 20 cm working length scope to the 0°–30° 2 mm 26 cm working length scopes (Storz endoscopy, USA). Corresponding operating sheaths are the 1.3-mm maximum diameter (6.5 Fr), pointed tip, sheath with laser fiber or puncture needle working channel, and the 9 Fr (Storz endoscopy, USA). Operating sheath with laser fiber working channel or more versatile 11, 11.5, or 14.5 Fr. operating sheaths (Storz endoscopy, USA) that include larger working channels or channels for working inserts.

6. For intravascular injections, the straight forward 0° 1 mm, 20 cm working length scope (Storz Endoscopy, USA), the 1.3-mm maximum diameter (6.5 Fr), pointed tip, sheath with laser fiber or puncture needle working channel, and the 0.6-mm 26-cm long puncture needle are used (Storz endoscopy, USA).

7. For fetal balloon tracheal occlusion, the 1.3-mm diameter, 30.6 cm working length, fetal tracheoscope and the 3.3-mm tracheoscopic sheath with three side ports (30.6 cm working length, precurved 30°) are ideal (Storz endoscopy, USA).

8. 10 Fr Flexor Check-Flo cannula (Check-Flo Introducer set).

9. BALTACCI—BDPE—detachable balloon catheter (outer diameter max. 0.9 mm, Balt, Montmorency, France).

10. 3.0 Fr Slip-cath infusion catheter (Cook, USA).

11. Warm Lactated Ringers solution.

12. 3-0 Vicryl.

2.6. Laparotomy Procedures

1. Povidone iodine antiseptic cleaning solution (Grampian Pharmaceuticals Ltd, UK).

2. Chlorhexidine gluconate cleaning scrub (Hibiscrub, AstraZeneca Ltd, UK).

3. Povidone iodine antiseptic solution (Grampian Pharmaceuticals Ltd, UK).

4. Sterile drapes.

5. An electrosurgical unit (ERBE VIO 300D Electrosurgical Generator, Tubingen, Germany).

6. A no. 21 blade and handle (Swann-Morton, Sheffield, UK).

7. Surgical instruments: Mayo and Metzenbaum scissors, surgical forceps set (atraumatic, Debackey, and grasping forceps), large and small needle holders (Fine Science Tools, InterFocus Ltd, Cambridge, UK).

8. Surgical sutures: 5-mm Mersiline tape (Ethicon, St. Stevens-Woluwe, Belgium), 2–0 PDS (Ethicon, St. Stevens-Woluwe, Belgium), 2-0 Monocryl (Ethicon, St. Stevens-Woluwe, Belgium) and 2-0 and 3-0 prolene (Ethicon, Norderstedt, Germany).

9. Surgical vessel loops (Bard Ltd, Crawley, West Sussex, UK).

10. 23- and 25-G butterfly winged perfusion sets (Terumo Europe NV, Leuven, Belgium).

11. 2-, 5-, and 10-ml syringes.

12. Blood collection tubes: EDTA, 9NC, and SST tubes (BD, Plymouth, UK).

13. Injectable antibiotics: Penstrep (200 mg/ml procaine penicillin and 250 mg/ml dihydrostreptomycin; Norbrook Laboratories Ltd, County Down, UK), sodium benzylpenicillin G (Crystapen, Schering-Plough, Uxbridge, UK), and Gentamicin (Genticin Injectable, Roche Products Ltd, Hertfordshire, UK).

14. Injectable analgesics: Buprenorphine (Vetergesic, Alstoe Animal Health, York, UK).

3. Methods

3.1. Generating Time-Mated Sheep

1. Once the sheep for experiments have been selected, a vaccination schedule against *Chlamydia abortus* and toxoplasmosis is drawn up and given. See Note 1.

2. To time sheep ovulation, Chronogest® sponges (Flugestone acetate) containing 30 mg progesterone is placed in the vagina of ewes for 2 weeks to induce ovulation.

3. Two days after removal of the progesterone sponges, ewes are put in a pen with the ram overnight. The ram is marked on its belly with Ram Raddle®, a colored powder mixed with liquid paraffin which is transferred and marks the back of the ewe once she has been served. Marked ewes are presumed to have been tupped.

4. Ewes that are unmarked or not marked clearly are put back for a "return service" with the ram 2 weeks later.

5. A spreadsheet is used to check on regular progress with tupping and to schedule experiments at specific gestational ages, with close liaison with the farmer.

6. When the sheep breeding season is advanced (March onwards) superovulation may help. Chronogest® sponges are placed vaginally for 12 days followed by an intramuscular injection of 1,000 IU Folligon® (Pregnant Mares Serum Gonadotrophin).

The ewe is placed in with the ram 2 days later and returned as before 14 days later. Superovulation often results in higher order multiple pregnancies such as triplets and quadruplets, and the conception rate is not as high as that achieved using progesterone alone.

7. Sheep should be moved from the farm to the experimental animal facility (if they are separate), at least 1 week before the scheduled procedure. See Note 2.

3.2. Confirmation of Pregnancy and Gestational Age by Ultrasound

1. All ewes that have been tupped are scanned to confirm pregnancy, fetal number, and gestation age at approximately 40 days of age to allow enough time to move them to the experimental facility. See Note 3.

2. Ewes are caught, turned up, and held in a sitting position for scanning using a standard B-mode gray-scale ultrasound scanner with an abdominal probe. This is the same position as is used by farmers to clip the toenails and for shearing sheep. Sheep are generally placid and will happily remain in this position for at least 15 min.

3. An area of the fleece is clipped close to the skin just above the udders extending 10 cm up the abdomen, to allow a clear view of the whole uterus.

4. The sonographer applies ultrasound coupling medium to the clipped area and scans the uterus using a 3.0–5.0-MHz probe held just above the udders. It is important to scan across both horns of the uterus to examine for twins and triplet pregnancies.

5. Still images of the occipito-snout length and biparietal diameter are obtained and the dimensions measured on the machine. Gestational age is confirmed according to standard tables.

3.3. Sheep Anesthesia

1. Before surgery ewes are starved for 12 h with free access to water, and bedded on wood chips. This is to prevent bloating which can occur when sheep under anesthesia are unable to belch (eructate) normally and release methane gas which is a by-product of fermentation of their food (9).

2. Animals are weighed to allow calculation of the correct dose for drugs.

3. On the morning of surgery sheep are walked to theater. This should be a dedicated theater space if possible that is kept clean, has anesthetic gas extraction facilities and air-conditioning to maintain the temperature at 37°C.

4. The sheep are turned up with an assistant sitting comfortably behind. The wool is clipped from their necks, the jugular vein is cannulated using a 19-Gauge butterfly winged perfusion set, and the cannula is securely taped in.

5. Blood can be taken at this point for baseline investigations of antibodies, hematology, and biochemistry parameters. Blood is collected into BD Vacutainer tubes containing potassium EDTA for plasma and into plain tubes for serum (SST).

6. 15 ml of thiopental sodium (10% w/v) is given intravenously over 1 min for induction of anesthesia.

7. Once asleep the ewes are rolled onto a mat with poles or handles to attach to a hoist and they are then lifted mechanically onto the operating table and laid supine. With the sheep supine, the neck is extended and the jaw is lifted up using tape applied across the tongue and teeth. A laryngoscope with a 30-cm blade is passed down the trachea to enable visualization of the vocal cords. A 9.0-mm cuffed endotracheal tube containing a stylet is passed through the vocal cords. The stylet is removed and the cuff is inflated with the required volume of air to prevent inhalation of regurgitated ruminal contents. The tube is secured with tape tied around the lower jaw.

8. Alternatively, the ewe may also be intubated while she is held sitting up by an assistant behind her. Using a tape the upper jaw is lifted up to straighten the trachea, and while the tongue is gently pulled to one side with some gauze, the endotracheal tube is passed through the vocal cords.

9. The endotracheal tube is connected to the anesthetic gas machine and the correct placement of the endotracheal tube is checked by a variety of methods:

   Auscultate both lungs while the bag and valve mask is being compressed to force air into the lungs.

   Observe the bag and valve mask inflating and deflating in the closed system.

   Feel movement of air within the endotracheal tube as the ewe breathes.

10. Once the sheep is supine on the operating table and anesthetized the head is lowered to allow drainage of saliva and ruminal fluid (10). Because access to the whole abdomen is needed for procedures the animals are kept in dorsal recumbency. Over long periods the rumen can compress the lungs causing dyspnoea and the vena cava reducing venous return to the heart. This is not felt to be a problem in the case of in utero ultrasound-guided procedures because they are usually of sufficiently short duration.

11. Anesthesia is maintained by inhalation on 2–2.5% isoflurane in 100% oxygen run at 2–3 l/min for spontaneously breathing animals and at 6–8 l/min for animals on intermittent positive pressure ventilation (IPPV) to maintain an end-tidal CO_2 <5%, O_2 saturation >95%, and a respiratory rate of 12–14/min using a semiclosed anesthetic circuit.

12. For lengthy experiments (over 20 min), it is advisable to pass an oral gastric tube into the sheep's stomach to allow the release of methane gas. A capnograph should also be connected to the endotracheal tube to monitor carbon dioxide saturation. We aim to keep end-tidal CO_2 <5% if possible.

13. For experiments that are expected to last over 30 min (complicated ultrasound-guided procedures or laparotomy), the ewe should be ventilated to prevent CO_2 retention. If the venous return or oxygen saturation is poor, and the position of the animal is compromised, the animal's head may be tilted too far down to rely on spontaneous breathing, in which case the ewe is then ventilated electively.

14. A pulse oximeter is placed on the ewe's ear or tongue and is used to monitor oxygen saturation and pulse rate. This data is documented every 15 min together with respiration rate, carbon dioxide concentration, anesthetic gas concentration, and level of anesthesia on an anesthetic sheet.

15. The anesthetic gas is turned off 1–2 min before the end of the experiment to allow recovery. Analgesia is not usually given for minimally invasive ultrasound-guided procedures. See Note 4.

16. The ewe is hoisted off the theater table to the floor and rolled onto her sternum. Her legs are placed underneath to ensure she can get upright as soon as awake. She is extubated once strong swallowing reflexes return and she is then returned to the sheep pen.

17. The ewe is watched carefully to ensure she stands up within a few minutes of the procedure, and she is helped to stand if she does not do so spontaneously after 20 min. Standing allows recovery of ruminal function, which can be affected by a long general anesthetic.

18. The day after the procedure, the well-being of the ewe is assessed and fetal survival is determined on ultrasound scan by turning the ewe up for ultrasound assessment. Ultrasound-guided injection sites are examined carefully for infection.

19. For experiments where the fetus will be delivered, serial fetal ultrasound assessment should be made during gestation and a plan made for care of the ewe during labor (see Chapter 14 and Note 5).

3.4. Ultrasound-Guided Procedures

1. Once anesthetized, the wool is clipped from the ewe's abdomen from the udders up to the sternum and round to the sides.

2. The skin is scrubbed for 5 min using povidone iodine antiseptic cleaning solution. See Note 6.

3. A detailed ultrasound examination of the uterus and its contents is performed using a 3–5-MHz probe on the ultrasound scanner. Fetal measurements such as fetal femur length, abdominal

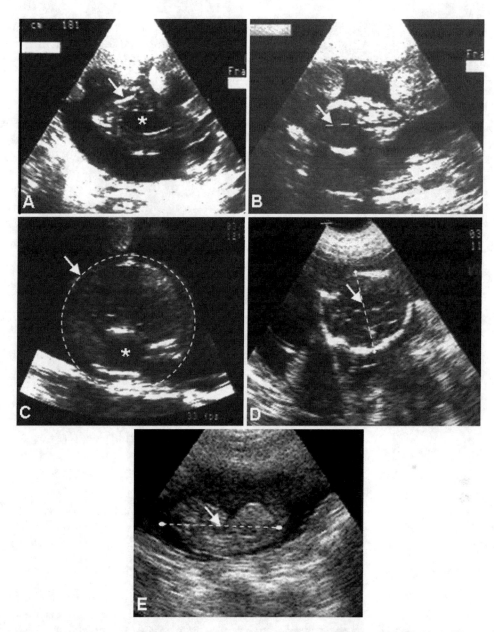

Fig. 1. Ultrasound measurements of fetal sheep. (**a**) Femur length (10.1 mm, *white arrow*) and fetal bladder (*white asterisk*) (**b**) Occipito-snout length (40.2 mm, *white arrow*) at 60 days of gestation. (**c**) Abdominal circumference (162 mm, *white arrow*) and fetal stomach (*white asterisk*) at 83 days of gestation. (**d**) Bi-parietal diameter (22 mm, *white arrow*) at 60 days of gestation. (**e**) Crown-rump length (27.4 mm, *white arrow*) at 36 days of gestation.

circumference, biparietal diameter, and occipito-snout length are measured for accurate assessment of gestational age (see Fig. 1). The appropriate site of injection is identified.

4. Ultrasound coupling medium (ultrasound gel) is used during this scan. At this stage nonsterile gel can be used.

5. The abdomen is then scrubbed again for a further 5 min using chlorhexidine gluconate cleaning solution and a final clean with povidone iodine solution is made over the injection site.

6. The ultrasound probe is wiped over with 70% ethanol and then sprayed again and the ethanol is allowed to evaporate.

7. The operators (usually two) scrub up and don sterile gowns and gloves. All instruments (needles, syringes containing the vector, and saline) are laid out on a small operating table draped with a sterile drape. The vector is drawn up into syringes ready to be administered immediately to the fetus.

8. The abdomen is now draped with sterile drapes to reveal a window for the ultrasound guided injection sit. See Note 7.

9. Sterile ultrasound coupling medium (ultrasound gel) is now used for ultrasound scanning. See Note 8.

10. A small cut in the skin, using a no. 11 blade may facilitate needle insertion and avoid blunting the sharp needle tip. This is only necessary when a larger needle (18 Gauge and higher) is used.

11. Under ultrasound guidance the needle is inserted (with an angle of 45° relative to the ultrasound beam) through the maternal skin, the uterus, and into the amniotic cavity avoiding passage through the placentomes if possible. Care should be taken to keep the whole needle length and particularly the needle tip under ultrasound view as this view most accurately reflects the needle position, otherwise it is easy to be mislead about the true needle orientation.

12. For each injection route a freehand technique is used. Needle-guides can be used but reduce the flexibility of the approach, especially if the fetal position moves relative to the needle as it is inserted. The fetus does not move around once the ewe has been under general anesthesia for at least 10 min.

13. The needle is then inserted through the fetal skin into the target by flicking it through the uterine wall and into the fetus. Once the needle tip is thought to be in the correct position, the stylet is removed and, the vector is injected followed by 40 μl 0.9% saline to flush the vector from the dead space of the spinal needle. This is important because in many experiments, small volumes of vector are often applied.

14. If the position of the needle tip is not certain, a small volume (40 μl 0.9% saline) can be introduced to confirm it. A white flash will be seen on the screen as turbulence was created by microbubbles.

15. After removal of the needle from the uterus, the fetus is observed for signs of fetal well-being such as a normal fetal heart rate, hematoma formation, or bleeding See Note 9.

16. Procedures are always video-taped so that after-injection review can be done to confirm the injection route.

17. The fetus should be scanned the day after injection to check on fetal well-being, by turning up the ewe. This allows examination of the injection site on the ewe's abdomen to be inspected.

3.4.1. Intra-amniotic Injection (see Note 10)

1. A 22-G Quincke type spinal needle (Becton Dickinson, UK) is used for procedures performed at any gestational age.

2. A good-sized pocket of amniotic fluid is identified by ultrasound.

3. The needle is aimed towards the center of the pocket of amniotic fluid while avoiding fetal parts, and passage through one of the placentomes.

4. The needle tip is advanced through the amniotic membrane using a rapid thrust since the amniotic membrane in sheep is relatively flaccid especially during early gestation.

5. The needle tip can be easily identified as a hyperechoic point inside fluid-filled anechoic amniotic cavity (even when using needles without an echotip). The needle stylet is removed and a small volume of amniotic fluid is aspirated for confirmation.

6. A small volume of 0.9% saline is injected and correct needle placement is reconfirmed by observing microbubbles moving around the fetus in the amniotic sac.

7. The vector is injected.

3.4.2. Intraperitoneal Injection

1. A 20- or 22-Gauge Echotip Lancet needle is used, depending on gestational age. 20 Gauge is used after 80 days of gestation.

2. A midsagittal ultrasonographic plane of the fetal abdomen is obtained. The needle tip is targeted below the level and to the side of the cord insertion and just superior to the fetal bladder. Care should be taken to prevent damage to umbilical cord insertion or the umbilical arteries as they flow down medio-laterally along the posterior aspect of the anterior fetal abdominal wall. Color or power Doppler is helpful in delineating the vascular anatomy if the operator is in doubt.

3. The position of the needle is viewed in two planes to confirm correct placement. Needle placement can also be verified by injecting 100 μl saline and observing turbulence as microbubbles spreading throughout the peritoneal cavity. In addition, an echodense area surrounding the fetal bowel usually develops at the injection site.

4. The vector is delivered in volumes of 100–500 μl before 60 days, up to 1 ml between 60 and 70 days and 1–2 ml after 70 days of gestation.

3.4.3. Umbilical Vein
Injection (See Note 11)

1. A 22-Gauge Echotip Lancet needle is used for all procedures up to 85 days of gestation. A 20-Gauge Echotip Lancet needle is used for experiments at later gestational ages.

2. The intrahepatic portion of the umbilical vein is the preferred site for intravascular injection in the sheep fetus. This is best visualized using a cross-sectional plane of the abdomen with a slight tilt towards the fetal head. Attempts should be made to puncture the vessel as close as possible to 90° as more parallel angles are associated with failed punctures. Injection of the fetal intrahepatic umbilical vein can be achieved reliably from approximately 80 days of gestation in the fetal sheep (unpublished observation).

3. As the needle hits the vessel wall, the wall tends to tent and a sharp but limited forward movement is needed to gain access into the lumen at that moment. The echodense needle tip can be seen inside the vessel lumen (Fig. 2). The stylet is removed and a small 1–2 ml syringe is connected to the needle. A tight seal of the 1–2 ml syringe to the end of the needle within the fetal UV is needed. After applying negative pressure a small volume of blood is usually withdrawn.

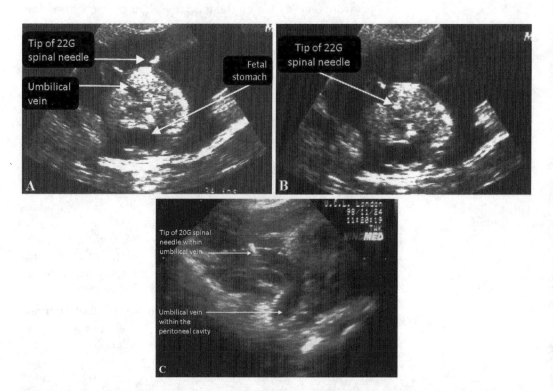

Fig. 2. Umbilical vein injection in the late first trimester (**a, b**) and late gestation (**c**). A sheep fetus (60 days of gestation) is shown in cross-section during umbilical vein delivery of AdhFIX vector. (**a**) A 22-G spinal needle approaches the fetus from the right side, passes through the hepatic parenchyma and (**b**) the tip can be seen within the UV as it traverses the fetal liver. (**c**) In late gestation (140 days of gestation), the UV can be injected as it enters the peritoneal cavity of the fetus.

4. If the blood does not come immediately, the needle is carefully and very slightly withdrawn while negative pressure is applied with the syringe, since sometimes the needle is touching the opposite vessel wall. As the needle moves away from the vessel side wall opposite, the blood should be easily withdrawn.

5. Once correct needle placement is confirmed, the syringe is then swopped to that containing the vector, which is then injected in a volume of 100–1,000 μl slowly over 1 min. Turbulence can be seen within the umbilical vein as microbubbles passing along the vessel. Great care must be taken during this change to the syringe containing the vector to avoid dislodging the needle from the umbilical vein.

6. Inadvertent intrahepatic injection can be seen as echogenic white material in the vessel wall and surrounding hepatic parenchyma.

3.4.4. Intracardiac Injection (see Note 12)

1. A 20-Gauge Echotip Lancet needle is used for this procedure to ensure the needle passes through the chest wall into the heart.

2. The left or right ventricle is the preferred site of injection. A four-chamber view of the fetal heart should be obtained if possible with the anterior chest wall facing anterior or laterally.

3. The heart should be approached from the inferior aspect so that the needle avoids the cardiac conduction system. This route has been found to reduce the mortality rate (11).

4. The needle should be passed through the fetal chest wall, between the fetal ribs and aiming for the ventricle in the direction of the AV valves. Once the needle tip is within the ventricle, the stylet is removed and fetal blood is aspirated to check correct needle position.

5. Once correct needle placement is confirmed, the syringe containing the vector is applied to the needle and a volume of 100–1,000 μl is injected slowly over 1 min. Turbulence can be seen within the ventricle as microbubbles pass through the chamber.

6. After removal of the needle, the fetal heart should be observed under ultrasound to confirm normal heart rate and to check for pericardial hemorrhage.

3.4.5. Intrahepatic Injection

1. A 22-Gauge Echotip Lancet needle is used for early gestation procedures. A 20-G needle may be used for procedures performed after midgestation.

2. Under ultrasound guidance the needle is aimed through the anterior fetal abdominal wall towards a point superior to the intrahepatic portion of the umbilical vein.

3. The position of the needle is viewed in two planes to confirm correct placement before delivery of the viral vector.

4. The vector is delivered in a volume of 100–200 µl in early gestation. A higher volume may be delivered after midgestation.

5. During intrahepatic vector administration, an echodense area develops within the hepatic parenchyma at the injection site.

3.4.6. Intratracheal Injection (see Note 13)

1. 20-Gauge Echotip Lancet spinal needles are used for midgestation sheep (80–115/145 days).

2. A coronal view of the fetal chest is obtained.

3. When the right fetal side faces the ultrasound beam, the pulmonary artery lies below the trachea in the field of view. The needle is inserted into the fetal thorax between the third and fouth ribs, penetrates the lung parenchyma, and enters the fetal trachea just proximal to the carina (Fig. 3).

4. When the left fetal side faces the ultrasound beam, the pulmonary artery and aorta lie above the trachea in the field of view. In this case, the needle is inserted into the fetal thorax between the second and third ribs, penetrates the lung parenchyma, and enters the fetal trachea at a more superior level.

5. Once the needle is thought to be in the trachea, the needle stylet is removed and tracheal fluid is withdrawn into a 2-ml syringe to confirm correct needle placement.

6. The gene therapy vector and transduction enhancing agents are delivered in a total volume of 1–10 ml of phosphate-buffered saline (PBS) (at midgestation) and are then instilled into the fetal trachea and can be seen flowing down the bronchii as microbubbles.

7. The needle is carefully removed and the fetus is observed for a few minutes afterwards to check on fetal well-being and for any sign of hematoma formation.

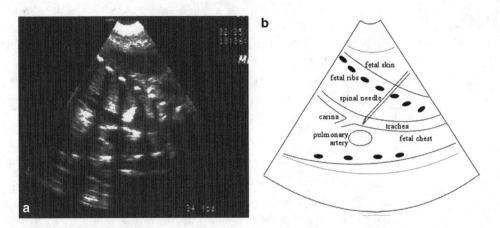

Fig. 3. Tracheal injection. An ultrasonogram (**a**) and diagram (**b**) showing injection of the trachea using the transthoracic route in a sheep fetus (114 days of gestation). A 20-G spinal needle is inserted into the fetal thorax between the third and fourth ribs, penetrates the lung parenchyma and enters the fetal trachea just proximal to the carina.

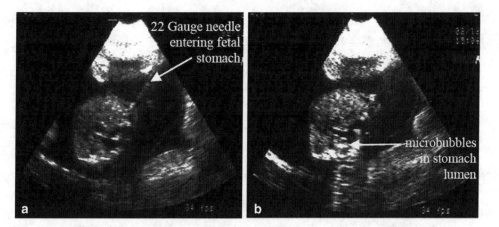

Fig. 4. Intragastric injection in a sheep fetus at 61 days of gestation. With the fetal abdomen in cross-section (**a**) the needle was inserted into the fetal stomach through the anterior abdominal wall and 100 μl gastric fluid was removed to confirm correct needle placement. (**b**)The vector was injected and microbubbles were observed in the stomach lumen.

3.4.7. Intragastric Injection (See Note 14)

1. A 22-Gauge Echotip Lancet needle (Cook Medical, UK) is used in early gestation as early as 60 days of gestation.

2. The fetal stomach is viewed in a transverse plane and the needle is inserted through the anterior abdominal wall or occasionally via the abdominal ribs.

3. The position of the needle is viewed in two planes to confirm correct placement and 100 μl of clear gastric fluid is withdrawn (Fig. 4).

4. The vector is administered in a volume of 100–200 μl of PBS and microbubbles can be seen as turbulence within the stomach lumen, to confirm correct injection. The needle is then flushed with 40 μl of saline to clear the dead space before removal.

3.4.8. Intraventricular Injection

1. A 22-Gauge Echotip Lancet needle is used and the procedure is performed at 50–60 days of gestation to make use of the soft cranium and wide sutures at this stage.

2. The lateral ventricles of the brain are visualized through the biparietal plane and the coronal suture identified.

3. The needle is inserted through the coronal suture in a posterior direction into the posterior horn of the lateral ventricle.

4. The position of the needle is viewed in two planes to confirm correct placement before delivery of the viral vector.

5. The vector is delivered in a volume of 100 μl.

6. Microbubbles are seen within the lateral ventricles on vector instillation.

3.4.9. Intramuscular Injection (see Note 15)

1. A 20- or 22-Gauge Echotip Lancet needle is inserted along the length of the femur and/or the buttock. 22 Gauge is used before 70 days of gestation and 20 Gauge used after 70 days of gestation. Echotip needles are used because the tip is easier to see while next to the echogenic femur bone, when compared with other needles.

2. Prior to injection, a sagittal ultrasonographic plane of the fetal thigh is obtained. The fetal thigh is best observed by obtaining a cross-sectional view of the fetal bladder and the femora are seen extending either side.

3. The fetal thigh (quadriceps or hamstring) and buttock muscles are chosen for injection because these are the largest accessible muscle groups at this gestational age in the sheep fetus (12)

4. The cross-sectional diameter of the fetal thigh or buttock in early gestation muscle is as wide as the bevel of the Echotip needle. Thus, insertion of the needle tip into the muscle in a transverse section may result in immediate loss of the vector into the amniotic fluid from the needle tip. To avoid this and to ensure a longer needle track, the needle is placed deeply within the muscle parallel to the bone.

5. In later gestations (100 days of gestation and beyond), the muscle bulk is larger and a more angled approach to the fetal bone is available without risk of vector leakage.

6. The needle position is now checked in longitudinal and transverse sections.

7. Once the correct needle position is confirmed, the stylet is removed. With a longitudinal view of the needle within the fetal muscle, the syringe containing the vector is then attached to the needle and as the needle is slowly withdrawn along the length of the muscle, the vector is injected. Echogenic foci can be observed within the muscle parenchyma confirming vector placement.

8. If necessary to achieve the required vector dose, a number of intramuscular injections on the same fetus can be done, on the same leg or using the opposite leg, if accessible.

9. The vector is delivered in a total volume of 100–500 μl by preferably 1 (up to 4) injection depending on the viral titre.

10. After needle removal, the fetal injection site is checked for bleeding.

11. Injection sites include the fetal thigh muscles (quadriceps), buttocks, and biceps/triceps.

3.4.10. Intrapleural Injection

1. A view of the fetal chest in longitudinal section is achieved.

2. A hydrothorax is created by inserting a 20-Gauge Echotip Lancet needle through the thoracic musculature between the ribs and up to the margin of the lung parenchyma.

3. The needle position is checked in longitudinal and transverse sections of the fetal chest.

4. 500 µl of PBS is then injected until a pool of fluid is seen by ultrasound. This not only confirms the correct needle position but also creates a hydrothorax into which the vector is delivered.

5. The vector is then injected into the pool of fluid in a volume of 100 µl.

6. After needle removal, the fetal injection site is checked for bleeding.

3.5. Fetoscopic Procedures

Fetoscopic procedures are required when higher resolution is required than can be provided by ultrasound guidance alone, or when the procedure requires real-time visualization that is beyond the capacity of ultrasound to provide. Fetoscopic procedures are often combined with ultrasound guidance to give the "big picture" and guide the fetoscope to the area of the intervention. In addition, if fetal positioning is an issue, as it often is in sheep with multiple fetuses, laparotomy may be the simplest way to improve access to the various amniotic compartments, to position the fetus or fetuses, and accurately place the fetoscope. We have employed this technique to perform intravascular injections as early as 55 days in the sheep fetus.

3.5.1. Fetoscopically Assisted Early Gestation Intravascular Injection

1. Sheep anesthesia and abdominal skin preparation/scrubbing are performed according to that described earlier for ultrasound-guided procedures.

2. Drape the sheep abdomen with sterile drapes.

3. If ultrasound guidance is utilized the ultrasound probe is prepared as above and the sheep abdomen is scanned to assess the location, orientation, and number of amniotic cavities, and fetuses within the maternal abdomen.

4. If the convergence of umbilical veins from each cotyledon (i.e., the root of the umbilical cord) can be visualized and is tangentially accessible from the maternal abdominal wall, the procedure can be performed percutaneously without laparotomy. This is the exception, however, and in most cases obtaining the optimal orientation of the fetoscope to the vessels as they emerge from the intercotyledonary membranes requires a maternal laparotomy.

5. If a maternal laparotomy is required, perform a laparotomy as described below and expose the uterine horn of interest. With the uterus exposed, the root of the umbilical cord can be relatively easily identified by either ultrasound, or transillumination of the uterine wall. See Note 16.

6. The insertion site of the fetoscope is chosen to achieve a 45° angle to the root of the cord in alignment with a major tributary from multiple placentomes.

7. Depending upon the sharpness of the fetoscope (1.3-mm pointed operating sheath) and the gestational age, the fetoscope may be directly inserted into the amniotic space, or a disposable cannula and fetoscopic trocar may be utilized for initial entry into the amniotic space and the fetoscope may then be directly introduced through the cannula. See Note 17.

8. The fetoscope is advanced under ultrasound guidance to the root of the umbilical cord.

9. The vessels are surveyed and a suitable vessel is chosen at the point of its emergence from the intercotyledenary membrane as it rises towards the confluence forming the cord. At this point the vessel is relatively tethered rather than freely floating which facilitates its puncture. See Note 18.

10. With the trocar in proper alignment to the chosen vessel and the vessel in view, the needle is advanced down the working channel of the fetoscope until it comes into view in the lower visual field and extends a few millimeter from the end of the scope.

11. The entire fetoscope and needle are advanced until the tip of the needle just approximates the vessel. The obturator is withdrawn from the needle and the vector-containing syringe is loaded. As the needle contains minimal dead space and the manipulations, once the needle is in the vessel risks loss of placement, no attempt is made to clear the deadspace in the needle.

12. The fetoscope is then advanced to the point where the needle indents the center of the vessel with the bevel directed inferiorly. A quick, very short thrust is made with the needle to penetrate the vein wall. If the needle penetrates the opposite wall it is slowly withdrawn until blood return is obtained. The vector is then injected and followed by a flush of 0.05 ml of saline. See Note 19.

13. The needle is withdrawn. A small amount of bleeding will occur but it will rapidly stop forming a clot on the vein.

14. The fetoscope is then withdrawn and the procedure repeated if a second fetus is targeted.

3.5.2. Fetoscopic Balloon Tracheal Occlusion to Facilitate Gene Transfer to Pulmonary Epithelium

Injection of vector directly into the larynx or trachea can result in significant loss of vector through the epiglottis depending upon a number of factors. To prevent this loss, one approach is reversible tracheal occlusion using a deployable balloon device. In addition to preventing loss of vector containing solution, balloon tracheal occlusion has been shown to increase pulmonary epithelial proliferation, which might theoretically further enhance expression, either through transduction of dividing cells or expansion of transduced cells. The technique can be successfully employed after 80 days gestation in the lamb fetus.

1. Sheep anesthesia, aseptic abdominal wall preparation and draping, and ultrasound guidance with or without laparotomy are as previously described.

2. Expose the uterine horn as previously described.

3. Position the fetus in the uterine horn upright with mouth facing the operator.

4. Insert the 10 Fr. Flexor Check-Flo cannula using seldinger technique through the uterine wall 5–10 cm from the fetal mouth.

5. Insert the tracheoscopic sheath with three side ports (3.3 mm outer diameter, 30.6 cm working length, and precurved 30°) containing the fetal tracheoscope (1.3 mm diameter, 30.6 cm working length) and the prefilled BALTACCI—BDPE—detachable balloon catheter (outer diameter max. 0.9 mm) in one of the side ports. See Note 20.

6. Maneuver the scope into the fetal mouth maintaining midline axial orientation with the curvature of the scope facing downward. Identify the tongue and follow it posteriorly to the epiglottis and cords. Advance the sheath into the trachea (clearly identified by the tracheal rings). See Note 21.

7. Advance a 3.0-Fr Slip-cath infusion catheter into the distal trachea through the empty side port for vector injection.

8. Advance the Balt balloon catheter to a position in the proximal trachea below the cords. Inflate the Balt balloon to the predetermined volume to approximately 1.5× the diameter of the trachea and withdraw the mandrel of the catheter to detach the balloon. See Note 22.

9. Withdraw any excess lung fluid distal to the balloon by aspiration through the Slip cath. Replace the syringe with the vector-containing syringe, and inject the vector into the trachea distal to the balloon, chasing the vector injection with an injection of saline equal to the catheter deadspace.

10. Withdraw the Slip-cath out of the trachea and withdraw the fetoscope and sheath.

11. Refill the amniotic space with warm Lactated Ringers solution if significant amniotic fluid leakage occurred.

12. Place a full thickness 3-0 Vicryl pursestring suture through the myometrium and membranes around the cannula and tighten and tie the pursestring as the cannula is removed.

3.6. Laparotomy for Local Delivery of Gene Therapy into the Uterine Artery

Laparotomy is chosen over ultrasound-guided or fetoscopic delivery techniques, when using minimally invasive routes of delivery into the fetus are not possible, for example, intraspinal delivery, or when the risk of ewe or fetal morbidity or mortality is high, for example, bleeding after injection of the uterine artery.

1. Sheep anesthesia and abdominal skin preparation/scrubbing are performed according to that described earlier for ultrasound-guided delivery.

2. Drape the sheep abdomen with sterile drapes to expose the site of incision.

3. Make a midline lower abdominal incision in the skin from just above the pubic bone to just below the umbilicus using a no. 11 blade.

4. Carefully diathermy any blood vessels within the fat.

5. Incise the rectus sheath and open the peritoneal cavity carefully.

6. Check the orientation of the uterus and identify the uterine arteries bilaterally. The bowel is then tucked away using two or three wet lap sponges to expose one of the uterine arteries.

7. The visceral peritoneum overlying the main uterine artery is incised 2–3 cm proximal to its first bifurcation and the underlying vessel is exposed.

8. Dissect it bluntly from it fascial attachment to the uterine wall using a right angle Kelly clamp.

9. A vessel loop is passed around the artery and held by a small Kelly clamp to stabilize the vessel and elevate it slightly.

10. The vessel is occluded digitally and the vector is diluted in 10 ml PBS and injected over 1 min distal to the occlusion site using a 23-G butterfly perfusion set connected to a 10-ml syringe.

11. The needle is removed and the injection site is digitally clamped while occlusion is kept on proximally as before for a further 4 min to minimize vector washout.

12. Hemostasis is secured and the visceral peritoneum overlying the artery is closed using 3-0 Prolene. See Note 23.

13. The procedure is repeated on the contralateral uterine artery.

14. Remove the lap sponges and instil Crystapen 3 g (sodium benzylpenicillin G)+gentamicin 80 mg into the peritoneal cavity before closure for antibiotic prophylaxis.

15. Close the ewe's abdomen in layers. The rectus sheath is closed with continuous 6-mm nylon tape to prevent herniation of the abdominal contents. In adolescent ewes where the rectus sheath can be fragile, an alternative suture material is interrupted mattress sutures with 0 Prolene. The subcutaneous tissue and the skin are closed with continuous 1-0 PDS and 1-0 Vicryl, respectively.

16. Intramuscular Penstrep 3 ml (200 mg/ml procaine penicillin and 250 mg/ml dihydrostreptomycin) is given for infection prophylaxis.

3.6.2. Uterine Vein Blood Sampling

1. After vector injection, the vein accompanying the main uterine artery is identified.

2. A 25-G butterfly perfusion set connected to a 2-ml syringe is inserted into the vein through its covering visceral peritoneum and blood is collected at predetermined intervals from vector injection. See Note 24.

3. Once blood collection is complete, the needle is withdrawn and hand pressure is exerted using a gauze to secure hemostasis.

3.7. Vector Dose

1. An adenoviral vector concentration of around 5×10^{11} to 5×10^{12} particles/ml is advised since higher doses may be toxic (5, 13). AAV concentration of 1×10^{12} vector genomes/kg is not toxic in early and late gestation, and gave up to 6 months of transgenic protein expression (14). See Notes 25–28.

2. For local delivery, the adenoviral particles may be complexed with diethylamnioethyl (DEAE) dextran for better transfection efficiency (15). The adenovirus polycation complexes are prepared by addition of adenovirus particles to PBS containing DEAE dextran (5 mg/mL) and allowed to form for 30 min at room temperature before injection.

3. All vectors are aliquoted freshly using pipettes, diluted in the needed volume, and delivered to the fetus within 10–15 min.

4. After vector delivery the contaminated needle is immediately placed into a sharps bin to avoid sharps injury. The syringe is flushed with a viricide such as Virkon for 30 min to destroy the vector particles before being discarded.

4. Notes

1. Research in pregnant sheep requires considerable planning to ensure time-mated ewes that are presented to the researcher at the correct gestational age in the best of health. A dedicated animal research unit is required as well as good communication between the animal technicians looking after the flock and the researcher themselves. Time-mating or tupping of sheep is usually begun in early autumn at the beginning of the sheep breeding season. A few breeds can be tupped all year round (e.g., Pol Dorset sheep) but they tend to be more expensive and have more multiple pregnancies.

2. Large animals need to be moved to the experimental facility at least 1 week before experiments to ensure that they acclimatize. Care needs to be taken with animal welfare so as to reduce stress, for example, housing sheep in pairs, or close by so that they can see each other is recommended. The diet of pregnant sheep is tailored to whether they have a multiple pregnancy.

Those that have a twin and other higher order multiple conception may require a concentrated source of oil, protein, and vitamins found in special feed which is fed at increasing amounts as gestation advances. This ensures adequate nutrition for the ewe and reduces the risk of toxemia caused by the large late gestation fetus compressing the rumen and reducing the ability of the ewe to digest enough hay for nutrition. Attention to this level of detail will improve animal well-being and thereby reduce the miscarriage rate.

3. For experiments at very early gestational ages, the ewe can be scanned to confirm pregnancy from 20 days of gestation. A view of the placentomes and gestational sac(s) can be easily achieved by careful sonography of the maternal abdomen on either side of her bladder. Assessment of multiple pregnancy is unreliable at this early gestation, since fetal resorption can occur.

4. Analgesia is not considered necessary for ultrasound-guided procedures since usually the technique involves a single injection of a fine needle (20 Gauge or less) and all are done under general anesthesia. These procedures are routinely carried out in humans without the need for analgesia. If analgesia is considered necessary, buprenorphine (0.1 mg/ml IM) can be given.

5. Ewes that are allowed to come to birth should be vaccinated with Heptavac-P plus® (Hoechst Roussel Vet Limited, Dublin) 1 month before their delivery date to prevent *Clostridium* and *Pasteurella* infection in the lambs. A further two doses are given to the lambs at 10 and 16 weeks of age.

6. It is vital to clean the skin/fleece of sheep adequately before embarking on surgery or ultrasound-guided procedures. Close clipping of the fleece helps remove dirt and a source of bacteria. Care is taken to ensure that even the finest wool is removed. We routinely clean the abdomen three times prior to surgery by percutaneous technique or laparotomy while the animal is under general anesthesia. The first scrub is performed before the animal is scanned to remove the majority of fleece and dirt, and we use povidone iodine antiseptic cleaning solution, which is worked into the clipped skin using a scrubbing brush and then wiped clean. The second scrub is done after the ultrasound scan assessment of fetal size and gestational age. For this we use chlorhexidine gluconate cleaning solution, which is again worked into the clipped skin using a scrubbing brush and then wiped clean. Finally, we dribble povidone iodine antiseptic cleaning solution onto the skin site over the incision or planned ultrasound-guided injection site. We have found judicious attention to this aspect of care reduces the rate of miscarriage.

7. The operator performing the ultrasound-guided injection must have a comfortable view of the ultrasound screen since procedures may take some time to perform. Injecting the sheep fetus requires a lighter touch in comparison to the human fetus, since the rectus muscle are more lax, and the fetus is more superficial. Care must be taken not to compress the fetus within the uterus by leaning on the probe too much.

8. Small aliquots of ultrasound coupling gel can be sterilized by irradiation for use during ultrasound-guided injection procedures although afterwards they become quite liquid. Liquid paraffin is a good alternative that is used for human fetal medicine procedures. It can be sterilized by irradiation for use. A further alternative for short term use is 70% ethanol but this must be reapplied underneath the scan probe every few minutes because it rapidly evaporates.

9. Following fetal injections the fetal heart rate generally does not alter significantly, although it may occasionally speed up for a minute temporarily. Hematomas can be seen as an enlarging echogenic area at the site of the injection. Bleeding can sometimes be observed as echogenic drops coming from the skin site of injection, which usually lasts a few seconds. The fetus is remarkably tolerant of such minor complications, but it is important to scan the fetus the day after procedures to check for viability.

10. For intra-amniotic injection, needles of gauge larger than 22 G may cause the fetal membranes to tent, especially when the procedure is performed during early gestation. During early gestation (less than 50 days of gestation), the allantoic cavity still occupies a large part of the uterine cavity, and the amniotic cavity is relatively smaller in volume. Care must be taken that the needle traverses the allantoic cavity and enters the amniotic cavity, which closely surrounds the fetus.

11. When considering ultrasound-guided intravascular delivery it is important to consider the differences between sheep and human development. The sheep umbilical cord contains two umbilical veins and two umbilical arteries and has very little Wharton's jelly compared to human fetuses. Injection of the free loop of cord causes vasoconstriction of the umbilical veins because of contained hemorrhage within the cord, and results in fetal death (unpublished observations). The placental cord insertion in the sheep is also different to that in humans. As the cord reaches the uterine wall, the umbilical vessels split and course within the fetal membranes along the uterine wall and over the placentomes. This makes it difficult to determine the optimum injection site. The diameter of each vessel becomes narrower and the more distal it is from the umbilical cord. Injection of the placental cord insertion has been achieved after

68 days of gestation (unpublished observations) but depends on good visualization, which can be improved using Doppler analysis of fetal blood flow within the vessel. Injection of the fetal cord insertion has not been tested but is likely to result in hemorrhage and fetal compromise as this is the position where the two umbilical veins join to become one as they enter the fetal liver.

12. The intracardiac injection route is limited by gestational age. Injection before 60 days of gestation led to miscarriage in all cases (5) due to pericardial hemorrhage. When the procedure was done on awake, on restrained ewes, the fetal loss rate from 100 days of gestation is reported to be 4.5% (16). Pericardial hemorrhage was the main cause of death. Fetoscopic intracardiac injection had an unacceptable 30% failure rate with 80% mortality (17). For systemic delivery, the intrahepatic umbilical vein route of injection is probably more useful because vector can be delivered at earlier gestational ages and with a lower miscarriage rate.

13. To improve transduction of the airways, NaCaprate (100 mM/ml, 5 ml) can be injected 5 min before delivery of the adenovirus vector. NaCaprate transiently opens the tight junctions between the airway epithelia, allowing the virus to reach the viral receptors on the basolateral side of the cell membrane. Complexing of the adenovirus with DEAE dextran (5 μg/ml) up to 20 min before vector use can also improve gene transfer (2). Injection of perflubron (5 ml) following vector injection can be used to flush the vector distally and enhance transduction of the distal airways. The small airways and lung parenchyma become very echogenic once perflubron is injected and this obscures the ultrasound view.

14. The fetal stomach can be injected in the sheep from 60 days of gestation. Earlier injection is associated with a very high mortality rate, and there are difficulties visualizing it before 50 days of gestation (3). To improve transduction of the fetal gut, NaCaprate (100–200 μl, 100 mM/ml) can be injected into the stomach lumen 5 min before adenovirus vector injection. Complexing the adenovirus vector with DEAE dextran (5 μg/ml) for 20 min before delivery improves gene transfer also (3). Flushing the vector with perflubron in a volume of 1,000–1,500 μl may be used to transduce distal bowel segments.

15. Gestational age at injection is an important consideration for intramuscular injection. Before approximately 70–80 days of gestation the fetal thigh musculature is relatively small compared to the size of the needle, making the injection technically more challenging. On the other hand, the relatively smaller muscle bulk means more is exposed to the vector, leading to better levels of gene transfer, and transduced muscle tissue is more easily identified on postmortem analysis.

16. The sheep placenta consists of between 70 and 100 placentomes each of which is made up of the uterine caruncle (maternal side) and fetal cotyledon. Each placentome has a vascular pedicle containing a single artery and vein. These converge in an arborized fashion from the two poles of the elongated chorionic sac to form the umbilical veins at the root of the umbilical vein. The area of the root of the umbilical vein can be imaged by ultrasound by following the umbilical vein to the uterine wall, or can be seen by transillumination of the uterus or by both techniques. It is important to approach the vessels at approximately a 45° angle to the plane of the intercotyledonary membrane for successful venous puncture so trocar position is critical and generally requires the freedom afforded by maternal laparotomy.

17. The sharpness of the fetoscope or trocar is critical. The membranes are poorly fixed and will tent and detach easily. The Storz pyramidal tip trocars designed for fetoscopy work reasonably well but must be inserted with a short rapid thrust. An alternative is a 14-Gauge or larger angiocath that will accommodate the scope. At younger gestations the chorionic sac is prominent and must be traversed to enter the amnion. We have utilized amnioinfusion to distend the amniotic space and provide internal counter-resistance for the amniotic membrane and reduce tenting of the membranes by the trocar or scope.

18. It can be exceedingly difficult to puncture the freely floating umbilical vein. While sheep umbilical vessels have less Wharton's jelly than human vessels, the vessel wall is nevertheless elastic and the vessel will roll unless tethered by membranes. We have found that it is easiest to puncture the vein at the point of emergence from the intercotyledonary membrane while it still has the orientation of the uterine wall rather than as it ascends into the umbilical cord. The other point of tethering, which we have not used but would be feasible is the fetal umbilical insertion of the cord.

19. The fetus is relatively tolerant of intravascular volume compared to intracavitary volume (intraperitoneal for instance) in early gestation likely due to the increased capacitance of the vascular system related to the placenta. Thus, volumes of up to 3 ml can be given to a 50-day lamb fetus without adverse consequence. In practice the vector dose is usually concentrated in 1 ml or less for administration.

20. The fetoscopic equipment described is optimal and is what is currently used in the clinical fetoscopic tracheal occlusion experience for prenatal treatment of congenital diaphragmatic hernia. Storz Endoscope currently makes state-of-the-art fetoscopy equipment and has little competition in the clinical marketplace. Other scopes with similar specifications can be used with the critical elements being three channels of adequate size, for

a scope with at least 1,000 pixel resolution (generally 1.2 mm or greater), a working channel that can accommodate the Balt detachable balloon catheter, and a working channel that will admit a catheter or needle for vector injection. The Balt balloon must currently be prefilled with a predetermined volume of saline using a blunt tipped needle that is inserted through the valve. The balloon is then guided over the mandrel and the catheter tip, which is already inserted through the fetoscopic sheath. The balloon then empties into the catheter, filling the dead space. When in the correct position, the same volume is injected into the catheter lumen, refilling the balloon. Withdrawal of the catheter out of the valve using the scope tip for counter pressure deploys the balloon.

21. A familiarity with airway anatomy is helpful and tracheal intubation can be practiced on fetal or neonatal sheep that are being harvested for other purposes. The 30° curvature of the scope is helpful in cannulating the airway without needing marked hyperextension of the neck, but not critical. Care must be taken with straight scopes not to perforate the trachea posteriorly.

22. Tracheal diameter will depend on gestational age. The trachea is reasonably compliant and will accommodate balloon inflation up to approximately two times its diameter without tearing. Overdistension of the balloon keeps it in place for an adequate time for vector transduction and for the physiologic effect on lung growth. The balloon will generally subsequently dislodge with fetal growth and be coughed out. Alternatively, the balloon can be removed at birth, or can be deflated with an ultrasound-guided needle when it will then be coughed out.

23. After uterine artery injection if bleeding persists from the puncture site despite continued pressure, a 6-0 Prolene suture can be used to close the defect in the vessel wall.

24. When sampling blood from the uterine vein, the parietal peritoneum should not be removed since it reduces blood loss from the injection site. Any venous bleeding can be easily controlled using hand pressure with sterile gauze. A larger volume syringe (e.g., 5 ml) may be used, however, care should be taken not to exert extra negative pressure in order to prevent vein collapse.

25. The lethal toxic level for adenovirus vectors in fetal sheep is approximately 8×10^{12} particles/kg (13). The dose given to the fetus is calculated according to the approximate weight of the fetal sheep determined from standard charts of fetal size (7, 8). Signs of toxicity can be observed in the fetus by serial ultrasound examination and include hydrops (e.g., skin edema, pleural effusion, ascites, or pericardial effusions), abnormal fetal heart rate (i.e., increased or reduced), abnormal umbilical artery Doppler examination (e.g., absent or reversed end diastolic flow), and reduced or absent fetal movements.

26. We have found adenovirus vectors to provide short term and highly efficient muscular gene transfer to fetal sheep. The vectors are tolerated to a high dose with minimal observed muscular damage on histological analysis (3, 5, 15).

27. We observed that adeno-associated virus vectors gave long-term gene transfer (up to 6 months) to fetal sheep after intraperitoneal delivery in early and late gestation (14).

28. Other vector systems have not proved so useful. Gene transfer to fetal sheep muscle in vitro is achievable using retrovirus vectors (18). We have found, however, that retrovirus vectors based on the Moloney Leukemia Virus (MLV) give poor gene transfer to the quadriceps of late gestation fetal sheep (120 days of gestation). This may reflect the relatively low dose that was applied (1×10^8 particles/kg fetus) to a relatively large muscle bulk at that gestation. Lentivirus vectors also seem to give poor gene transfer in the fetal sheep. Vectors based on human immunodeficiency virus (HIV) or equine immune anemia virus (EIAV) and pseudotyped with VSV-G or Mokola gave no significant gene transfer to the fetal sheep muscle in early gestation when injected at a dose of 1×10^8 particles/kg fetus.

References

1. David AL, Peebles DM, Gregory L et al (2003) Percutaneous ultrasound-guided injection of the trachea in fetal sheep: a novel technique to target the fetal airways. Fetal Diagn Ther 18: 385–390

2. Peebles D, Gregory LG, David A et al (2004) Widespread and efficient marker gene expression in the airway epithelia of fetal sheep after minimally invasive tracheal application of recombinant adenovirus in utero. Gene Ther 11:70–78

3. David AL, Peebles DM, Gregory L et al (2006) Clinically applicable procedure for gene delivery to fetal gut by ultrasound-guided gastric injection: toward prenatal prevention of early-onset intestinal diseases. Hum Gene Ther 17: 767–779

4. Davey MG, Hedrick HL, Bouchard S et al (2003) Temporary tracheal occlusion in fetal sheep with lung hypoplasia does not improve postnatal lung function. J Appl Physiol 94: 1054–1062

5. David AL, Cook T, Waddington S et al (2003) Ultrasound guided percutaneous delivery of adenoviral vectors encoding the beta-galactosidase and human factor IX genes to early gestation fetal sheep in utero. Hum Gene Ther 14: 353–364

6. David AL, Weisz B, Gregory L et al (2006) Ultrasound-guided injection and occlusion of the trachea in fetal sheep. Ultrasound Obstet Gynaecol 28:82–88

7. Barbera A, Jones OW, Zerbe GO et al (1995) Ultrasonographic assessment of fetal growth: comparison between human and ovine fetus. Am J Obstet Gynecol 173:1765–1769

8. Kelly RW, Newnham JP (1989) Estimation of gestational age in Merino ewes by ultrasound measurement of fetal head size. Aust J Agr Res 40:1293–1299

9. Wolfensohn S, Lloyd M (1998) Handbook of laboratory animal management and welfare. Blackwell, Oxford, pp 257–277

10. Taylor PM (1991) Anaesthesia in Sheep and Goats. In: Melling M, Alder M (eds) Sheep and goat practice. Saunders, London, pp 99–116

11. Newnham JP, Kelly RW, Boyne P, Reid SE (1989) Ultrasound guided blood sampling from fetal sheep. Aust J Agr Res 40:401–407

12. Joubert DM (1956) A study of prenatal growth and development in the sheep. J Agric Sci 47: 382–427

13. Themis M, Schneider H, Kiserud T et al (1999) Successful expression of galactosidase and factor IX transgenes in fetal and neonatal sheep after ultrasound-guided percutaneous adenovirus vector administration into the umbilical vein. Gene Ther 6:1239–1248

14. David AL, McIntosh J, Peebles DM et al (2011) rAAV mediated *in utero* gene transfer gives therapeutic transgene expression in the sheep. Hum Gene Ther 22(4):419–426

15. Weisz B, David AL, Gregory LG et al (2005) Targeting the respiratory muscles of fetal sheep for prenatal gene therapy for Duchenne muscular dystrophy. Am J Obstet Gynecol 193: 1105–1109

16. Newnham JP, Kelly RW (1993) Ultrasound for research with foetal sheep. In: Neilson JP, Chambers SE (eds) Obstetric Ultrasound I. Oxford Medical Publications, pp 203–222

17. Kohl T, Strumper D, Witteler R et al (2000) Fetoscopic direct fetal cardiac access in sheep: an important experimental milestone along the route to human fetal cardiac intervention. Circulation 102:1602–1604

18. John HA (1994) Variable efficiency of retroviral-mediated gene transfer into early-passage cultures of fetal lamb epithelial, mesenchymal, and neuroectodermal tissues. Hum Gene Ther 5:283–293

Chapter 12

Animal Models for Prenatal Gene Therapy:
The Nonhuman Primate Model

Citra N. Mattar, Arijit Biswas, Mahesh Choolani, and Jerry K.Y. Chan

Abstract

Intrauterine gene therapy (IUGT) potentially enables the treatment and possible cure of monogenic diseases that cause severe fetal damage. The main benefits of this approach will be the ability to correct the disorder before the onset of irreversible pathology and inducing central immune tolerance to the vector and transgene if treatment is instituted in early gestation. Cure has been demonstrated in small animal models, but because of the significant differences in immune ontogeny and the much shorter gestation compared to humans, it is unlikely that questions of long-term efficacy and safety will be adequately addressed in rodents. The nonhuman primate (NHP) allows investigation of key issues, in particular, the different outcomes in early and late-gestation IUGT associated with different stages of immune maturity, longevity of transgene expression, and delayed-onset adverse events in treated offspring and mothers including insertional mutagenesis. Here, we describe a model based on the *Macaca fascicularis* using ultrasound and fetoscopic approaches to systemic vector delivery and the processes involved in vector administration and longitudinal analyses.

Key words: Nonhuman primate, Monogenic disease, Intrauterine gene therapy, Immune response, Time-mating, Ultrasound guidance, Fetoscopy, Cesarean delivery, Hand-rearing, Safety, Efficacy, Germ-line transmission

1. Introduction

A number of inherited monogenic disorders can cause neurological, hematological, or metabolic pathology in fetal life that may ultimately lead to intrauterine demise, stillbirth, or irreversible neurological disability, reviewed extensively by Coutelle et al. (1) and Waddington et al. (2). There are several potential advantages for adopting an intrauterine gene transfer (IUGT) approach to the treatment of these candidate diseases (3, 4). The first and most cogent reason is the opportunity to treat an illness before the onset

Charles Coutelle and Simon N. Waddington (eds.), *Prenatal Gene Therapy: Concepts, Methods, and Protocols*,
Methods in Molecular Biology, vol. 891, DOI 10.1007/978-1-61779-873-3_12, © Springer Science+Business Media, LLC 2012

of irreversible pathology, particularly important for disorders in which postnatal therapy has limited usefulness. Second to this is the possibility of inducing central tolerance to the vector and transgene by introducing foreign molecules or cells during the window of thymic processing that exists in the late first trimester (5–7). Third, the greater migratory and differentiation potential of fetal progenitor and stem cells in comparison to adult stem cells makes them attractive targets for gene correction (8), permitting the lifelong production of the missing protein. Finally, treating the fetus that weighs between 30 and 60 g at 12–14 weeks of gestation allows the use of a smaller quantity of vectors or genetically manipulated cells to achieve a therapeutic effect, compared to the dose that may be required to achieve the same outcomes in a larger postnatal recipient (9).

1.1. Proof of Principle Studies of IUGT in Rodent Models of Human Diseases

Over the past decade, a substantial body of evidence has been amassed supporting the effectiveness of IUGT for the correction of monogenic diseases in small animal models such as Leber congenital amaurosis in mice (10) and chickens (11), Crigler–Najjar syndrome in rats (12), hemophilia B (13), and mucopolysaccaridosis (14) in mice, while limited success has been achieved in murine models of cystic fibrosis (15–17) and Duchenne muscular dystrophy (18, 19) with this approach. While small animal models are essential to the initial assessment of a new vector delivery system, these models may not fully recapitulate human disease pathogenesis and long-term outcomes of IUGT because of significant differences in fetal physiology and development, placentation (see ref. 20, Chapter 9), and life-span.

1.2. Limitations in the Use of Rodents in IUGT Experiments

There are certain limitations to drawing parallels between the physiological responses to experimental interventions in mice and humans. Differences in body size, blood volume, and cardiac output can influence the efficacy with which vectors are delivered to target organs. In addition, the processes by which foreign antigens are presented to the immune system may ultimately produce divergent responses between rodents and humans. Because of the shorter life-span of rodents (1.5–3 years) (21), any assessment of insertional tumorigenesis, abnormal physical development, organ-specific toxicity, and maternal complications may not be observable over a time period long enough to guide clinical application (22), particularly if persistent biological activity of the vector is expected or desired.

One of the main determinants in the successful outcome of gene transfer is in achieving immune tolerance toward the transgene in question. The importance of immune naiveté to the vector and transgene in facilitating therapeutic long-term expression is demonstrated elegantly in a series of mouse experiments comparing outcomes in fetal, neonatal, and adult mice injected with the same vector (23). These data showed an inverse relationship between anti-vector immunoglobulin (Ig) response and the level and longevity of transgene expression. Mice injected in utero demonstrated sustained transgene activity in the absence of an Ig response to the capsid, a consistent finding replicated by other laboratories (24, 25), while neonatal and adult mice showed a progressively earlier onset of Ig production which consequently nullified transgene expression. The importance of delivering the vector to a subject with naive or suppressed immunity is demonstrated in other preclinical (26–29) and clinical trials (30). The immune naiveté in early gestation should facilitate the acceptance of vector and transgene while avoiding the morbidity associated with myeloablative conditioning treatments associated with postnatal hemopoietic stem cell transplantations.

An important question particular to fetal gene therapy is the effect that timing of intervention will have on the outcome, which will be influenced by the temporal pattern of immune system development during gestation. In animals with shorter gestations, the immune system completes maturation in the postnatal period, as seen in the mouse which has a 21-day gestational period. Human fetuses are developmentally more mature than mouse fetuses at the corresponding gestational time-points (31). Mitogen-responsive lymphocytes are detectable in peripheral blood by 0.3–0.4 G (32–34), and functional natural killer (NK) cells develop at 0.7 G (35). By contrast, the mouse fetus demonstrates a two- to threefold lag in immune maturity. T and B lymphocytes appear at 0.85 G (6, 36–38) and become fully functional only after a few weeks postdelivery. Similarly, functional NK cell activity is only detected 3 weeks after birth (39). In addition to the dissimilar ontogeny, there are several important discrepancies between mouse and human in terms of the repertoire of expressed Ig isotypes, B and T cell development and regulation, and tolerance induction to grafts, as comprehensively reviewed by Mestas and Hughes (40). These differences limit the parallels that can be drawn from experimental mouse data to predict possible human clinical outcomes, particularly in gene therapy.

To interrogate the effects of immune maturity on cellular transduction and transgene expression, it would be necessary to obtain access to the fetus in the first trimester. The size of the mouse fetus at early gestation presents the main challenge to parenteral vector delivery. Intracerebral (14), intracardiac (41), intraperitoneal (42), intrahepatic (43), intramuscular (44), and intravenous (via yolk sac vessels) (45) routes of delivery are achievable from around mid-gestation onwards and most murine fetal injections have been performed after E13. However, the only practical route of vector delivery at the equivalent of the first trimester (<E9) is intraamniotic (46) as the thoracic and peritoneal cavities are too underdeveloped to visualize.

Because of the phylogenetic relatedness and consequent physiological similarities (47) between nonhuman primates (NHPs) and humans, the need to answer questions of clinical relevance specific to the timing and routes of fetal gene delivery, related pregnancy complications, and transplacental trafficking of vectors or cells can be better met with the NHP model than with any other animal model. The ability to monitor a pregnancy and gain access to the fetus with techniques already established in the clinic is an added advantage.

1.3. NHPs as Models for IUGT

NHPs have mainly monotocous pregnancies, with a similar prevalence of spontaneous pregnancy loss rates compared with humans (48). Embryo implantation into the decidua, trophoblast invasion, modification of maternal vessels, and evolution of the uteroplacental circulation follow a similar course in NHP and humans (49–51). Macroscopically and histologically, the NHP placenta is very similar to the human placenta, and undergoes a similar process of differentiation from cytotrophoblast to syncytiotrophoblast, forming an endocrine unit during gestation as the human placenta does (52). NHP embryonic development closely follows the human embryonic stages with regard to the formation and differentiation of major organ systems (53, 54). Therefore, preclinical data on the fetal and maternal effects of early intervention may be expected to closely predict clinical outcomes.

Primate immune ontogeny parallels the human temporal pattern (55). Lymphocytes appear in lymphoid tissue at 0.4–0.5 G (gestation 155–165 days) and expression of immunoglobulins is detected around the same period. The macaque fetus has a fully competent immune system by 0.8–0.9 G, thus gene transfers performed at late (0.9 G) and early gestation (0.4 G) are expected to be informative of the outcomes expected from human fetal gene transfers.

The long life-span compared to other animal models (35 years), anatomical and biological similarities (56) to humans and accessibility to the late and early gestation fetuses using established clinical procedures allows the interrogation of key questions, including the efficacy of tissue transduction, longevity of transgene expression, exploration of the dose–effect relationship and attainability of therapeutic target protein levels, long-term maternal outcomes resulting from transplacental trafficking of the vector, integration events, delayed adverse clinical effects, and analysis of gamete transduction to determine the potential for germ-line transmission.

We and others have demonstrated intravenous or intraperitoneal access in human fetuses via ultrasound (US) and thin-gauge fetoscopic methods (57), and by adapting these procedures we can gain similar access to the early gestation NHP fetus (see Fig. 1a–h), enabling the investigation of specific organ targeting via reporter transgene expression following different modes of vector delivery,

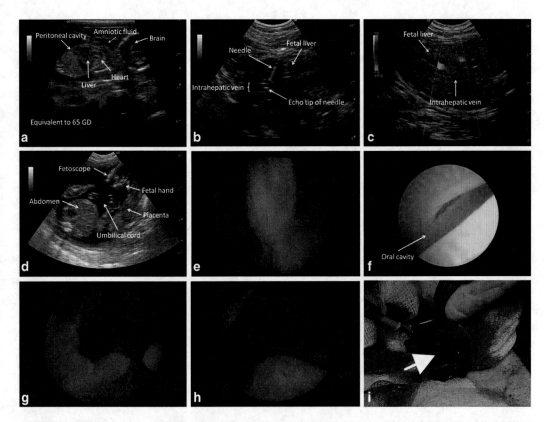

Fig. 1. Ultrasound, fetoscopy, and laparotomy images of IUGT in the NHP model. (a) Pre-IUGT US assessment aids in defining the fetal cavities that can be targeted for vector delivery. Possible routes for vector delivery include intraamniotic, intraperitoneal, intrahepatic, and intracardiac. (b) A 22-G amniocentesis needle is inserted percutaneously into the fetal intrahepatic vein with the tip of the needle seen in the lumen. (c) Doppler flow confirms correct intravenous placement of the needle and turbulence is observed when the vector is injected. (d) Fetoscopy is performed with a 1 mm endoscope inserted percutaneously into the amniotic cavity with US visualization. (e) The umbilical cord attachment at the fetal abdomen is a stable site for blood sampling or injection with an appropriate needle inserted through a side port. (f) Other features can be easily examined, such as the oral cavity, a target for fetal tracheal occlusion (to treat congenital diaphragmatic hernia), (g) hand, and (h) face (nose seen in foreground). (i) The maternal liver (*white arrow*) is accessible via a midline incision at the epigastrium, and hemostasis at the biopsy site is achieved with electrocautery.

including intraamniotic (58), intracardiac (59, 60), intrathoracic (59), intrapulmonary (60, 61), intraperitoneal (62, 63), and intrahepatic routes (62, 64) (see Fig. 1a). We have demonstrated that a therapeutic and sustained level of human Factor IX can arise from a single intravenous injection (see Fig. 1b, c) of a self-complementary AAV vector delivered to macaque fetuses at 0.9 G (65). To date, the intramuscular and intracerebral routes for IUGT in NHP has not been reported, although with the current technical capabilities this could be readily achieved.

Despite the successful use of the macaque as models for the study of human disease ranging from reproductive biology to atherosclerosis (66–68), one distinct disadvantage is the current lack of appropriate NHP disease models of relevance to IUGT that can demonstrate proof of cure. However, the development of a transgenic

NHP in the last decade (69) and the recent generation of a transgenic NHP model of Huntington disease (70) suggests that proof of cure in an NHP disease model with IUGT may soon be possible. For now, the possibility of asking many more pertinent questions with the NHP which, for biological limitations, cannot be addressed with the mouse model, strengthens the scientific and ethical argument for the use of the NHP model to interrogate the IUGT paradigm.

Here, we describe a clinically applicable high-resolution US-guided approach to intravenously delivered gene transfer using the *Macaca fasicularis* as a model, which can be adapted to deliver vectors directly to the hepatic parenchyma, peritoneal cavity, thoracic cavity, or to specific organs including the heart, trachea, lung parenchyma, muscle, and brain. Fetoscopic-assisted tissue biopsies (e.g., to diagnose epidermolysis bullosa) and blood samples can be obtained to permit intermittent prenatal analysis of transgene expression, and to visually guide access to body cavities (see Fig. 1e–h). In this chapter, we describe procedures used with this model to interrogate the IUGT paradigm.

2. Materials

The following describes the basic surgical set used in the described procedures. It is not exhaustive but includes the most useful instruments and accessories.

2.1. Time-Mating and Pregnancy Monitoring

1. Reproductive-age male and female macaques screened for simian retroviruses, herpes B virus, and anti-vector serotype antibodies if applicable.
2. Logiq P5 ultrasound (US) machine with the 12 L probe set at 7 MHz (General Electric), or equivalent.
3. Menstrual calendar.
4. Fetal growth chart (71).

2.2. Intravenous Vector Delivery

1. Logiq P5 ultrasound machine with an 8 C probe with a small foot-print at 10 MHz (General Electric), or equivalent.
2. 22 G Cook EchoTip® Amniocentesis needles (Cook medical, product number G16309) and/or 27 G Whitacre spinal needles (BD, product number 405144).
3. 1- and 5-mL syringes for aspirating fetal blood and flushing with saline postinjection.
4. Vector on ice.
5. EDTA tubes for blood collection.
6. Sterile 10% povidone iodine, for skin antisepsis.

7. Sterile gel for ultrasound.

8. Sterile normal saline for postvector injection flush.

2.3. Fetoscopy

1. Small surgical blade to make initial skin incision, e.g., size 10 or 11 with a handle no. 3.

2. 1 mm 0° fiber-optic fetoscope (Karl Storz, product code 11510A).

3. 2 mm 0° endoscope with Hopkins lens (Karl Storz, Hopkins II telescopes, product code 26008AA).

4. 26-G puncture needles for sampling (Karl Storz, product code 11510KC).

5. Camera unit, monitor, and recorder.

6. Xenon light source.

2.4. Cesarean Section, Laparotomy, Infant Resuscitation, and Hand-Rearing

1. Laparotomy set: sterile drapes, towel clips, gallipots, gauze, a range of tissue forceps for skin (rat tooth forceps) and viscera (Allis forceps), tissue and suture scissors, electrocautery wand with disposable flat rounded tip or ball tip and insulation pad.

2. A range of surgical blades, e.g., sizes 11, 18, 22, and appropriate blade handles (e.g., handles nos. 3 and 4).

3. Sterile sutures 3/0–5/0, e.g., Polyglactin 910 (Vicryl, Ethicon 360, product code VCP305H) or Poliglecaprone 25 (Monocryl, Ethicon 360, product code MCP944H), for soft tissue approximation and subcuticular closure of the skin, respectively.

4. Polydioxanone (PDS, Ethicon 360, product code Z258H) suture, 3/0, for approximation of the rectus sheath.

5. Silk ties (without needle) and silk sutures with needle for hemostatic sutures.

6. Sterile saline (37°C) for irrigation of peritoneal cavity.

7. Surgicel® (Ethicon 360, product code 1953), absorbable hemostatic agent.

8. Interceed® (Ethicon 360, product code M4350), absorbable adhesion barrier.

9. Xylocaine for local skin infiltration (2% solution).

10. PBS to clean biopsied tissues.

11. Liquid N_2 to snap-freeze samples and dry ice for slow cryopreservation.

12. Eppendorf tubes and tissue forceps.

13. Size 2 Portex® endotracheal tube (Smiths Medical, product code 100/111/020), neonatal size 0 laryngoscope (Smiths Medical, product code DS.2940.185.05), Micropore® medical tape (3 M, product code 1532-1).

14. Neonatal warmer blanket, infrared light warmer.

15. Veterinarian feeding bottles and feeders.

16. Soy-based infant formula for newborns.

17. Veterinarian incubators.

2.5. Necropsy

1. Laparotomy set, blades, and handles as described above.

2. Skull saw for retrieving brain and brain derivatives.

3. Large bore (18- or 16-G needles and large heparinized syringe for cardiac puncture).

4. PBS to wash tissues.

5. Liquid N_2 to snap-freeze tissues; dry ice and aluminum foil in ice tray for slow-freezing whole organs.

6. Eppendorf tubes for samples; plates for washing tissues before freezing.

2.6. Oocyte Harvest and Analysis of Germ-Line Transmission

1. Follicular stimulating hormone (e.g., Gonal-F, Merck; or Puregon, Organon/MSD).

2. Gonadotrophin releasing hormone antagonist (e.g., Ganirelix acetate, MSD).

3. US machine for monitoring follicular response, as above.

4. Laparotomy set for ovarian harvest, as above.

5. 22-G needle and dissecting microscope to aspirate follicles.

6. Flame-pulled fine-bore glass pipettes for dissociation of oocytes from oocyte–cumulus complexes (OCC).

7. Hyaluronidase (10 IU/mL, Sigma, product code H3506), prepared fresh as needed, stored at 4°C.

8. Sterile phosphate-buffered saline (PBS, 1×).

9. Wash buffer (0.4% bovine serum albumin in 1× PBS), filter-sterilized and stored at 4°C.

10. Lysis buffer (0.2 M potassium hydroxide), filter-sterilized and stored at 4°C.

11. Neutralizing buffer (0.2 M Tricine, Sigma, product code T9784), filter-sterilized and stored at 4°C.

12. DNA-Away (Molecular BioProducts, product code 7010).

3. Methods

3.1. Time-Mating of Macaques

Sexually mature adult female and male macaques (*M. fasicularis*) between the ages of 4 and 6 years undergo a clinical review and serological screening for specific pathogens (Herpes B, simian retroviruses, simian HIV) and for preexisting immunity to AAV 5, 8, and 9 before entering the IUGT program. Females are observed for menstruation by performing daily vaginal swabs (see Note 1). When the first day of the menstrual cycle is detected, an individual menstrual chart is commenced, with the cycle duration readily established after monitoring over 3–4 menstrual cycles. Because a menstrual cycle in the macaque species is approximately 32 days with ovulation occurring around mid-cycle (72), females are individually time-mated with proven male breeders from days 10 to 16 of the menstrual cycle, for a few hours each day, with the middle day of the multiple mating designated as day zero of the resulting pregnancy (53).

1. Perform a trans-abdominal pelvic ultrasound (US) with a Logiq P5 US machine (General Electric) or similar with the 12 L probe set at 7 MHz, a week after the next expected menstruation that was not observed.

2. Diagnose a viable pregnancy by observing the gestational sac, yolk sac, and fetus with a heartbeat.

3. Measure the crown-rump length (CRL, longest longitudinal length of the embryo) and the approximate gestational age derived from published growth charts derived for rhesus and cynomolgus macaques (71).

3.2. Pregnancy Monitoring

Pregnancies are monitored under sedation (intramuscular ketamine at 10 mg/kg body weight) by monthly US to determine fetal viability and growth.

1. After shaving the abdominal skin, perform an US with a 12-L probe as above.

2. Parameters recorded are the CRL until the skull becomes more calcified and well-defined, at which point the biparietal diameter (BPD) is measured. Femur length (FL), head, and abdominal circumferences can also be measured although gestational age and growth parameters are confirmed primarily using BPD and FL (71).

3. Evidence of pregnancy loss includes vaginal bleeding, a fetus without a heartbeat where viability was confirmed previously, and an empty uterus or evidence of pregnancy resorption.

3.3. Intravenous Vector Delivery

1. Prepare the pregnant dam for the procedure by feeding her a normal vitamin-enriched diet the day before and fasting her overnight.

2. After sedation with IM ketamine (10 mg/kg), shave the abdomen from nipple line to symphysis pubis.

3. US assessment is performed prior to fetal injection using the 12-L probe and fetal heart rate, BPD, FL, placental location, liquor volume, and the lie and position of the fetus are recorded

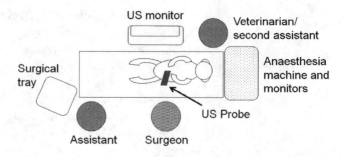

Fig. 2. Schema of operating theater set-up for IUGT. This diagram illustrates a suggested arrangement of personnel and equipment for IUGT, with the surgeon primarily performing the US and inserting the needle and the assistant delivering the vector dose. Another assistant continuously monitors the dam.

(see Note 2), following which general anesthesia (GA) is induced and the dam intubated to protect the maternal airway if possible difficulty is anticipated due to fetal position (see Note 3).

Surgical antisepsis of the maternal abdomen is achieved with povidone iodine paint. The dam is positioned supine and lightly restrained with ties around the upper and lower limbs (see Note 4, Fig. 2).

4. Apply sterile US gel for needle insertion and switch to a sterilized 8 C probe with a small foot-print at 10 MHz to visualize the intrahepatic portion of the umbilical vein. The target vessel is confirmed with application of color Doppler.

5. Insert a 22-G siliconized needle (EchoTip®, Cook Medical) percutaneously through the uterine wall and into the fetal intrahepatic vein under direct US guidance, taking special precautions to avoid the placenta (see Fig. 1b). A smaller 25-G spinal needle may be used as per operator preference.

6. Remove the stylet and confirm correct intravascular placement of the needle tip by gentle aspiration of fetal blood.

7. Inject the pre-warmed vector in an appropriate volume of carrier solution as a slow bolus dose, which can be observed as turbulence under ultrasound imaging (see Fig. 1c).

8. Remove the needle and syringe as a single unit quickly and immediately after the injection to minimize vector spillage into maternal tissue. Reverse maternal GA.

9. Observe the fetus throughout and after IUGT under US for evidence of bradycardia. Confirm normal fetal heart activity at 5 min post-IUGT.

10. Administer prophylactic antibiotic to the dam (enrofloxacin 5%, 5 mg/kg body weight, intramuscularly (IM) given as a single dose). Analgesia is usually not required after IUGT and a normal diet is reintroduced after full recovery from GA.

3.4. Fetoscopy

The aim of this procedure is to visualize the fetus, placenta, and umbilical cord during gestation in order to plan sampling of trophoblast or fetal blood, or to plan and execute delivery of vector and/or cells to the fetus.

1. Prepare the dam for surgery as above.

2. Perform an US with the 12-L probe to map the location of the bi-lobed placenta and plan the route of access into the amniotic cavity.

3. After surgical preparation of the abdomen, make a 1 mm skin incision over the maternal abdomen to accommodate the fetoscopic trocar.

4. Insert a 21-G fetoscopic trocar into the amniotic cavity under US guidance (see Fig. 1d) followed by a 1-mm diameter rigid fiber-optic fetoscope (11510, Karl Storz) comprising 10,000 optical fibers with a 0° viewing angle, and illuminated by a xenon light source (150 W) as previously described (57).

5. Adjust the focus and inspect the amniotic cavity, fetus, placental discs, communicating inter-lobe vessels and umbilical cord insertion (see Note 5).

6. At this juncture, a sampling needle (e.g., 26-G hypodermic puncture needle, 11510KC, Karl Storz) can be used to puncture the umbilical vein (see Fig. 1e) for fetal blood collection and/or for delivery of vector or cells, as described in human first trimester pregnancies by Chan et al. (57). Typically, the placental or the fetal insertion of the umbilical cord is targeted as these points are relatively immobile compared to the free portion of the cord. Withdrawal or infusion of fluids is done slowly, with constant monitoring of the fetal heart rate. Transient bradycardia may be observed during or immediately following the procedure; this normally resolves spontaneously.

7. Upon withdrawal of the needle, observe the venepuncture site for bleeding for up to a minute before withdrawing the fetoscope.

8. Close the skin incision with a subcutaneous suture and reverse GA. Give the dam a course of antibiotics (IM enrofloxacin 5%, 5 mg/kg daily for 5 days) and analgesia (SC carprofen 2 mg/kg once daily for 3 days) and allow her to consume a normal diet as tolerated when she has fully recovered from GA. A larger 2 mm endoscope (Hopkins lens, Karl Storz) may be employed in a similar manner to obtain higher resolution images (see Fig. 1f–h).

3.5. Cesarean Section

Where possible, injected offspring are delivered surgically, as spontaneous vaginal delivery is associated with stillbirth and early perinatal demise. The aim of this surgery is twofold: to ensure safe delivery of the injected offspring and to collect biopsies of maternal and extra-fetal birth tissues (placenta, membranes, and cord blood) for molecular analyses. Deliveries are scheduled to take place around 150–155 days of gestation.

1. Perform a physical review of the dam, screen for hemoglobin, creatinine, urea, and transaminase levels, and fast her overnight.

2. On the day of surgery, prepare the dam as described above and secure peripheral IV access. Induce GA and maintain with isofluorane. Intubate the dam and monitor end-tidal CO_2 and oxygen saturation.

3. Make a midline incision extending from the umbilicus to the symphysis pubis. Expose the peritoneal cavity by dissecting the subcutaneous tissue and abdominal fascia in layers and dividing the rectus muscle (see Fig. 3a, b).

4. Inspection of the gravid uterus often reveals large turgid uterine veins coursing over the anterior surface. Make a classical

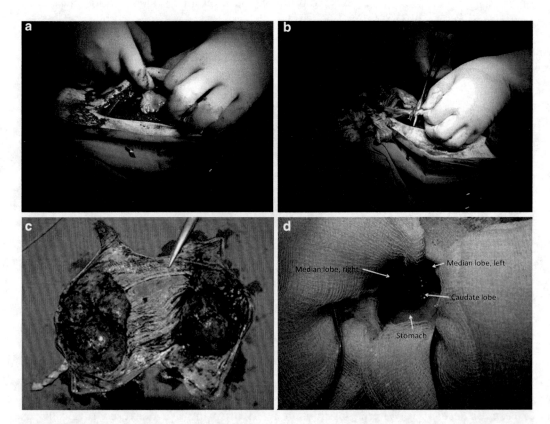

Fig. 3. Cesarean section and liver biopsy. A classical Cesarean section is performed at approximately 150 days gestation. A lower midline abdominal incision facilitates delivery of the infant (**a** and **b**). The bi-discoid placenta features two distinct placental discs connected by communicating vessels (**c**). Subsequent access to the various maternal tissues of interest such as the liver (**d**) is achieved through a second midline incision at the epigastrium.

vertical uterine incision over the anterior uterine surface so as to avoid dissecting through these vessels, and enter the amniotic sac.

5. Deliver the fetus and clamp and cut the umbilical cord. Hand the neonate to the veterinarian team for resuscitation with a combination of manual stimulation, warming, and iv doxapram or caffeine where required to stimulate the onset of respiration (see Note 6).

6. Collect umbilical cord blood and remove the placenta (see Fig. 3c) carefully. Administer IV 1 U oxytocin to the mother to accelerate uterine contraction and arrest placental bed hemorrhage. If necessary give a second dose of oxytocin and perform uterine massage to reduce blood loss from uterine atony.

7. Clean the uterine cavity with surgical gauze and close the myometrium in one continuous layer with 3/0 polyglactin braided sutures (Vicryl, Ethicon 360).

8. The maternal liver can be biopsied at this stage through a second midline incision in the epigastrium from either the left or right lobe avoiding the underlying gallbladder (see Fig. 3d). It is best to place two vertical mattress sutures running from the apex of the wedge toward the periphery, demarcating a wedge biopsy region with 3/0 polyglactin sutures to achieve hemostasis prior to taking the biopsy. Alternatively, take a small biopsy with a scalpel and apply electrocautery for hemostasis (see Note 7).

9. Finally, irrigate the pelvic cavity and apply anti-adhesion barrier (Interceed®, Johnson and Johnson) across the cavity. Close the rectus sheath in both incisions with 3/0 PDS and the skin with 3/0 or 4/0 subcuticular poliglecaprone 25 (Monocryl, Ethicon 360) using a "knotless" technique to bury the distal ends of the suture in the subcutaneous layer to avoid it being picked at by the very dextrous animal.

10. Administer antibiotics and analgesia as described. Allow the dam to consume a normal diet once she has recovered from anesthesia.

3.6. Infant Resuscitation and Hand-Rearing

1. Following cesarean delivery, dry and warm the infant, and perform manual suction of mucus from the oral and nasal passages, gentle rubbing of the chest and back, and gentle rocking help to clear the respiratory tract and stimulate breathing.

2. Administer oxygen via a face mask adapted to fit the newborn macaque by stretching a latex glove across the opening and making a 2–3-cm slit in the middle through which to insert the neonatal head.

3. Attach the pulse oximeter to the infant's paw or foot for heart rate and oxygen saturation monitoring. If the infant is unresponsive to these initial measures (no spontaneous respiration or movement, bradycardia) perform bag and mask ventilation in addition to tactile stimulation.

4. Obtain IV access and give a slow infusion of crystalloid. If these measures fail, attempt endotracheal intubation with a size 2 endotracheal tube (Portex®, Smiths Medical) and a neonatal size 0 (straight blade) laryngoscope (see Note 8).

5. When the neonate is stable and has been thoroughly examined, feed it a soy-based infant formula with a syringe, applying a few drops onto the tongue or into the cheek pouch while holding the infant securely in an upright position. On the first day of life, infants are fed a mean total daily volume of 30 mL of formula in eight divided doses. The total daily volume fed is 60–80 mL by the end of the first week.

6. Feed the infant every 3 h by syringe and then by bottle when it is able to suckle, and by 6 weeks a daily total volume of 120 mL of formula should be reached, divided over 5–6 feeds.

7. By 3 months when the infant is able to drink from a suspended animal feeder, decrease the frequency of milk feeds to 2–4 times daily while increasing the volume proportionately to reach approximately 100–200 mL a day and introduce a solid diet. The infant is weaned off formula by 6 months of age.

3.7. Laparotomy for Liver and Other Organ Biopsy (Survival Surgery)

Liver biopsy is performed on injected offspring and dams at 6 month intervals where there is a need to assess the efficacy of liver-directed gene therapy.

1. Surgical preparation is performed as previously described. GA is induced, the animal is intubated to protect the airway, and monitoring of parameters is performed as described above.

2. The liver is accessed through an upper abdominal midline incision extending 2–3 cm from the xiphisternum. If adhesions are present from previous surgery, these may impair visualization of or impede access to the liver and will thus require adhesiolysis. Palpate the liver carefully for parenchymal and surface tumors.

3. Obtain a wedge biopsy with sharp dissection from the inferior margin of either the left or right lobe. If there was a previous biopsy taken from one lobe, the new sample is taken from opposite lobe as described.

4. Other tissues which can be readily biopsied include the omentum, subcutaneous fat, rectus muscle, and skin.

5. Perform mass closure of the abdominal wall and repair the skin as previously described.

6. Antibiotics and analgesia are administered as previously described. The animal is placed on a normal diet once it has recovered from anesthesia. Movement, drinking, and eating habits are observed for evidence of adequate postsurgical analgesia.

3.8. Necropsy

At predetermined time-points, the injected offspring or dam is euthanized and a full organ harvest is performed for molecular and other analyses.

1. Sedate the animal with IM ketamine and induce and maintain GA with isofluorane. Provide oxygen by face-mask, secure IV access, and maintain patency with an infusion of crystalloids.

2. Position and restrain the animal before performing a thoracotomy. Expose the thoracic and peritoneal cavities by extending the incision caudally. Dissect the pericardium to expose the myocardium, identifying the left ventricle.

3. Use a large bore 16- or 18-G cannula to perform a cardiac puncture via the left ventricle and slowly withdraw a large volume of blood into a heparinized syringe. Remove the syringe while the cannula remains intraventricular and attach the cannula to a primed infusion set. We infuse normal saline (0.9%) or Ringer's lactate followed by 1% paraformaldehyde where histological staining is required.

4. Make an incision in the right atrium with surgical scissors or blade to allow exsanguination and harvest organs when the effluent is clear. All organs are removed (brain, cerebellum, and heart) or sampled (skin, fat, and muscle) and stored in 1% paraformaldehyde for tissues in which there is a need to preserve a reporter protein for gene marking analysis, or in 4% PFA for other purposes.

3.9. Oocyte Harvest and Analysis of Germ-Line Transmission in Dams

When viral vectors are delivered in utero there is a risk of germ-cell transduction in the fetal recipient. We have demonstrated transplacental trafficking of AAV vectors and consequent maternal ovarian transduction, as well as fetal gonadal transduction after late-gestation AAV-mediated IUGT. However, analysis of purified maternal oocytes confirmed that there was no germ-cell transduction despite the presence of vector in bulk ovarian tissues (65). The analysis of gametes for vector sequences can be done upon sexual maturity at 4–6 years of age (73). In order to harvest pure oocytes for analysis, a strategy for ovarian hyperstimulation with a modified antagonist-mediated cycle can be pursued.

1. Commence ovarian stimulation on day 2 of menstruation. Administer subcutaneous (sc) 30 IU of recombinant human follicular stimulating hormone (FSH Gonal-F, Merck; or Puregon, Organon/MSD) twice daily.

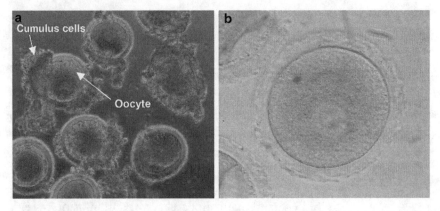

Fig. 4. Oocyte–cumulus complexes (OCC) and purified oocyte. (a) Aspiration of ovarian follicles yields OCC which need to be processed to yield pure oocytes for accurate analysis of germ-line vector transmission. (b) The purified oocyte after treatment with hyaluronidase and manual removal of cumulus cells.

2. From day 6, give sc 125 μg of gonadotrophin releasing hormone antagonist (GnRH antagonist Ganirelix acetate, Orgalutron, MSD) once daily to prevent premature ovulation prior to oocyte harvest.

3. Continue both medications until day 12 when the ovaries are harvested at necropsy.

4. Working in a laminar flow hood that has been wiped thoroughly with ethanol and DNA-Away and under a dissecting microscope, use a 22-G needle to aspirate large and small follicles over the surface of the ovary. Bisect the ovary to aspirate remaining internal follicles. The material collected will consist of OCC which should be pooled and stored in a small volume of sterile PBS on ice (see Fig. 4a).

5. Prepare hyaluronidase (10 IU/mL) at room temperature and place a thin layer of this solution in a Petri dish. Pipette the OCC into the dish and gently agitate to disperse the complexes. Incubate at room temperature for 10 min.

6. Under the microscope, gently dissociate oocytes from cumulus cells by teasing the latter off the oocytes using a fine-bore glass pipette pulled over a flame. Confirm visually that the oocytes have been completely denuded (see Fig. 4b).

7. Pool the purified oocytes in 20 μL of PBS. With a p100 micropipette and aerosol-resistant tip, place 5 drops (50 μL each) of wash buffer on a clean Petri dish. Ensure that each droplet is sufficiently far away from the other to prevent the droplets from merging together.

8. Transfer one or two oocytes into the first droplet of wash buffer.

9. With a p2 micropipette and new aerosol-resistant tip, transfer one oocyte from the first droplet to the second droplet. Wash oocyte and discard the tip.

10. Repeat the previous step across all five droplets of wash buffer till the oocyte is washed in the final fifth droplet. Perform this procedure for each oocyte.

11. With a p2 micropipette and new aerosol-resistant tip, transfer the oocyte (suspended in 0.5 µL of wash buffer) into a tube containing 5 µL of lysis buffer. Quickly spin to bring down the tube's contents.

12. Heat-denature the oocytes in lysis buffer at 65°C for 10 min in a thermal cycler.

Once completed, neutralize the contents by adding an equivalent volume of neutralizing buffer to obtain cell-free DNA. Use 1 µL to determine vector copy number by quantitative PCR.

3.10. Monitoring of Safety and Long-Term Outcomes

One of the main advantages of interrogating vector delivery technology in a NHP model is the ability to monitor the injected offspring and their mothers over several years. This allows the assessment of short-term safety issues, such as acute complications of trans-abdominal intravascular injection, fetal or maternal loss related to the surgical procedure, acute immune- or hepatotoxicity, and mid- to long-term adverse effects, such as transaminitis, chronic hepatotoxicity, neutralizing antibody production that nullifies transgene expression, loss of transgene expression (gene silencing), localized immune response to the intracellular vector causing site-specific inflammatory effects (e.g., meningoencephalitis, myocarditis) and tumor development, particularly if vector integration is a concern.

In addition, because of the similarities in placental development, downstream maternal effects of transplacental vector trafficking can be observed longitudinally. We have performed serial analyses of maternal viremia following IUGT to determine the rapidity of vector clearance from the maternal circulation, maternal immune response to vector capsid and transgene, tissue transduction and transgene expression, and are studying the risk of germ-line transmission by harvesting oocytes for molecular analyses. The importance of studying maternal safety and well-being cannot be underestimated as this is one of the major ethical issues to be addressed before IUGT is approved for clinical trials.

We monitor the safety aspects of all interventions by creating individual charts for each animal with which to document parameters, changes in surgical protocol, and outcomes. Each adult and infant animal undergoes monthly clinical screening which includes a physical review, a full blood count, and a panel of hepatic transaminases (AST and ALT), urea, and creatinine, in addition to immunological and functional assays. If an animal is ill, it is isolated from the others in the colony and the appropriate diagnostic tests are performed.

During laparotomy or necropsy, organs biopsied or harvested are inspected for macroscopic tumors. Fixed tissues are sectioned and a review of the histology is done after H&E staining. Obvious inflammatory infiltration and pleomorphic changes are documented. Immunohistochemistry or in situ hybridization can be performed to determine the exact location of the vector within certain tissues, for example, to establish the presence or absence of the vector within the gametes.

4. Conclusions

The NHP model is a critical development in IUGT research as it allows exploration of clinically relevant issues. The data this model has the potential to yield, particularly with respect to long-term safety and efficacy of transgene expression in injected offspring, the ability to target different organs of interest by varying vector type or route of delivery, and the risk of vector transmission to future offspring through germ-line transmission, may be useful in guiding the first clinical trials of IUGT in humans.

5. Notes

1. Monitoring the menstrual cycle: It is important in a time-mating strategy to perform daily perineal/vaginal swabs to determine the start of menstruation so that female macaques can be mated during the fertile period, around the mid-cycle. The females are mated between days 10 and 16 (counting the first recorded day of menstruation as day 1). To facilitate the diagnosis of early pregnancy, the females undergo a trans-abdominal ultrasound scan (TAUS) 2 weeks after mating. A gestational sac may be observed at this time. If there is a thickened endometrial lining but no gestational sac the US scan is repeated in 2 weeks, until either a pregnancy is confirmed or menstruation begins again.

2. Pre-IUGT assessment: Shortly before the IUGT, after the mother has been sedated, it is important to assess fetal viability by TAUS before thawing and preparing the vector. A biometric assessment can be performed to assess growth and confirm the gestational age. This will be useful in planning the route of needle entry should the fetus be in an abnormal lie or position. If needle insertion is anticipated to be difficult because of this, we prefer to anesthetize, intubate, and ventilate the mother in the event that sedation wears out before the end of the procedure and to prevent aspiration which can be catastrophic.

3. Difficult IUGT: If the needle is not successfully inserted intravenously on the first pass, even with careful adjustment of the tip, it is withdrawn, the angle of approach reassessed and a second attempt at cannulation made. If the intrahepatic vein is inaccessible, an alternative route of delivery is intracardiac. The heart is visualized by US, and the needle is inserted smoothly, aiming for one of the ventricles and withdrawn quickly after vector delivery. We have observed fetal bradycardia (normal heart rate 150–180 bpm) in 10–15% of injected fetuses, sometimes lasting beyond 1 h postinjection, although this often resolves after 30 min.

4. Arrangement of equipment and personnel in the operating theater: Surgical instruments and equipment should be arranged to facilitate smooth flow of events, such as the change between aspirating fetal blood and injecting the vector. The surgeon will often operate the US probe in the left/nondominant hand while inserting the needle with right/dominant hand. The US monitor should be located in front of the surgeon, with the animal's head on the surgeon's right so that the needle can be inserted comfortably in a direction toward the animal's pelvis. A surgical tray should be prepped and contain cleaning solutions for skin antisepsis, sterile gel, and two syringes, one loaded with vector in the correct dose and the other with sterile saline. It is helpful to have an assistant present to aid the surgeon performing the injection. The assistant's task is to remove the introducer when the needle has been inserted, aspirate fetal blood to confirm correct needle placement, and attach the loaded syringe containing the vector, without disturbing the position of the needle. The assistant will also deliver the vector as a slow bolus.

5. Fetoscopy: Visual inspection can be improved by infusion of warmed normal saline to increase the distension of the amniotic cavity as low levels of amniotic fluid is normal, in contrast to human pregnancy.

6. Neonatal resuscitation with IV caffeine or doxapram: Occasionally, the neonate exhibits signs of respiratory depression after delivery, possibly contributed to by general anesthesia used on the mother. This may lead to hypoxia and bradycardia, and must be treated immediately. We find that the administration of IV caffeine (20 mg/kg or 5 mg for a 250 g neonate) as a bolus dose at delivery is very effective in stimulating the initiation of respiration (74, 75). Upon delivering the fetus, the cord is clamped and caffeine is injected through the umbilical vein. We have observed a marked reduction in the time taken to resuscitate the neonate with the institution of this practice.

7. Liver biopsy, hemostasis, and cautery: When a liver biopsy is performed as described, there may be brisk bleeding from the apex of the wedge. It is important to cut out the wedge biopsy cleanly and to take care with tissue retraction so that an unintended tear does not cut through the apex. Haemostasis must be secured at the apex first with electrocautery and the two sides of the lobe flanking the wedge may need to be gently retracted to expose the bleeding apical vessels. Pressure applied with surgical gauze on the biopsy site for a few minutes may help reduce blood flow to facilitate hemostasis. When active bleeding has been arrested, a small sheet of Surgicel® can be folded into the wedge to prevent further bleeding. After irrigation of the peritoneal cavity, we place a small sheet of Interceed® superior to the liver to minimize adhesions forming in anticipation of future surgery.

8. Intubating the neonate: The small bore ETT is uncuffed so an improvised cuff is made with Micropore™ (3 M) medical tape wound several times in a tapered fashion around the distal end of the tube to improve the laryngeal seal. In our research facility, this procedure is undertaken only for the purpose of acute resuscitation and, on occasion where prematurity is significant enough to cause concerns about surfactant insufficiency, to administer intratracheal Survanta (Abbott) when there is evidence of improved cardiorespiratory stability.

References

1. Coutelle C, Themis M, Waddington S et al (2003) The hopes and fears of in utero gene therapy for genetic disease—a review. Placenta 24:S114–S121

2. Waddington SN, Kramer MG, Hernandez-Alcoceba R et al (2005) In utero gene therapy: current challenges and perspectives. Mol Ther 11:661–676

3. Waddington S, Buckley S, David A et al (2007) Fetal gene transfer. Curr Opin Mol Ther 9:432–438

4. Surbek D, Schoeberlein A, Wagner A (2008) Perinatal stem-cell and gene therapy for hemoglobinopathies. Semin Fetal Neonatal Med 13(4):282–290

5. Haynes B, Scearce R, Lobach D et al (1984) Phenotypic characterization and ontogeny of mesodermal-derived and endocrine epithelial components of the human thymic microenvironment. J Exp Med 159:1149–1168

6. Billingham R, Brent L, Medawar P (2003) 'Actively Acquired Tolerance' of foreign cells. Transplantation 76:1409–1412

7. Takahama Y (2006) Journey through the thymus: stromal guides for T-cell development and selection. Nat Rev Immunol 6:127–135

8. Guillot PV, Gotherstrom C, Chan J et al (2007) Human first-trimester fetal MSC express pluripotency markers and grow faster and have longer telomeres than adult MSC. Stem Cells (Dayton Ohio) 25:646–654

9. Benirschke K, Kaufmann P (eds) (2003) Pathology of the human placenta. Springer, New York

10. Dejneka NS, Surace EM, Aleman TS et al (2004) In utero gene therapy rescues vision in a murine model of congenital blindness. Mol Ther 9:182–188

11. Williams M, Coleman J, Haire S et al (2006) Lentiviral expression of retinal guanylate cyclase-1 (RetGC1) restores vision in an avian model of childhood blindness. PLoS Med 3:e201

12. Seppen J, van der Rijt R, Looije N et al (2003) Long-term correction of bilirubin UDP glucuronyltransferase deficiency in rats by in utero lentiviral gene transfer. Mol Ther 8:593–599

13. Waddington SN, Nivsarkar MS, Mistry AR et al (2004) Permanent phenotypic correction of hemophilia B in immunocompetent mice by prenatal gene therapy. Blood 104:2714–2721

14. Karolewski BA, Wolfe JH (2006) Genetic correction of the fetal brain increases the lifespan of mice with the severe multisystemic disease mucopolysaccharidosis type VII. Mol Ther 14:14–24

15. Buckley SMK, Waddington SN, Jezzard S et al (2008) Intra-amniotic delivery of CFTR-expressing adenovirus does not reverse cystic fibrosis phenotype in inbred CFTR-knockout mice. Mol Ther 16:819–824

16. Davies L, Varathalingam A, Painter H et al (2008) Adenovirus-mediated in utero expression of CFTR does not improve survival of CFTR knockout mice. Mol Ther 16:812–818

17. Larson J, Morrow S, Happel L et al (1997) Reversal of cystic fibrosis phenotype in mice by gene therapy in utero. Lancet 349:619–620

18. Reay D, Bilbao R, Koppanati B et al (2008) Full-length dystrophin gene transfer to the mdx mouse in utero. Gene Ther 15:531–536

19. Koppanati BM, Li J, Reay DP et al (2010) Improvement of the mdx mouse dystrophic phenotype by systemic in utero AAV8 delivery of a minidystrophin gene. Gene Ther 17(11):1355–1362

20. Georgiadesa P, Ferguson-Smith A, Burton G (2002) Comparative developmental anatomy of the murine and human definitive placentae. Placenta 23:3–19

21. Harkness J, Wagner J (1995) The biology and medicine of rabbits and rodents, 4th edn. Williams & Wilkins, Philadelphia

22. Williams DA (2009) Recombinant DNA Advisory Committee updates recommendations on gene transfer for X-linked severe combined immunodeficiency. Mol Ther 17:751–752

23. Sabatino DE, Mackenzie TC, Peranteau W et al (2007) Persistent expression of hF.IX after tolerance induction by in utero or neonatal administration of AAV-1-F.IX in hemophilia B mice. Mol Ther 15:1677–1685

24. Waddington SN, Buckley SMK, Nivsarkar M et al (2003) In utero gene transfer of human factor IX to fetal mice can induce postnatal tolerance of the exogenous clotting factor. Blood 101:1359–1366

25. Waddington SN, Mitrophanous KA, Ellard F et al (2003) Long-term transgene expression by administration of a lentivirus-based vector to the fetal circulation of immuno-competent mice. Gene Ther 10:1234–1240

26. Tran ND, Porada CD, Almeida-Porada Ga et al (2001) Induction of stable prenatal tolerance to b-galactosidase by in utero gene transfer into preimmune sheep fetuses. Blood 97:3417–3423

27. Kafri T, Morgan D, Krahl T et al (1998) Cellular immune response to adenoviral vector infected cells does not require de novo viral gene expression: implications for gene therapy. Proc Natl Acad Sci U S A 95:11377–11382

28. van Ginkel F, McGhee J, Liu C et al (1997) Adenoviral gene delivery elicits distinct pulmonary-associated T helper cell responses to the vector and to its transgene. J Immunol 159:685–693

29. Liberatore C, Capanni M, Albi N et al (1999) Natural killer cell-mediated lysis of autologous cells modified by gene therapy. J Exp Med 189:1855–1862

30. Manno CS, Pierce GF, Arruda VR et al (2006) Successful transduction of liver in hemophilia by AAV-Factor IX and limitations imposed by the host immune response. Nat Med 12:342–347

31. Holladay SD, Smialowicz RJ (2000) Development of the murine and human immune system: differential effects of immunotoxicants depend on time of exposure. Environ Health Perspect 108(Suppl 3):463–473

32. Stites D, Carr M, Fundenberg H (1974) Ontogeny of cellular immunity in the human fetus. Development of responses to phytohaemmagglutinin and to allogeneic cells. Cell Immunol 11:257–271

33. Mumford D, Sung J, Wallis J et al (1978) The lymphocyte transformation response of fetal hemolymphatic tissue to mitogens and antigens. Pediatr Res 12:171–175

34. Ohama K, Kaji T (1974) Mixed culture of fetal and adult lymphocytes. Am J Obstet Gynecol 119:552–560

35. Toivanen P, Uksila J, Leino A et al (1981) Development of mitogen responding T cells and natural killer cells in the human fetus. Immunol Rev 57:89–105

36. Mosier D (1977) Ontogeny of T cell function in the neonatal mouse. In: Cooper M, Dayton D (eds) Development of host defenses. Raven Press, New York

37. Verlarde A, Cooper M (1984) An immunofluorescence analysis of the ontogeny of myeloid, T, and B lineage cells in mouse hemopoietic tissues. J Immunol 133:672–677

38. Tyan ML, Herzenberg LA (1968) Studies on the ontogeny of the mouse immune system: II.

Immunoglobulin-producing cells. J Immunol 101:446–450

39. Santoni A, Riccardi C, Barlozzari T et al (1982) Natural suppressor cells for murine NK activity. In: Herberman R (ed) NK cells and other natural effector cells. Academic Press, New York, p 443

40. Mestas J, Hughes CCW (2004) Of mice and not men: differences between mouse and human immunology. J Immunol 172:2731–2738

41. Christensen G, Minamisawa S, Gruber PJ et al (2000) High-efficiency, long-term cardiac expression of foreign genes in living mouse embryos and neonates. Circulation 101: 178–184

42. Lipshutz GS, Gruber CA, Cao Y-a et al (2001) In utero delivery of adeno-associated viral vectors: intraperitoneal gene transfer produces long-term expression. Mol Ther 3:284–292

43. Lipshutz GS, Flebbe-Rehwaldt L, Gaensler KM (1999) Adenovirus-mediated gene transfer in the midgestation fetal mouse. J Surg Res 84:150–156

44. Gregory LG, Waddington SN, Holder MV et al (2004) Highly efficient EIAV-mediated in utero gene transfer and expression in the major muscle groups affected by Duchenne muscular dystrophy. Gene Ther 11:1117–1125

45. Waddington SN, Buckley SM, Bernloehr C et al (2004) Reduced toxicity of F-deficient Sendai virus vector in the mouse fetus. Gene Ther 11:599–608

46. Endo M, Henriques-Coelho T, Zoltick PW et al (2010) The developmental stage determines the distribution and duration of gene expression after early intra-amniotic gene transfer using lentiviral vectors. Gene Ther 17:61–71

47. Magness CL, Fellin PC, Thomas MJ et al (2005) Analysis of the *Macaca mulatta* transcriptome and the sequence divergence between macaca and human. Genome Biol 6:R60

48. Hendrie T, Peterson P, Short J et al (1996) Frequency of prenatal loss in a macaque breeding colony. Am J Primatol 40:41–53

49. Enders A, Lantz K, Peterson P et al (1997) Symposium: reproduction in baboons. From blastocyst to placenta: the morphology of implantation in the baboon. Hum Reprod Update 3:561–573

50. Blankenship TN, Enders AC, King BF (1993) Trophoblastic invasion and modification of uterine veins during placental development in macaques. Cell Tissue Res 274:135–144

51. Blankenship TN, Enders AC, King BF (1993) Trophoblastic invasion and the development of uteroplacental arteries in the macaque:

immunohistochemical localization of cytokeratins, desmin, type IV collagen, laminin, and fibronectin. Cell Tissue Res 272:227–236

52. Musicki B, Pepe G, Albrecht E (2003) Functional differentiation of the placental syncytiotrophoblast: effect of estrogen on chorionic somatomammotropin expression during early primate pregnancy. J Clin Endocrinol Metab 88:4316–4323

53. Makori N, Rodriguez C, Cukierski M et al (1996) Development of the brain in staged embryos of the long-tailed monkey (*Macaca fascicularis*). Primates 37:351–361

54. Hendrickx A, Peterson P (1997) Symposium: reproduction in baboons. Perspectives on the use of the baboon in embryology and teratology research. Hum Reprod Update 3:575–592

55. Hendrickx A, Makori N, Peterson P (2002) The nonhuman primate as a model of developmental immunotoxicity. Hum Exp Toxicol 21:537–542

56. Plopper C, Alley J, Weir A (1986) Differentiation of tracheal epithelium during fetal maturation in the rhesus monkey *Macaca mulatta*. Am J Anat 175:59–71

57. Chan J, Kumar S, Fisk NM (2008) First trimester embryo-fetoscopic and ultrasound-guided fetal blood sampling for ex vivo viral transduction of cultured human fetal mesenchymal stem cells. Hum Reprod 23:2427–2437

58. Garrett D, Larson J, Dunn D et al (2003) In utero recombinant adeno-associated virus gene transfer in mice, rats, and primates. BMC Biotechnol 3:16–23

59. Tarantal A, Lee C (2010) Long-term luciferase expression monitored by bioluminescence imaging after adeno-associated virus-mediated fetal gene delivery in rhesus monkeys (*Macaca mulatta*). Hum Gene Ther 21:1–6

60. Tarantal AF, McDonald RJ, Jimenez DF et al (2005) Intrapulmonary and intramyocardial gene transfer in rhesus monkeys (*Macaca mulatta*): safety and efficiency of HIV-1-derived lentiviral vectors for fetal gene delivery. Mol Ther 12:87–98

61. Tarantal AF, Lee CI, Ekert JE et al (2001) Lentiviral vector gene transfer into fetal rhesus monkeys (*Macaca mulatta*): lung-targeting approaches. Mol Ther 4:614–621

62. Tarantal AF, O'Rourke JP, Case SS et al (2001) Rhesus monkey model for fetal gene transfer: studies with retroviral-based vector systems. Mol Ther 3:128–138

63. Tarantal AF, Lee CC, Jimenez DF et al (2006) Fetal gene transfer using lentiviral vectors: in vivo detection of gene expression by microPET and

optical imaging in fetal and infant monkeys. Hum Gene Ther 17:1254–1261

64. Lai L, Davison BB, Veazey RS et al (2002) A preliminary evaluation of recombinant adeno-associated virus biodistribution in rhesus monkeys after intrahepatic inoculation in utero. Hum Gene Ther 13:2027–2039

65. Mattar CN, Nathwani AC, Waddington SN et al (2011) Stable human FIX expression after 0.9G intrauterine gene transfer of self-complementary adeno-associated viral vector 5 and 8 in macaques. Mol Ther 19(11):1950–1960

66. Bishop CV, Sparman ML, Stanley JE et al (2009) Evaluation of antral follicle growth in the macaque ovary during the menstrual cycle and controlled ovarian stimulation by high-resolution ultrasonography. Am J Primatol 71:384–392

67. Adams M, Kaplan J, Manuck S et al (1990) Inhibition of coronary artery atherosclerosis by 17-beta estradiol in ovariectomized monkeys. Lack of an effect of added progesterone. Arteriosclerosis 10:1051–1057

68. Gallagher M, Rapp PR (1997) The use of animal models to study the effects of aging on cognition. Annu Rev Psychol 48:339

69. Chan AW, Chong KY, Martinovich C et al (2001) Transgenic monkeys produced by retroviral gene transfer into mature oocytes. Science (New York, NY) 291:309–312

70. Yang SH, Cheng PH, Banta H et al (2008) Towards a transgenic model of Huntington's disease in a non-human primate. Nature 453:921–924

71. Tarantal A, Hendrickx A (1988) Prenatal growth in the cynomolgus and rhesus macaque (*Macaca fascicularis* and *Macaca mulatta*): a comparison by ultrasonography. Am J Primatol 15:309–323

72. Blakely G, Beamer T, Dukelow W (1981) Characteristics of the menstrual cycle in non-human primates. IV. Timed mating in *Macaca nemestrina*. Lab Anim 15:351–353

73. Bonadio C (2000) *Macaca fascicularis* (On-line). Animal Diversity Web

74. Henderson-Smart DJ, Steer PA (2010) Caffeine versus theophylline for apnea in preterm infants. Cochrane Database Syst Rev (1):CD000273

75. Schmidt B, Roberts RS, Davis P et al (2006) Caffeine therapy for apnea of prematurity. N Engl J Med 354:2112–2121

Chapter 13

Choice of Surrogate and Physiological Markers for Prenatal Gene Therapy

Juliette M.K.M. Delhove, Ahad A. Rahim, Tristan R. McKay, Simon N. Waddington, and Suzanne M.K. Buckley

Abstract

Surrogate genetically encoded markers have been utilized in order to analyze gene transfer efficacy, location, and persistence. These marker genes have greatly accelerated the development of gene transfer vectors for the ultimate application of gene therapy using therapeutic genes. They have also been used in many other applications, such as gene marking in order to study developmental cell lineages, to track cell migration, and to study tumor growth and metastasis. This chapter aims to describe the analysis of several commonly used marker genes: green fluorescent protein (GFP), β-galactosidase, firefly luciferase, human factor IX, and alkaline phosphatase. The merits and disadvantages of each are briefly discussed. In addition a few short examples are provided for continual and endpoint analysis in different disease models including hemophilia, cystic fibrosis, ornithine transcarbamylase deficiency and Gaucher disease.

Key words: Green fluorescent protein, GFP, Luciferase, β-galactosidase, Bioluminescence, Alkaline phosphatase, Quantitative PCR, Immunohistochemistry

1. Introduction

Gene therapy of monogenic diseases is perfected by achieving tissue-targeted expression resulting in appropriate levels of gene expression whilst using the smallest amount of gene transfer vector possible. Moreover, the gene expression is ideally regulated in a physiologically appropriate fashion. Although other gene therapy targets, such as cancer, depend upon expression of genes which may be toxic and which may be expressed in extremely non-physiological concentrations, the concept of expression targeting and regulation remains valid. To interrogate the fundamental mechanisms of vector function and as a precursor to application of therapeutic transgenes, surrogate marker genes are widely used. These have been used extensively in pre and postnatal gene transfer and

Charles Coutelle and Simon N. Waddington (eds.), *Prenatal Gene Therapy: Concepts, Methods, and Protocols*,
Methods in Molecular Biology, vol. 891, DOI 10.1007/978-1-61779-873-3_13, © Springer Science+Business Media, LLC 2012

gene therapy in order to analyze the efficacy, duration, and location of gene expression following administration of gene transfer vectors. In addition, marker genes may also permit monitoring of transduced cell lineages (ideally if the marker gene has integrated into the cell genome) and to monitor tumor growth and metastasis. More recently, marker genes have been used to monitor intracellular signaling pathways and have been placed downstream of known regulatory sequences (including promoters and enhancers) in order to act as surrogate markers of endogenous gene expression.

One of the first examples in preclinical studies was the delivery of the neomycin-resistance gene (neo[R]) into blood and bone marrow cells. The phosphotransferase activity of the gene product was determined in tissue extracts (1). Since then, several other enzymes have been used as genetically-encoded marker proteins, including human α-1-antitrypsin (2) and chloramphenicol acetyltransferase, although β-galactosidase (β-Gal) and human placental secreted alkaline phosphatase (SEAP) (3, 4) have been more widely used (and are thus described in this chapter). β-Galactosidase can be assayed in tissue homogenates by several means but one of the preferable assays is by ELISA. It can also be located by histochemical staining of whole or sectioned tissues using a substrate which forms a blue precipitate after reaction with the enzyme (5). SEAP, being a secreted enzyme, can be measured in plasma and also in tissue culture medium. Similarly, human factor IX (hFIX) can be detected in plasma and is particularly useful for long-term, noninvasive expression studies. It can also be detected by immunohistochemistry in tissues (6).

Green fluorescent protein (GFP) is derived from the jellyfish *Aequorea victoria* (7). It emits a green light when excited by light in the long wave ultraviolet region of the spectrum. It can be visualized directly in whole tissues, can be detected by immunohistochemistry, and can be quantified by ELISA in tissue homogenates. Subsequently, several homologs of jellyfish GFP have been developed which emit light at different wavelengths, such as blue, yellow and red fluorescent proteins. More recently, fluorescent proteins from other organisms, such as the sea anemone *Entacmaea quadricolor*, have been identified and are subject to further modifications to improve their brightness and stability. Moreover, it is now feasible to detect these proteins by whole-body imaging and the efficiency of this technology has been improved by modification of these fluorescent proteins so that they emit light towards the infrared end of the spectrum as light at these wavelengths is least absorbed by mammalian tissue. Over the past 10 years, whole-body imaging has been revolutionized by the use of the enzyme luciferase from the firefly *Photinus pyralis* (8, 9). This enzyme converts D-luciferin into oxyluciferin, consuming ATP and oxygen in the process but, most importantly, also emitting light (10). This light is emitted at 557 nm and is therefore yellow-green in color, however, red-shifted variants have been developed which are absorbed less by mammalian tissue and are, therefore, more suited for whole-body bioimaging (11).

Each marker gene has advantages and disadvantages, therefore, the use of more than one marker gene should be considered. For example, there is some evidence that GFP is toxic in certain cell types (12) and that expression levels of different colored fluorescent proteins depend upon the tissue type (13). Secreted marker genes such as SEAP or hFIX are particularly useful for long-term, noninvasive, and nonterminal monitoring of gene expression, however, efficiency of release may depend upon the cell type which has been transduced and therefore may not be the same for all tissues. Luciferase has proven immensely useful of late for performance of whole-body bioluminescence and can, therefore, be used for longitudinal monitoring and localization of gene expression. However, immunohistochemical detection of luciferase is poor; nevertheless, this has been overcome by the addition of a peptide tag (e.g. FLAG, his, myc), generation of a GFP fusion construct or incorporation of GFP in a bicistronic cassette.

Physiological endpoints indicating therapeutic success of gene therapy will naturally vary from disease to disease. For many pre-clinical gene therapy studies, survival and weight gain/loss would be useful generic markers. However, most of the longitudinal and endpoint analyses in these cases will very specifically depend on the symptoms of the disease being studied and details of their determination would be beyond the scope of this book. To illustrate the range of different specific physiological endpoints that may apply, we present examples for five diseases, without going in to methodological details in the following paragraphs.

Gene therapy for hemophilia, for which several animal models exist, would entail continual blood collection at, say, monthly time points. Blood would be analyzed by functional tests to assess blood coagulation and by measuring expression of coagulation factors VIII or IX (for hemophilias A or B, respectively) by antigen ELISA. In addition, the blood could be analyzed by chromogenic coagulation factor activity assay. Finally, endpoint analysis could be performed to detect the coagulation factor in cells by immunohistochemistry, bearing in mind that secreted coagulation factor may result in background in the vascular lumen unless tissue is perfused (6).

Detection of the coagulation factors, in particularly factor IX, can also serve as a surrogate endpoint in preparatory studies requiring long-term repeated monitoring of a secreted protein (see Subheading 3.5).

For ornithine transcarbamylase deficiency, the sparse fur-abnormal skin and hair mouse model (spf-ash) acts as a useful model—it remains relatively well but lacks hair until approximately a month of age. The first readout for successful gene expression in this model would be early (i.e. normal) acquisition of hair.

For longitudinal analysis measurement of urinary orotic acid is standard. For endpoint analysis histochemical (rather than immunohistochemical) analysis of enzyme expression is a standard assay (14).

For cystic fibrosis, various mouse models exist, several of which exhibit fatal gut blockage around birth or weaning. Survival would therefore be one of the most striking consequences of successful gene expression in the gut. However, even in the absence of improved survival, electrophysiological properties of the airways and intestine would be useful phenotypic markers of expression of cystic fibrosis transmembrane conductance regulating protein (CFTR). There are also several antibodies directed against mouse and human CFTR, which are available from the CFTR folding consortium (http://www.cftrfolding.org/CFFTReagents.htm), which could be used in immunohistochemistry.

For neuronopathic Gaucher disease, the mouse model generated by the Karlsson laboratory is one of extreme severity, with all mice succumbing to motor neuron degeneration and death before 20 days of age (15). Again, survival and weight gain would be the most salient features of successful gene delivery. In addition, direct measurement of glucocerebrosidase, the enzyme deficient in this disease, is possible by the use of a fluorometric enzyme activity assay. The enzyme can also be detected by western blot and immunohistochemistry. Enzyme expression could also be detected indirectly by assay for glucosylceramide. Alternatively, surrogate markers may also be examined. For example, microglial activation is a hallmark of the brain inflammation which precedes death, therefore, immunohistochemical detection of activated microglia is useful and can be performed using an anti-CD68 antibody (15).

2. Materials

The majority of *marker gene constructs* for performance of the following assays are readily available from commercial or academic sequence repositories such as Origene (http://www.origene.com/) and the DF/HCC DNA Resource Core (http://plasmid.med.harvard.edu/PLASMID/).

2.1. Total Protein Determination

1. Homogenizing pestles (disposable).
2. Lysis buffer (from β-gal ELISA kit, Roche, Indianapolis, USA; #11539426001).
3. Bicinchoninic acid (BCA) reagents A and B (Pierce, Rockford, USA; #23223 and 23224, respectively).
4. Albumin protein standard 2.0 mg/ml (Pierce; #23209).
5. Microtiter plate.
6. Plate reader.

2.2. Green Fluorescent Protein

2.2.1. GFP ELISA

1. Monoclonal anti-GFP primary antibody (Abcam, Cambridge, UK; #ab1218).
2. Biotin conjugated secondary antibody (Abcam; #ab6658).
3. Streptavidin–horse radish peroxidase (HRP) (Invitrogen, Paisley, UK; #SNN2004).
4. 3,3′,5,5′-Tetramethylbenzemidine (TMB) (Sigma, Gillingham, UK; #T0440).
5. 2.5 M H_2SO_4.
6. Bicarbonate buffer: 0.8 g Na_2CO_3 + 1.465 g $NaHCO_3$ in 500 ml.
7. Wash buffer: 200 µl Tween-20 in 400 ml phosphate-buffered saline (PBS).
8. Block solution: 4 g bovine serum albumin (BSA) in 400 ml PBS.
9. Microtiter plate.
10. Plate reader.

2.2.2. Direct Immunofluoresence Detection

1. Isofluorane.
2. Heparinized PBS.
3. Stereoscopic fluorescence microscope.
4. Digital microscope camera.
5. Image analysis software.

2.2.3. Immuno-histochemical Detection

1. Histoclear (National Diagnostics, Atlanta, GO, USA; #HS-200).
2. Industrial methylated spirits (IMS).
3. Deionized water.
4. 0.01 M citrate buffer, pH 6.
5. Tris-buffered saline (TBS).
6. 1% H_2O_2 (Sigma; #H1009) made up in TBS.
7. Blocking solution: TBS containing 0.3% Triton X-100 (TBS-T) and 15% normal goat serum (NGS) (Vector Laboratories, Burlingame, USA; #S-1000).
8. Rabbit polyclonal anti-GFP antibody (Abcam, Cambridge, UK; #ab290).
9. Biotinylated goat anti-rabbit IgG (Vector Laboratories; #BA-1000).
10. VECTASTAIN Elite ABC kit (Vector Laboratories; #PK-1000).
11. 0.05% 3,3-Diaminobenzidine (DAB) solution (Sigma; #D8001) containing 0.01% H_2O_2.
12. DPX mounting medium (Sigma; #44581).
13. Microscope.

2.3. Firefly Luciferase (Luc)

2.3.1. Whole-Body Bioimaging

1. Isofluorane.
2. D-Luciferin (Goldbio, St. Louis, USA; #LUCK-1G).
3. Light-tight chamber: specifically a Caliper Life Sciences IVIS machine (see Note 1).
4. Cooled charge-coupled (CCCD) camera (as included in the IVIS machine above).
5. LivingImage (Xenogen) software.

2.3.2. Luminometry of Tissue Lysates

1. Lysis buffer (5×) (from Luciferase Assay System, Promega, Madison, USA; #E1500).
2. Luciferase assay reagent (from Luciferase Assay System).
3. Luminometer tubes.
4. Vortex.
5. Centrifuge.
6. Luminometer.

2.4. β-Galactosidase

2.4.1. β-Gal ELISA (from β-Gal ELISA Kit)

1. β-Gal standards.
2. β-Gal enzyme stock solution.
3. Anti-β-gal-digoxigenin (DIG).
4. Anti-DIG-peroxidase (POD).
5. POD substrate, 2,2′-azino-bis(3-ethylbenzthiazoline-6-sulfonic acid) (ABTS).
6. ABTS substrate solution with enhancer.
7. Washing buffer: containing 10× PBS.
8. Sample buffer: containing PBS and blocking reagents.
9. Anti-β-gal-coated microtitre plate.
10. Plate cover.
11. Plate reader.

2.4.2. X-Gal Staining of Tissues

1. 70% Ethanol.
2. Stain 1: 0.169 g potassium ferrocyanide (Sigma; #P9387), 0.132 g potassium ferricyanide (Sigma; #P8131), 0.01 g $MgCl_2$ made up in 100 ml PBS—protect solution from light and store at room temperature.
3. Stain 2: 40 mg 5-bromo-4-chloro-3-indolyl-β-D-galactopyranoside (X-gal) (Goldbio, St. Louis, USA; #X4281C) in 1 ml DMSO.
4. PBS.
5. 10% Formalin.
6. Shaker.

2.5. Human Factor IX

2.5.1. hFIX Antigen ELISA

1. Capture antibody: polyclonal affinity purified anti-FIX antibody (Affinity Biologicals, Ancaster, USA; #EIA-FIX).

2. Detecting antibody: peroxidase-conjugated polyclonal anti-FIX antibody (Affinity Biologicals; #EIA-FIX).

3. Coating buffer: 1.59 g Na_2CO_3 and 2.93 g $NaHCO_3$ in 1 l distilled water, adjust buffer to pH 9.6, store for up to 1 month at 2–8°C.

4. PBS.

5. hFIX protein standard (Cambridge Biosciences, Cambridge, UK; #HCIX-0040).

6. Wash buffer: 1 ml Tween-20 in 1 l PBS, adjust to pH 7.4.

7. Sample diluent (HBS–BSA–EDTA–T20): 5.95 g HEPES, 1.46 g NaCl, 0.93 g Na_2EDTA, 2.5 g BSA, dissolved in 200 ml H_2O, add 250 μl Tween-20 and adjust solution to pH 7.2—make solution up to a final volume of 250 ml with distilled water, aliquot, and store at –20°C.

8. Substrate buffer: 2.6 g citric acid and 6.9 g Na_2HPO_4—make solution up to a final volume of 500 ml with distilled water, store for up to 1 month at 2–8°C.

9. *o*-Phenylenediamine 2HCl (OPD) substrate: 5 mg OPD in 12 ml substrate buffer, add 12 μl 30% H_2O_2—make fresh for each assay.

10. Stopping solution: 2.5 M H_2SO_4.

11. Microtitre plate.

12. Plate reader.

2.5.2. hFIX Chromogenic Activity Assay

1. Reagent 1: human factor X and FVIII:C (from Biophen Factor IX chromogenic assay kit, Hyphen Biomed, Neuville-sur-Oise, France; #221802).

2. Reagent 2: XIa-thrombin-calcium-phospholipids (from Biophen Factor IX chromogenic assay kit).

3. Reagent 3: chromogenic substrate SXa-11 (from Biophen Factor IX chromogenic assay kit).

4. Reagent 4: Tris–BSA buffer (from Biophen Factor IX chromogenic assay kit).

5. Abnormal plasma (internal control) (Hyphen Biomed; #223301).

6. Normal plasma (standards) (Hyphen Biomed; #223201).

7. Citric acid (2%).

8. Microtitre plate.

9. Plate reader.

2.6. Secreted Alkaline Phosphatase

2.6.1. SEAP Activity Assay (See Note 2)

1. SEAP assay buffer: 1.0 M diethanolamine pH 9.8, 0.5 mM $MgCl_2$, 10 mM L-homoarginine.
2. Diluted *p*-nitrophenylphosphate in SEAP assay buffer (to a final concentration of 120 mM).
3. Microcentrifuge.
4. Microtitre plate.
5. Plate reader.

2.7. Vector Detection

2.7.1. Isolation of DNA from Mammalian Tissues

1. Mammalian Genomic DNA Preparation Kit (Sigma–Aldrich).
2. Water bath or heating block.
3. Scalpel.
4. Ice-cold ethanol (95–100%).
5. Microcentrifuge.

2.7.2. Vector Detection and Quantitation by PCR

1. Quantitative PCR machine.
2. PCR primers.
3. qPCR reagents: SYBR Green Mastermix (Roche), deoxyNTPs (2.5 mM) (NEB).

---

3. Methods

3.1. Total Protein Determination

It is often useful to quantify the total amount of protein present in tissue samples as the amount of transgene protein expression can then be expressed per mg of total protein. This allows the degree of transgene expression in different tissues to be compared directly.

1. Dissect the tissues and freeze in lysis buffer.
2. Defrost and homogenize the tissue with a homogenizing pestle (see Note 3).
3. Spin for 2 min at $13,000 \times g$, keep the supernatant, and remove cell debris.
4. Dilute the supernatant (depending on the concentration of the sample) with final volume >20 μl.
5. Add 10 μl of each protein standard (2, 1.5, 1.0, 0.75, 0.5, 0.25, 0.125, 0 mg/ml) and each sample to a microtitre plate (in duplicate).
6. Mix 50 parts BCA reagent A to 1 part BCA reagent B and add 200 μl of this to each well.
7. Leave at room temperature for 10–15 min.
8. Read on a plate reader at 570 nm.

9. Use the BSA standard samples of known concentration to generate a standard curve.

10. Using the standard curve, determine the total protein concentration in the samples in mg/ml (see Notes 4 and 5).

3.2. Green Fluorescent Protein

GFP has become one of the more popular marker genes in recent years and there have been numerous modifications to increase expression and to achieve fluorescence at different wavelengths. It can be detected readily by ELISA, direct fluorescence, and by immunohistochemical detection of tissue sections and is therefore very useful for endpoint analysis.

3.2.1. GFP ELISA

1. Dilute the primary antibody in bicarbonate buffer (1:10,000 monoclonal anti-GFP).

2. Add 100 µl of the diluted primary antibody per well, cover the plate, and incubate at 4°C overnight.

3. Discard the antibody solution and wash 3× with 300 µl wash buffer per well.

4. Add 300 µl of block solution to each well, cover the plate, and incubate at 37°C for a minimum of 1 h.

5. Discard the block solution and wash 3× with 300 µl wash buffer per well.

6. Add 100 µl of standards to each well (in duplicate). To make up the standards, dilute standard GFP protein to a concentration of approximately 4,000 pg/well. Do serial dilutions from here (300 µl of dilution in 300 µl wash buffer, etc.) down to 5–10 pg/well. Also include a buffer blank sample.

7. Dilute the protein samples so they contain between 2 and 200 µg/well.

8. Add 100 µl of protein sample per well (in duplicate).

9. Cover the plate and incubate at 37°C for 1 h.

10. Discard the standards and samples and wash 3× with 300 µl wash buffer per well.

11. Dilute biotin conjugated secondary antibody in block solution (1:5,000 polyclonal anti-GFP).

12. Add 100 µl of diluted secondary antibody to each well, cover the plate, and incubate at 37°C for 1 h.

13. Discard the diluted secondary antibody and wash 3× with 300 µl wash buffer per well.

14. Dilute the streptavidin–HRP in block solution (1:20,000).

15. Add 100 µl diluted streptavidin–HRP to each well, cover the plate, and incubate at 37°C for 1 h.

16. Discard the diluted streptavidin–HRP solution and wash 3× with 300 µl wash buffer per well.

17. Add 100 μl of peroxidase substrate (TMB) to each well. Incubate for 10 min at room temperature. Avoid placing the plate in direct light.

18. Add 100 μl of 2.5 M H_2SO_4 to each well, in the same order as the other reagents were added, to stop the enzymatic reaction.

19. Read the plate using an ELISA plate reader at 450 nm within 30 min.

20. Use the GFP standard samples of known concentration to generate a standard curve.

21. Using the standard curve, determine the amount of GFP protein in the samples and express the value as pg GFP protein/mg of total protein (see Notes 4 and 5).

3.2.2. Direct Fluorescent Detection

1. Anesthetize the mice using isofluorane.

2. Perform whole-body perfusion using heparinized PBS.

3. Dissect tissues of interest.

4. Place sample under stereoscopic fluorescence microscope for GFP visualization by means of a digital microscope camera.

5. Analyze the images using image analysis software.

3.2.3. Immuno-histochemical Detection

1. Dewax paraffin embedded sections by immersing slide in Histoclear for 5 min. Repeat once.

2. Rinse with 100% industrial methylated spirit for 5 min. Repeat once.

3. Rinse the slides using deionized water.

4. Perform antigen retrieval by boiling sections for 10 min in 0.01 M citrate buffer.

5. Once the sections have cooled, rinse the slides in deionized water and subsequently treat with 1% H_2O_2 made up in TBS for 30 min.

6. Place sections in TBS-T/15% NGS blocking solution for 30 min.

7. Dilute rabbit polyclonal anti-GFP antibodies (1:1,000) in TBS-T/10%NGS.

8. Using a grease pen, draw a perimeter around the tissue and then add rabbit anti-GFP antibodies. Incubate overnight at 4°C.

9. Wash sections in TBS for 5 min. Repeat twice.

10. Dilute goat anti-rabbit IgG (1:1,000) in TBS-T/10% NGS.

11. Incubate sections in diluted goat anti-rabbit IgG for 2 h.

12. Dilute Vectastain avidin–biotin solution (1:200) in TBS. Make this solution up to 30 min prior to use.

13. Rinse sections in TBS for 5 min. Repeat twice.

14. Incubate sections in diluted Vectastain avidin–biotin solution for 2 h.

15. Rinse sections in TBS for 5 min. Repeat twice.

16. Apply 0.05% DAB solution containing 0.01% H_2O_2 to the sections to visualize immunoreactivity.

17. Rinse the sections twice in ice-cold TBS in order to stop the reaction.

18. Dehydrate the sections by submerging in rising concentrations of industrial methylated spirits.

19. Clear the sections by immersing in Histoclear for 20 min. Repeat once.

20. Add DPX mounting medium and overlay section with a coverslip. Allow to dry overnight in a flow hood.

21. Microscopically visualize DAB staining.

3.3. Firefly Luciferase

Luciferase describes a family of light-emitting compounds, however, firefly luciferase has proven to be the most popular as there is little background light emission and the substrate is relatively cheap. It is ideal for longitudinal monitoring of gene expression through whole-body bioimaging and it is also possible to perform endpoint analysis by luminometry of tissue lysates. Unlike with GFP however, immunohistochemical detection in tissue sections and analysis by ELISA are relatively poor.

3.3.1. Whole-Body Bioimaging

1. Anesthetize the mice using isofluorane (see Note 6).

2. Administer D-luciferin through intranasal, intravenous, intraperitoneal, or subcutaneous routes (see Note 7).

3. Place the anesthetized mouse in the light-tight chamber.

4. Take an image 5 min after administration of D-luciferin using a charge-coupled device (CCD) or CCCD camera (see Note 8).

5. Perform data acquisition and analysis with LivingImage (Xenogen) software.

3.3.2. Luminometry of Tissue Lysates

1. The luciferase assay system should be stored at –20°C, and must be equilibrated at room temperature (20–25°C) for 30 min before use.

2. Perfuse the tissues with PBS to remove as much blood as possible as hemoglobin may interfere with the assay.

3. Immerse approximately 1/8 of the liver in 200 μl of 1× reporter lysis buffer.

4. Homogenize tissue using a pestle and mortar or an electronic homogenizer.

5. Centrifuge sample at $13,000 \times g$ for 2 min at room temperature to pellet the cellular debris.

6. Transfer supernatant to clean eppendorf tube.

7. Using a nontransparent 96-well plate, pipette 20 µl supernatant into each well.

8. Add 100 µl luciferase assay reagent.

9. Place plate in the luminometer and initiate reading at the appropriate wavelength depending on the type of luciferase being assayed.

3.4. β-Galactosidase

β-Gal has been used extensively and so detection by both ELISA and by histochemical staining of whole tissues are well established. The motivation for choice of this marker gene would likely be the availability of preexisting constructs.

3.4.1. β-Gal ELISA

1. Determine the total protein content of the samples (see Subheading 3.1).

2. Ensure that all reagents are at 15–25 °C before starting the test.

3. Dilute the protein samples in sample buffer to obtain 2–200 µg protein per well and add 200 µl to anti-β-gal-coated microtitre plate.

4. Dilute the β-gal standards (1.250, 0.625, 0.312, 0.156, 0.078, 0 ng/ml) and add to anti-β-gal-coated microtitre plate.

5. Cover the plate and incubate at 37 °C for 1 h.

6. Discard the standards and samples and wash 3× with 300 µl wash buffer per well.

7. Dilute reconstituted anti-β-gal-DIG (50 µg/ml) in sample buffer to a final concentration of 0.5 µg anti-β-gal-DIG/ml sample buffer.

8. Add 200 µl diluted anti-β-gal-DIG to each well, cover the plate, and incubate at 37 °C for 1 h.

9. Discard the solution and wash 3× with 300 µl wash buffer per well.

10. Dilute reconstituted anti-DIG-POD (20 U/ml) in sample buffer to a final concentration of 150 mU anti-β-gal-DIG/ml sample buffer.

11. Add 200 µl diluted anti-DIG-POD to each well, cover the plate, and incubate at 37 °C for 1 h.

12. Discard the solution and wash 3× with 300 µl wash buffer per well.

13. Add 200 µl substrate with enhancer into each well and incubate at 15–25 °C for 15–40 min (until a green color can be detected by eye).

14. Read the plate using an ELISA plate reader at 405 nm.

15. Use the β-gal standards of known concentration to generate a standard curve.

16. Using the standard curve, determine the amount of β-gal protein in the samples and express the value as pg β-gal protein/mg of total protein (see Notes 4 and 5).

3.4.2. X-Gal Staining of Tissues

1. Dissect the tissues of interest and rinse thoroughly in PBS.

2. Immerse the tissue in 100% ethanol to fix it for at least 1 h (see Note 9).

3. Wash tissue 3× in PBS.

4. Mix stains 1 and 2 and add to samples making sure the tissues are covered.

5. Incubate tissues on a shaking table at room temperature for between 15 min and overnight.

6. Remove staining solution and wash cells 3× in PBS.

7. Place sample under stereoscopic microscope for visualization of the blue precipitate (produced by the reaction of the X-gal substrate with the β-gal enzyme) by means of a digital microscope camera.

8. Analyze the images using image analysis software.

9. For permanent preservation, fix the tissues in formalin.

3.5. Human Factor IX

hFIX is both a physiological endpoint when applying gene therapy to animal models of Hemophilia B or to hemophilic patients and a useful surrogate marker for longitudinal analysis of functional activity of gene therapy constructs. This is most relevant as a simple model for gene expression of a secretory protein using long-term minimally invasive monitoring by repeated bleeds. There are a variety of commercial factor IX antibodies therefore ELISA can be readily performed.

3.5.1. hFIX Antigen ELISA

1. Dilute the capture antibody (1/100) with coating buffer and add 100 μl/well. Incubate at room temperature for 2 h.

2. Discard the solution and wash 3× with 300 μl wash buffer per well.

3. Dilute the hFIX protein standard so the highest standard would be equivalent to ~100% of normal hFIX levels then dilute this 1/50 in sample diluent. Do serial dilutions from here (300 μl of dilution in 300 μl wash buffer, etc.) down to <1% hFIX. Also include a buffer blank sample.

4. Dilute the plasma samples 1/50 in sample diluent (see Note 10).

5. Add 100 μl of standards and samples to wells (in duplicate) and incubate at room temperature for 90 min.

6. Discard the solution and wash 3× with 300 μl wash buffer per well.

7. Dilute the detecting antibody (1/100) in sample diluent.

8. Add 100 μl of diluted detecting antibody to each well and incubate the plate at room temperature for 90 min.

9. Discard the solution and wash 3× with 300 μl wash buffer per well.

10. Add 100 μl freshly prepared OPD substrate per well, allow 5–10 min for enzymatic reaction to occur and for color to develop.

11. Add 50 μl 2.5 M H_2SO_4 to each well to stop the enzymatic reaction.

12. Read the plate using an ELISA plate reader at 490 nm.

13. Use the hFIX standards of known concentration to generate a standard curve.

14. Using the standard curve, determine the amount of hFIX protein in the samples and express the value as % of normal hFIX levels (see Note 4).

3.5.2. hFIX Chromogenic Activity Assay

1. Reconstitute reagents R1, R2, and R3 with 2.5 ml distilled water per vial; reconstitute normal and abnormal plasma with 1 ml distilled water. Allow them to equilibrate at room temperature for 30 min, shaking occasionally.

2. Dilute the normal plasma 1/50 with Tris–BSA buffer (this is the highest standard and is equivalent to ~200% hFIX (see kit insert for precise concentration)). Perform serial dilutions to 1–5% hFIX. Also include a buffer blank sample.

3. Dilute samples (see Note 10) and abnormal plasma (see Note 11) 1/100 with Tris–BSA buffer.

4. Incubate samples, standards and reagents at 37°C for 15 min (see Note 12).

5. Add 50 μl of standards and samples to the plate (in duplicate).

6. Add 50 μl R1 to each well. Incubate at 37°C for 2 min.

7. Add 50 μl R2 to each well. Incubate at 37°C for 3 min.

8. Add 50 μl R3 to each well. Incubate at 37°C for 3 min.

9. Add 50 μl citric acid to each well to stop the reaction.

10. Read the plate using an ELISA plate reader at 405 nm.

11. Use the hFIX standards of known concentration to generate a standard curve.

12. Using the standard curve, determine the amount of hFIX activity in the samples and express the value as % of normal hFIX activity (see Note 4).

3.6. Secreted Alkaline Phosphatase

Human placental secreted alkaline phosphatase (hSEAP) is the most common form of this marker gene however alternative forms, such as murine SEAP, have shown promise, specifically in mice.

3.6.1. SEAP Activity Assay

1. Heat 250 µl plasma at 65°C for 5 min (see Note 13).

2. Clarify medium by centrifuging at $14,000 \times g$ for 2 min.

3. Add 100 µl SEAP assay buffer and 100 µl of clarified medium to each well of the microtiter plate and mix.

4. Warm the samples in the plate to 37°C for 10 min.

5. Pre-warm p-nitrophenylphosphate to 37°C.

6. Add 20 µl pre-warmed 120 mM p-nitrophenylphosphate to each well and incubate at 37°C.

7. Read plate in 1 min intervals at an absorbance of 405 nm using the luminescence plate reader.

8. Use the change in absorbance over time to determine the linear reaction rate.

9. One milliunit of SEAP enzyme is defined as the amount of enzyme required to catalyze the formation of 1 nmol p-nitrophenol per hour.

3.7. Vector Detection

3.7.1. Isolation of DNA from Mammalian Tissues

Mammalian cells and tissues are lysed with a chaotropic salt-containing buffer to ensure denaturation of macromolecules. DNA is bound to the spin column membrane and the remaining lysate is removed by centrifugation. A filtration column is used to remove cell debris. After washing to remove contaminants, the DNA is eluted with buffer into a collection tube.

Before beginning the procedure, do the following:

1. Preheat a water bath or shaking water bath to 55°C.

2. Thoroughly mix reagents (any precipitate can be dissolved at 55–65°C).

3. Dilute wash solution concentrate with ethanol (95–100%).

4. Dissolve Proteinase K in water to obtain a 20-mg/ml stock solution.

DNA isolation using the GenElut Kit (Sigma–Aldrich)

1. Cut and weigh approximately 25 mg of tissue and mince with a scalpel.

2. Add lysis solution T followed by the Proteinase K solution to the tissue. Mix by vortexing. Incubate the sample at 55°C until the tissue is completely digested and no particles remain. Vortex occasionally or use a shaking water bath. Digestion is usually complete in 2–4 h. After digestion is complete, vortex briefly.

3. Add RNase A solution and incubate for 2 min at room temperature.

4. Add lysis solution C to the sample and vortex thoroughly. Incubate at 70°C for 10 min.

5. Add column preparation solution to the GenElute Binding Column and centrifuge at $12,000 \times g$ for 1 min. Discard flow-through liquid.

6. Add 200 ml of ethanol (95–100%) to the lysate; mix thoroughly by vortexing 5–10 s. A homogeneous solution is essential.

7. Transfer the contents of the tube into the treated binding column from step 4. Centrifuge at $6,500 \times g$ for 1 min. Discard the collection tube containing the flow through liquid and place the binding column in a new 2 ml collection tube.

8. Add 500 ml of wash solution to the binding column and centrifuge for 1 min at $6,500 \times g$. Discard the collection tube containing the flow-through liquid and place the binding column in a new 2 ml collection tube.

9. Repeat the wash and centrifuge for 3 min at maximum speed ($12,000–16,000 \times g$) to dry the binding column. The binding column must be free of ethanol before eluting the DNA. Centrifuge the column for one additional minute at maximum speed if residual ethanol is seen. Finally, discard the collection tube containing the flow through liquid and place the binding column in a new 2 ml collection tube.

10. Pipette the elution solution directly into the center of the binding column, centrifuge for 1 min at $6,500 \times g$ to elute the DNA.

3.7.2. Vector Detection and Quantitation by PCR

Real time qPCR detects quantity of nucleotide sequence target by using specific oligonucleotides homologous to the target sequence along with the incorporation of fluorescent SYBR Green label to the double stranded product. SYBR Green (such as Power SYBR Green Master Mix—Applied Biosystems; P/N 4367659) is a dye that can bind to the minor groove of dsDNA, and become highly fluorescent on excitation. In general terms, the reaction mix (SYBR Green Master Mix, forward primer, reverse primer and PCR grade water is prepared in a MicroAmp Optical 96-well Reaction Plate (AB; P/N N801-0560)) with each reaction prepared in triplicate. The plates are sealed with Optical Adhesive Covers (AB; P/N 4360954) and the qPCR reaction run on a real-time qPCR machine (such as a Sequence Detection 7500 system—Applied Biosystems) following the preprogrammed amplification parameters. Data output is expressed as an algorithm output (ddCt value) which indicates the amplification threshold for each reaction. These can either be expressed in comparison to a known and stable control or quantified in comparison to amplification of a series of standards.

SYBR green kits and methodologies vary with respect to the real-time PCR machine used (Applied Biosystems 7500, Roche LightCycler, Stratagene Mx3000), the final quantitation process (amplification comparisons to a quantified standard curve or to a housekeeping gene such as GAPDH), and the primers used, therefore, the authors advise to follow manufacturer's instructions.

4. Notes

1. There are several whole-body bioluminescence imaging machines available from different companies, however, we prefer the IVIS machines from Caliper Life Sciences which have been extensively calibrated to permit comparison between machines.

2. There is both a chromagenic and a chemiluminescent assay available for SEAP; we have chosen the former as a standard spectrophotometer which is likely to be available in a wider range of labs. However, it is useful to bear in mind that the chemiluminescent assay is the most sensitive.

3. An electronic homogenizer can be used if the tissue samples size and lysis buffer volume are >500 μl. Take care to clean the homogenizer carefully between samples.

4. All sample values must fall within the limits of the standard curve (if samples lie outside the standard curve limits, they must be diluted and the assay repeated).

5. Variable background is detected in some tissues; therefore, controls with un-transfected tissues of the type being analyzed should always be included.

6. We have observed that prolonged anesthesia can reduce the light emission and there have been some publications describing this phenomenon (16).

7. Although intraperitoneal injection has been the most widely used route of injection, there is evidence that other routes are more suitable. Particularly for general application, the subcutaneous route is believed to result in more stable and prolonged light emission than other routes. In addition, we have observed that intranasal delivery is best for genes expressed within the lung (9).

8. Ideally, the kinetics of light emission and the concentration of luciferin administered to the animal should be determined for each experiment; this is likely a cost–benefit analysis as higher concentrations of luciferin are likely to give greater light output but may be prohibitively expensive.

9. We observe no problems with fixing overnight in ethanol at 4°C. Some protocols describe fixation with formalin, paraformaldehyde, or gluteraldehyde but we did not find them any better for tissue fixation.

10. Plasma samples should be used fresh or immediately after defrosting but should not be freeze/thawed more than once.

11. Abnormal plasma is pooled human plasma at a known but less than physiological concentration (see kit insert for precise concentration) so it can be used as an internal control for the assay if treated the same as a sample.

12. The plate should be placed in a water bath for the assay.

13. This is to eliminate background alkaline phosphatase activity.

References

1. Eglitis MA, Kantoff P, Gilboa E et al (1985) Gene expression in mice after high efficiency retroviral-mediated gene transfer. Science 230:1395–1398

2. Morral N, O'Neal W, Zhou H et al (1997) Immune responses to reporter proteins and high viral dose limit duration of expression with adenoviral vectors: comparison of E2a wild type and E2a deleted vectors. Hum Gene Ther 8:1275–1286

3. Berger J, Hauber J, Hauber R et al (1988) Secreted placental alkaline phosphatase: a powerful new quantitative indicator of gene expression in eukaryotic cells. Gene 66:1–10

4. Christou C, Parks RJ (2011) Rational design of murine secreted alkaline phosphatase for enhanced performance as a reporter gene in mouse gene therapy preclinical studies. Hum Gene Ther 22:499–506

5. Naylor LH (1999) Reporter gene technology: the future looks bright. Biochem Pharmacol 58:749–757

6. Waddington SN, Nivsarkar M, Mistry A et al (2004) Permanent phenotypic correction of Haemophilia B in immunocompetent mice by prenatal gene therapy. Blood 104:2714–2721

7. Misteli T, Spector DL (1997) Applications of the green fluorescent protein in cell biology and biotechnology. Nat Biotechnol 15:961–964

8. Wu JC, Sundaresan G, Iyer M et al (2001) Noninvasive optical imaging of firefly luciferase reporter gene expression in skeletal muscles of living mice. Mol Ther 4:297–306

9. Buckley SMK, Howe SJ, Rahim AA et al (2008) Luciferin detection after intra-nasal vector delivery is improved by intra-nasal rather than intra-peritoneal luciferin administration. Hum Gene Ther 19:1050–1056

10. Branchini BR, Ablamsky DM, Davis AL et al (2010) Red-emitting luciferases for bioluminescence reporter and imaging applications. Anal Biochem 396:290–297

11. Branchini BR, Ablamsky DM, Murtiashaw MH, et al (2006) Thermostable red and green light-producing firefly luciferase mutants for bioluminescent reporter applications. Anal Biochem 22:499–506

12. Liu HS, Jan MS, Chou CK et al (1999) Is green fluorescent protein toxic to the living cells? Biochem Biophys Res Commun 260:712–717

13. Bell P, Vandenberghe LH, Wu D et al (2007) A comparative analysis of novel fluorescent proteins as reporters for gene transfer studies. J Histochem Cytochem 55:931–939

14. Mian A, McCormack WM Jr, Mane V et al (2004) Long-term correction of ornithine transcarbamylase deficiency by WPRE-mediated overexpression using a helper-dependent adenovirus. Mol Ther 10:492–499

15. Enquist IB, Bianco CL, Ooka A et al (2007) Murine models of acute neuronopathic Gaucher disease. Proc Natl Acad Sci U S A 104:17483–17488

16. Franks NP, Jenkins A, Conti E et al (1998) Structural basis for the inhibition of firefly luciferase by a general anesthetic. Biophys J 75:2205–2211

Chapter 14

Monitoring for Potential Adverse Effects of Prenatal Gene Therapy: Use of Large Animal Models with Relevance to Human Application

Vedanta Mehta, Khalil N. Abi-Nader, David Carr, Jacqueline Wallace, Charles Coutelle, Simon N. Waddington, Donald Peebles, and Anna L. David

Abstract

Safety is an absolute prerequisite for introducing any new therapy, and the need to monitor the consequences of administration of both vector and transgene to the fetus is particularly important. The unique features of fetal development that make it an attractive target for gene therapy, such as its immature immune system and rapidly dividing populations of stem cells, also mean that small perturbations in pregnancy can have significant short- and long-term consequences. Certain features of the viral vectors used, the product of the delivered gene, and sometimes the invasive techniques necessary to deliver the construct to the fetus in utero have the potential to do harm.

An important goal of prenatal gene therapy research is to develop clinically relevant techniques that could be applied to cure or ameliorate human disease in utero on large animal models such as sheep or nonhuman primates. Equally important is the use of these models to monitor for potential adverse effects of such interventions. These large animal models provide good representation of individual patient-based investigations. However, analyses that require defined genetic backgrounds, high throughput, defined variability and statistical analyses, e.g. for initial studies on teratogenic and oncogenic effects, are best performed on larger groups of small animals, in particular mice.

This chapter gives an overview of the potential adverse effects in relation to prenatal gene therapy and describes the techniques that can be used experimentally in a large animal model to monitor the potential adverse consequences of prenatal gene therapy, with relevance to clinical application. The sheep model is particularly useful to allow serial monitoring of fetal growth and well-being after delivery of prenatal gene therapy. It is also amenable to serially sampling using minimally invasive and clinically relevant techniques such as ultrasound-guided blood sampling. For more invasive long-term monitoring, we describe telemetric techniques to measure the haemodynamics of the mother or fetus, for example, that interferes minimally with normal animal behaviour. Implanted catheters can also be used for serial fetal blood sampling during gestation. Finally, we describe methods to monitor events around birth and long-term neonatal follow-up that are important when considering human translation of this therapy.

Key words: Ultrasound, Doppler, Fetal growth velocity, Amniocentesis, Fetal blood sampling, Blood flow, Blood pressure, Heart rate, Neonate, Miscarriage, Liver biopsy, Bone marrow biopsy

Charles Coutelle and Simon N. Waddington (eds.), *Prenatal Gene Therapy: Concepts, Methods, and Protocols*, Methods in Molecular Biology, vol. 891, DOI 10.1007/978-1-61779-873-3_14, © Springer Science+Business Media, LLC 2012

1. Introduction

1.1. Potential Adverse Consequences of Fetal Gene Therapy

One of the most important goals of the extensive experimentation described in this book is to define clinically relevant techniques that could be applied to cure or ameliorate human disease. Safety is obviously a necessary prerequisite for introducing any new therapy and therefore the need to monitor the consequences of administration of both vector and transgene to the fetus is particularly important. The unique features of fetal development that make it an attractive target for gene therapy, such as its immature immune system and rapidly dividing populations of stem cells, also mean that small perturbations in pregnancy can have significant short- and long-term consequences. The potential to do harm is heightened further by features of the viral vectors used, the product of the delivered gene, and sometimes the invasive techniques necessary to deliver the construct to the fetus in utero. Some of the potential hazards are detailed below, evidenced from human observations and experimental fetal physiology.

Fetal access. The majority of studies of fetal gene therapy rely on injecting the fetus or amniotic cavity through the uterine and sometimes the maternal abdominal wall. There is a significant experience of performing these procedures in humans; the most important associated risk is of inducing a miscarriage (between 0.5 and 5% of procedures depending on the gauge of the needle used and the complexity of the procedure (1–3)) and more rarely of causing fetal death in utero. Interestingly, there are few reports of long-term fetal damage with these procedures, partly because of the accuracy of insertion using ultrasound and also because of the well-described potential for fetal healing.

Normal fetal growth and development results from a complex interaction between the genetic blueprint and the environment; of particular importance is a constant, balanced supply of nutrition provided by a healthy mother with a functional placenta, and a well-developed fetoplacental circulation. Viral vector-mediated gene delivery to mother or fetus could interfere with any stage of this sequence of events. Congenital viral infections such as rubella or cytomegalovirus are associated with fetal growth restriction and a number of developmental abnormalities such as sensorineural hearing loss, visual impairment, and cerebral palsy (4, 5). Early studies also commonly detected adenovirus DNA in the amniotic fluid of women with fetal growth restriction or a fetal abnormality, but more recent larger and more rigorous studies have found no increased risk of complications in pregnancies infected with adenovirus (6).

Congenital infection with some viruses can cause fetal growth restriction. Whilst this is thought to be highly unlikely using vectors that are replication deficient, it is important to be sure that delivery resulting in widespread tissue transfection does not cause even minor perturbations of fetal growth. The sensitivity of the system, particularly to interventions at specific "time windows" of development is illustrated by studies in fetal sheep, showing that maternal administration of steroids can lead to either lung maturation or long-term alteration of blood pressure, depending on the gestation at which it was given (7, 8). Small variations in fetal birthweight are associated with lifelong risk of cardiovascular disease and type 2 diabetes (9). Although fetal gene therapy interventions are yet to be tested in clinical practice, these data highlight their potential to influence development and the need for long-term follow-up.

Delivery of vector/gene product into either the maternal or fetal circulation could have a detrimental effect on placental function and blood supply, either of which could reduce fetal growth. For instance, the use of sildenafil, a potent vasodilator, to increase uteroplacental blood flow and improve fetal growth, actually led to fetal growth restriction and fetal death as a result of widespread vasodilation and maternal hypotension (10). Current studies on determining whether over-expression of VEGF in the uteroplacental circulation improves placental perfusion and fetal growth provide a good example of an intervention where either the vector (adenovirus) or the transgene product (VEGF) could have a detrimental effect on maternal or fetal blood flow (11). Alternatively, transfection of the syncytiotrophoblast or underlying cytotrophoblast cells might adversely affect the function of active transporter mechanisms for amino acids and lipids, leading to growth restriction.

Vector spread. Although vector leakage with transduction and gene expression in tissues other than those desired is a risk associated with most gene delivery systems, there are several aspects that are of particular relevance to pregnancy. One of these risks, the inadvertent transduction of the male germline has been assessed in sheep, mice, and monkeys. Following intraperitoneal injection of retroviral vector, PCR on purified sperm cells from the rams postnatally, and immunohistochemistry on testis, both revealed very low numbers of transduced germ cells. The authors estimated a transduction frequency of 1 in 6,250 germ cells and noted that this was several orders of magnitude below the calculated frequency of naturally occurring endogenous insertions and below the upper tolerable limit set by the US Federal Drug Administration (12).

The second risk, unique to pregnancy is that of transplacental passage of vector, leading to unwanted transfection in either mother or fetus. Evidence from large and small animal studies of fetal gene transfer suggests that there is only low level spread of vector to the mother after fetal injection. Spread of the vector to the fetus after delivery to the mother is probably limited by the placenta. In the pregnant rabbit, uterine artery infusion of adenovirus, plasmid/PEI or plasmid/liposome formulations efficiently transduced the trophoblast cells, but gene transfer to the fetal tissues was only detectable by sensitive PCR methods (13). The ability of adenoviruses to infect human trophoblast cells in vitro is related to the state of trophoblast differentiation (14, 15). Recombinant adenoviruses efficiently transduce the inner cytotrophoblast of the human placenta that is present on the fetal side throughout gestation, but there is a significant reduction in the transduction efficiency of these vectors after the terminal differentiation of the mononucleated cytotrophoblast cells into the multinucleated syncytiotrophoblast. This is probably due in part to the lack of coxsackie adenovirus receptor (CAR) expression in the syncytiotrophoblast (16), thus rendering it resistant to maternal adenovirus infection and limiting transplacental transmission in that direction.

Vector toxicity. Many viral vectors are toxic at high concentrations, mainly through their immunogenic properties. This can manifest as fetal death or tissue damage. The effects can be highly species specific, e.g. VSVG pseudotyped lentivirus, used successfully in mouse studies, caused fetal ascites and death, even at low doses when given to the early gestation fetal sheep (unpublished data, authors lab). More subtle signs of toxicity can be detected; for instance, VSV-G pseudotyped HIV delivering GFP cDNA to the fetal murine cochlear resulted in good expression in inner and outer hair cells of the cochlea several weeks after injection, but with evidence of mild hearing loss (17). Data showing that even transient fetal exposure to pro-inflammatory cytokines has deleterious effects on pulmonary and neurological development highlight the need for detailed histological and functional follow-up after experimental fetal exposure to viral vectors. For instance, some cases of autism or schizophrenia are thought to be caused by fetal exposure to maternally acquired viruses, such as influenza (18).

Insertional mutagenesis. It has been suggested that the high level of cellular proliferation in the fetus, the abundance of growth factors, and the transcriptionally active state of genes associated with the regulation of growth and differentiation may predispose the organism to increased risk of cancer from vectors which integrate into the host genome. Themis and colleagues demonstrated a high incidence of hepatocellular carcinoma after delivery of lacZ or human factor IX cDNA using an Equine Infectious Anaemia Virus (EIAV) vector, but not after use of an HIV vector (19). This suggests that the fetus may be uniquely sensitive to genetic perturbations arising from gene transfer, as EIAV has been used in adult animals for transduction of the nervous system with no adverse events reported (20). This issue is discussed in more detail in Chapter 16.

1.2. Benefits of Large Animals for Monitoring Adverse Effects

1. *Use of relevant delivery techniques.* The size of the two most commonly used experimental large animals, the sheep and the non-human primate, makes it possible to use techniques to deliver the vector that are identical to those in current use in human fetal medicine. Ultrasound imaging can be used to guide needle insertion into a variety of organs or vessels including muscle, liver, brain, lung, intraperitoneal, intratracheal, and intravascular (normally the intrahepatic section of the umbilical vein) (21). As described above, each of these procedures is associated with the potential for fetal death in utero or miscarriage.

2. *Fetal monitoring.* The ovine or non-human primate fetus is easily visualised from early gestation using ultrasound. Major technological improvements over the last 20 years have made ultrasonography a powerful method for monitoring fetal growth and development; it is the predominant method used in a clinical setting. Imaging can provide detailed information about structural changes, e.g. if the fetus develops ascites as a result of intraperitoneal administration of vector or bleeding. The fetus can also be measured (biometry); typically, this would include the head circumference, abdominal circumference, and femur length. Normal values are available for human and sheep fetuses. Finally, using Doppler ultrasonography, it is possible to measure blood flow velocity; the vessels most commonly investigated are the uterine, umbilical and middle cerebral arteries, and the ductus venosus. Analysis of the Doppler waveforms provides a direct measurement of downstream vascular resistance. These data can be informative in a number of ways: vascular resistance could increase as a result of vascular obstruction/constriction (e.g. in the umbilical artery as a result of poor fetoplacental circulation), alternatively resistance can fall as a result of vasodilation (e.g. as a result of vector mediated over production of VEGF in the uterine circulation or hypoxia-related dilation of fetal cerebral vessels).

   The well-described ability of the sheep myometrium to tolerate exteriorised fetal catheters without contractions and preterm birth means that it is possible to obtain an extensive, in vivo assessment of many aspects of fetal physiology over long periods of time (e.g. for 8 weeks). Most commonly performed monitoring procedures are placing catheters into the fetal carotid or femoral arteries to measure fetal blood pressure and heart rate (either telemetrically or using pressure transducers connected directly to the catheter) and/or to take samples of fetal blood.

   Finally, it is possible to measure vector presence, concentrations of product, antibody responses or markers of organ damage in blood or fluid aspirated from a needle inserted under ultrasound guidance.

3. *Length of gestation.* The longer gestation of large animals (145 days in sheep compared with 21 days in rats and mice) means that it is possible to monitor the in utero consequences of an earlier intervention for significant periods of time before birth. For example, we have injected vector into the peritoneal cavity as early as 50 days of gestation in the fetal sheep, allowing monitoring for adverse reactions to continue for at least 3 months before birth.

4. *Organ development.* A multitude of studies using the sheep fetus have provided the foundation for the current understanding of fetal physiology and our appreciation of how it differs from that of the adult. To a degree this is because the time course and sequence of organ maturation in the fetal sheep can be correlated with similar events in the human, e.g. lung maturation, myelination of the brain. A similar argument can be made for the developmental processes at birth and in the immediate postnatal period.

5. *Postnatal monitoring.* The size and relatively slow development of the offspring mean that it is possible to perform detailed monitoring following birth. Apart from routine blood sampling, studies have used a range of techniques including liver biopsy, bone marrow aspiration, and electro-ejaculation to study germline transmission. In addition, sampling of a wide range of tissues is possible post-mortem enabling detailed histological assessment and molecular analysis to determine vector spread and toxicity as well as gene expression.

1.3. Limitations to the Use of Large Animals

There are a number of factors that limit the degree to which data from large animal models can be extrapolated to the clinical situation, raising the issue that at some stage in the development of human fetal gene therapy it will be necessary to determine efficacy and safety in the human. To some extent, the use of non-human primates circumvents many of these problems, but raises ethical and financial issues.

Different placentation. The sheep has a different placental structure to the human (epithelio-chorial vs haemochorial) meaning that studies assessing transplacental vector spread should be interpreted with caution. This is described further in Chapter 9.

Different parturition. One of the pregnancy complications that most concerns clinicians is preterm birth; this could be a potential consequence of either invasive procedures or inflammation. However, the events leading to parturition in sheep are very different from those in the human; the tolerance of ovine myometrium to handling and surgery has been referred to. See also Chapters 9 and 11.

Long gestation. The long gestation and limited number of offspring compared with rodent pregnancy make large animals less suitable for monitoring transgenerational consequences of fetal gene therapy. In addition, the same factors mean that the sheep in particular is rarely used for genetics studies; this would make mice a more attractive species in which to study insertional mutagenesis (see Chapter 16).

Cost. In general, large animal studies are expensive and labour intensive, compared to smaller animals such as mice, which limits group numbers as well as the number of studies that include long-term follow-up.

The described methodology in this chapter is based on our sheep experiments; however, despite highlighted limitation, it is likely that most of these techniques can be transferred with only little modification to non-human primates and eventually, selectively, also to humans.

2. Materials

2.1. Monitoring Fetal Growth and Well-being by Ultrasound

1. Appropriate ultrasound machine (e.g. Logiq 9, GE Medical Systems, Milwaukee, USA) with real-time image display, cine loop facility, B-mode, pulsed wave and colour flow Doppler modes and preferably image storage function.

2. 3.0–5.0 MHz curvilinear abdominal probe.

3. Clippers (e.g. Masterclip Duo, Outlandish Items Ltd., Leicestershire, UK).

4. Cleaning solution, made up per litre using 750 ml 100% ethanol (Sigma-Aldrich), 150 ml deionised water, and 100 ml 5% chlorhexidine gluconate (Hibitane®, Zeneca Ltd., Macclesfield, Cheshire, UK).

5. Ultrasound coupling gel (e.g. Electro Medical Supplies Ltd., Oxon, UK).

6. Record sheets including three data-entry boxes for each ultrasound parameter to be measured.

2.2. Ultrasound-Guided Fetal Sampling

Refer to Chapter 11, Subheading 3.4 for Ultrasound-guided procedures, Intra-amniotic and Umbilical vein injection.

2.2.1. Amniocentesis

1. 22G, 9-cm long echotip needles with a lancet tip (Cook Medical, UK).

2. 2-ml syringe.

2.2.2. Fetal Blood Sampling

1. 18 or 20G, 9-cm long echotip needle with a lancet tip (Cook Medical, UK).

2. 1-ml syringe, 2-ml syringe.

3. Blood bottles (see Subheading 2.4.7).

2.3. Invasive Maternal and Fetal Monitoring Using Telemetry or Catheters to Allow Serial Blood Sampling

2.3.1. Long-Term Measurement of Maternal Blood Pressure and Heart Rate Using Telemetric Catheters

1. TA11 PA-D70 monitoring device (Data Sciences International, USA)—Blood pressure and heart rate monitoring device for large animals.

2. AM/FM Radio.

3. Magnet (Data Sciences International, USA).

4. Metallic receivers RMC-1 (Data Sciences International, USA).

5. Ambient pressure reference-1 (Data Sciences International, USA).

6. Data Exchange Matrix (Data Sciences International, USA).

7. Computer to receive and process data. Currently requires Windows XP professional or Windows Vista Business Edition, Core 2 Duo E6750/2.66 GHz processor, 4 GB of RAM, minimum 250 GB of hard drive storage, 256 MB ATI Radeon HD 2400 XT Graphics, Dual DVI or VGA (or equivalent) video card.

8. Dataquest ART Software (Data Sciences International, USA).

9. No. 11 blade (Swann-Morton, Sheffield, UK).

10. 1-0 silk ties (Ethicon, New Jersey USA), 2-0 Vicryl (Johnson & Johnson Intl., St. Stevens-Woluwe, Belgium).

11. Normal saline solution 0.9%.

12. Terramycin aerosol spray (3.92% cutaneous spray containing oxytetracycline hydrochloride) (Pfizer, Kent).

13. Povidone iodine antiseptic cleaning solution (1% w/w Iodine solution, Vetasept Animal Care Ltd., York, UK).

14. Clippers for fleece (Masterclip Duo, Outlandish Items Ltd., Leicestershire, UK).

2.3.2. Long-Term Measurement of Uterine Artery Blood Flow Using Telemetric Flow Probes

1. Transit time flow probe (Transonic System Inc., NY, USA).

2. Physiometer, battery pack and rechargeable batteries (Transonic System Inc., NY, USA).

3. Physioview Software (Transonic Systems Inc., NY, USA).

4. No. 11 blade (Swann-Morton, Sheffield, UK).

5. Diathermy and diathermy pad (ERBE VIO 300 D Electro-surgical Generator (Tübingen, Germany).

6. Trocar and cannula, length of trocar 20 cm, outer diameter of cannula 1.2 cm, inner diameter of cannula 1.0 cm (Medema Limited, London, UK).

7. Vetguard Cold sterilant (Steritech, Cheltenham, UK).

8. Lap sponges (National Veterinary Services, UK).

9. Penstrep (200 mg/ml procaine penicillin and 250 mg/ml dihydrostreptomycin; Norbrook Laboratories Ltd., County Down, UK), Crystapen (Sodium benzylpenicillin G,

Schering-Plough, Uxbridge, UK), and Gentamicin (Genticin Injectable, Roche Products Ltd., Hertfordshire, UK).

10. 4-0 Prolene suture (Ethicon, Edinburgh, UK), 6 mm nylon tape (Johnson and Johnson Intl., St. Stevens-Woluwe, Belgium), 1-0 PDS (Johnson and Johnson Intl., St. Stevens-Woluwe, Belgium) and 1-0 Vicryl (Ethicon, Norderstedt, Germany).

11. Terramycin aerosol spray (3.92% cutaneous spray containing oxytetracycline hydrochloride) (Pfizer, Kent).

12. Vetergesic (Buprenorphine, Alstoe Animal Health, York, UK).

13. Povidone (1% w/w Iodine solution, Vetasept Animal Care Ltd., York, UK).

14. "Hibiscrub" (Chlorhexidine gluconate 4% w/v, Regent Medical, Manchester, UK).

15. Clippers for fleece (Masterclip Duo, Outlandish Items Ltd., Leicestershire, UK).

16. Computer to acquire and analyse data, currently requires Windows XP or Windows 7 that includes XP, free USB port, Core 2 Duo E6750/2.66 GHz processor, minimum 250 GB of hard drive storage.

2.3.3. Long-Term Measurement of Fetal Blood Pressure and Heart Rate

1. D70-PCTP monitoring device (Data Sciences International, USA) for large animals attached to two haemodynamic catheters.

2. AM/FM Radio.

3. Magnet (Data Sciences International, USA).

4. Metallic receivers RMC-1 (Data Sciences International, USA).

5. Ambient pressure reference-1 (Data Sciences International, USA).

6. Data Exchange Matrix (Data Sciences International, USA).

7. Computer to receive and process data. Currently requires Windows XP professional or Windows Vista Business Edition, Core 2 Duo E6750/2.66 GHz processor, 4 GB of RAM, minimum 250 GB of hard drive storage, 256 MB ATI Radeon HD 2400 XT Graphics, Dual DVI or VGA (or equivalent) video card.

8. No. 21 blade (Swann-Morton, Sheffield, UK).

9. 3-0 silk ties (Ethicon, New Jersey USA), 2-0 Vicryl (Johnson & Johnson Intl., St. Stevens-Woluwe, Belgium), 2-0 PDS (Ethicon, St. Stevens-Woluwe, Belgium), and 3-0 Prolene (Ethicon, Norderstedt, Germany), 6 mm nylon tape (Johnson and Johnson Intl., St. Stevens-Woluwe, Belgium), 1-0 PDS (Johnson and Johnson Intl., St. Stevens-Woluwe, Belgium), and 1-0 Vicryl (Ethicon, Norderstedt, Germany).

10. Babcock and small Kelly surgical clamps.

11. Fine-tipped scissors and forceps.

12. 3 ml Penstrep (200 mg/ml procaine penicillin and 250 mg/ml dihydrostreptomycin; Norbrook Laboratories Ltd., County Down, UK), benzylpenicillin G 3 g (Crystapen, Schering-Plough, Uxbridge, UK), and gentamicin 80 mg (Genticin Injectable, Roche Products Ltd., Hertfordshire, UK).

13. Normal saline solution 0.9%.

14. Lap sponges (National Veterinary Services, UK).

15. Terramycin aerosol spray (3.92% cutaneous spray containing oxytetracycline hydrochloride) (Pfizer, Kent).

16. Povidone iodine antiseptic cleaning solution (1% w/w Iodine solution, Vetasept Animal Care Ltd., York, UK).

17. "Hibiscrub" (Chlorhexidine gluconate 4% w/v, Regent Medical, Manchester, UK).

18. Clippers for fleece (Masterclip Duo, Outlandish Items Ltd., Leicestershire, UK).

19. Diathermy and diathermy pad (ERBE VIO 300 D Electro-surgical Generator (Tübingen, Germany).

2.3.4. Catheters to Allow Serial Sampling

1. Povidone (1% w/w Iodine solution, Vetasept Animal Care Ltd., York, UK).

2. "Hibiscrub" (Chlorhexidine gluconate 4% w/v, Regent Medical, Manchester, UK).

3. Clippers for fleece (Masterclip Duo, Outlandish Items Ltd., Leicestershire, UK).

4. No. 11 blade and no. 21 blade (Swann-Morton, Sheffield, UK).

5. Diathermy and diathermy pad (ERBE VIO 300 D Electro-surgical Generator, Tübingen, Germany).

6. Trocar and cannula, length of trocar 20 cm, outer diameter of cannula 1.2 cm, inner diameter of cannula 1.0 cm (Medema Ltd., London, UK).

7. Vetguard Cold sterilant (Steritech, Cheltenham, UK).

8. Lap sponges (National Veterinary Services, UK).

9. Penstrep (200 mg/ml procaine penicillin and 250 mg/ml dihydrostreptomycin; Norbrook Laboratories Ltd., County Down, UK), Crystapen (Sodium benzylpenicillin G, Schering-Plough, Uxbridge, UK) and Gentamicin (Genticin Injectable, Roche Products Ltd., Hertfordshire, UK).

10. 3-0 silk ties (Ethicon, New Jersey USA), 2-0 Vicryl (Johnson & Johnson Intl., St. Stevens-Woluwe, Belgium), 2-0 PDS (Ethicon, St. Stevens-Woluwe, Belgium), and 3-0 Prolene (Ethicon, Norderstedt, Germany), 6 mm nylon tape (Johnson and Johnson Intl., St. Stevens-Woluwe, Belgium), 1-0 PDS

(Johnson and Johnson Intl., St. Stevens-Woluwe, Belgium), and 1-0 Vicryl (Ethicon, Norderstedt, Germany).

11. Babcock and small Kelly surgical clamps, fine-tipped scissors and forceps.

12. Polyvinyl tubing. Internal diameter (ID) 0.86 mm; outer diameter (OD) 1.52 mm (Critchly Electrical Products, NSW, Australia).

13. Terramycin aerosol spray (3.92% cutaneous spray containing oxytetracycline hydrochloride) (Pfizer, Kent).

14. Vetergesic (Buprenorphine, Alstoe Animal Health, York, UK).

15. Heparinised saline for flushing line (Hepsal, Heparin Sodium 10 I.U./ml, Wockhardt UK Ltd., Wrexham, UK).

2.4. Neonatal Monitoring and Follow-up

2.4.1. Neonatal Monitoring in the Immediate Postnatal Period

1. Stopwatch that reads in seconds.

2. Rectal thermometer (e.g. Becton Dickinson, Plymouth, Devon, UK).

3. Scales to measure up to 40 kg to the nearest gram (e.g. D-7470, Sauter, Ebingen, Germany).

4. Oxytocin (Intervet UK Ltd., Schering-Plough Animal Health, Milton Keynes, Bucks, UK).

5. 500 ml measuring jug (plastic or glass).

6. Standard baby bottles.

7. Steam steriliser (Lindam Ltd., Harrogate, North Yorkshire, UK).

8. Lamb feeding tubes (William Daniels UK Ltd., Withernsea, East Yorkshire, UK).

9. Enrofloxacin (Baytril®, Bayer Ltd., Newbury, Berks, UK).

10. Vitamin E/Selenium (Vitesel®, Norbrook Laboratories, Newry, Co. Down, Northern Ireland).

11. Piece of string.

12. Metre stick or long ruler.

13. Horse height measuring stick with an extending stem, folding arm, and spirit level (http://www.brighteyesandbobtails.co.uk/acatalog/Measuring_Tapes_and_Sticks.html).

2.4.2. Liver Biopsy in Lambs or Sheep

1. Povidone iodine antiseptic cleaning solution (Grampian Pharmaceuticals Ltd., UK).

2. Procaine hydrochloride (Arnolds Veterinary Products Ltd., UK).

3. A no.11 blade, (Swann-Morton, Sheffield, UK).

4. Veterinary Quick-Core Biopsy Needle (Cook, Australia).

5. 2-0 vicryl (Johnson & Johnson Intl., St. Stevens-Woluwe, Belgium) for skin sutures.

6. Terramycin aerosol spray (3.92% cutaneous spray containing oxytetracycline hydrochloride) (Pfizer, Kent).

7. Amoxycillin trihydrate (Norbrook Laboratories Ltd., Newry, Northern Ireland).

8. Clippers for fleece (Masterclip Duo, Outlandish Items Ltd., Leicestershire, UK).

2.4.3. Bone Marrow Biopsy in Lambs or Sheep

1. Povidone iodine antiseptic cleaning solution (Grampian Pharmaceuticals Ltd., UK).

2. Lidocaine chloride (Hameln Pharmaceuticals, Gloucester, UK).

3. Veterinary Quick-Core Biopsy Needle (Cook, Australia).

4. Terramycin aerosol spray (3.92% cutaneous spray containing oxytetracycline hydrochloride) (Pfizer, Kent).

5. Clippers for fleece (Masterclip Duo, Outlandish Items Ltd., Leicestershire, UK).

2.4.4. Immune Response to Challenge with Transgenic Protein

1. Freund's complete adjuvant (Sigma-Aldrich Chemie Gmbh, Steinheim, Germany) stored at 4°C.

2. Transgenic protein, e.g. hFIX protein, 1 mg, Replenine-VF 1000, Bio Products Laboratory, Elstree, UK.

3. 10-ml syringe and 19-gauge needle to collect blood.

4. Clippers for fleece (Masterclip Duo, Outlandish Items Ltd., Leicestershire, UK).

5. Povidone iodine antiseptic cleaning solution.

6. 1-ml syringe and 19-gauge needle.

2.4.5. Immune Response to Routine Vaccination

1. Clippers (e.g. Masterclip Duo, Outlandish Items Ltd., Leicestershire, UK).

2. Cleaning solution (made up as above).

3. Vacutainer® system with 20-gauge PrecisionGlide® needle and 3×10 ml tubes containing 18 mg Potassium EDTA (Becton Dickinson, Plymouth, Devon, UK).

4. Refrigerated centrifuge (e.g. ALC 4237R).

5. Storage vial and pipette.

6. Heptavac P Plus® 8 in 1 Clostridial and pasteurella vaccine for sheep (Intervet UK Ltd., Schering-Plough Animal Health, Milton Keynes, Bucks, UK).

7. Serum amyloid A testing kit—Phase™ Range Multispecies SAA ELISA kit (Tridelta Development Ltd., Bray, Co. Wicklow, Ireland).

2.4.6. Glucose Tolerance Test

1. Scales (e.g. CW-11, Ohaus Corporation, Pine Brook, NJ, USA).

2. Clippers (e.g. Masterclip Duo, Outlandish Items Ltd., Leicestershire, UK).

3. Cleaning solution (made up as above).

4. Two 16-gauge short (60 mm) intravenous cannulae (Vygon, Ecouen, France).

5. 2.0 Mersilk suture (Ethicon Inc., Somerville, NJ, USA).

6. Selection of 2.5-ml, 5-ml and 10-ml syringes.

7. Sterile 50% glucose solution (Baxter Healthcare Corporation, Deerfield, IL, USA).

8. Sterile normal saline (0.9% NaCl).

9. 5-ml assay tubes×13 (Sarstedt, Nümbrecht, Germany) each containing 50 µl heparin at a concentration of 10 IU per ml (Leo Laboratories Ltd., Princes Risborough, Bucks, UK).

10. Refrigerated centrifuge (e.g. ALC 4237R).

11. Storage vials and pipettes ×13.

2.4.7. Blood Sampling and Analysis in Lambs and Adult Sheep

1. 19-gauge needle or butterfly.

2. 10-ml syringe.

3. Blood bottles: EDTA bottle, sodium citrate, serum sample (Becton Dickinson, UK).

3. Methods

All methods described here are based on the sheep as experimental model

3.1. Monitoring Fetal Growth and Well-being by Ultrasound

Fetal growth velocity is a good indicator of fetal well-being. Normal fetal growth indicates the fetus and placenta are healthy long term, but it can be really only assessed adequately over a minimum of 2 weeks, because of errors inherent in the measurement. Measurement of umbilical artery Doppler provides a more acute assessment of fetal well-being, over days rather than weeks.

1. An appropriate set-up for scanning is used—either utilising an assistant to turn and hold the sheep supine, or a modified milking crate allowing ewe to stand upright whilst loosely restrained by head collar with access to hay.

2. Cleaning solution is sprayed generously onto the fleece anterior to the udder, and the abdominal wool is clipped carefully.

3. Ultrasound coupling gel is liberally applied to the probe prior to contact with the skin.

4. The gain and depth settings are adjusted to achieve an optimal ultrasound image.

3.1.1. Occipitosnout Length and Biparietal Diameter (See Note 1)

1. An axial view of the fetal head is obtained at the level of the thalami, taking care that the falx cerebri that splits the brain is visible anteriorly and posteriorly and is in the midline. The thalamic nuclei should be in the midline. The choroids plexus may be visible within the lateral ventricles in early pregnancy.

2. The biparietal diameter (BPD) is measured at the widest point of the outer table of the proximal skull to the outer table of the distal skull.

3. The occipitosnout length (OSL) is measured in the same plane as the BPD, from the outer table of the skull overlying the posterior fossa, to the anterior portion of the jaw.

4. See (22, 23) for further description, an illustration of these measurement and reference values.

3.1.2. Abdominal Circumference (See Note 1)

1. A transverse circular view of the fetal abdomen is obtained at the level of the lowermost rib, just above the umbilical cord insertion.

2. The abdominal circumference is measured directly using an ellipse or traced manually.

3. See (22) for further description, an illustration of this measurement and reference values.

3.1.3. Renal Volume (See Note 2)

1. A longitudinal view of the kidney is obtained in parasagittal section, taking care to visualise the full length of the organ (Fig. 1a) (24).

2. Renal length (RL) is measured by placing the callipers on the border of the renal tissue at the superior and inferior poles of the kidney.

3. The probe is then rotated to obtain a circular transverse view of the fetal abdomen and moved up and down to identify the widest mid-section of the kidney in this plane (Fig. 1b).

4. At this point, the anteroposterior and transverse renal diameters (APRD and TRD) are measured at right angles to each other.

5. Each measurement should be taken three times and averaged for accuracy.

6. The renal volume (RV) is calculated using the ellipsoid equation: $RV = RL \times TRD \times APRD \times \pi / 6$ (25).

3.1.4. Placentome Index

The placentome index provides an assessment of placental size during pregnancy without resorting to necropsy.

1. A quick sweep is made through the gravid uterine horn to roughly survey the number, size, and general distribution of the placentomes.

2. Thereafter, ten representative placentomes are randomly selected for measurement.

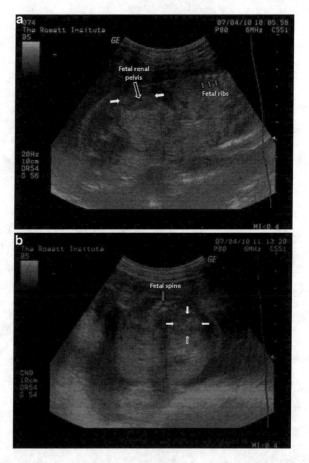

Fig. 1. (**a**) *Fetal renal length*. Sonogram demonstrating a parasagittal view of the ovine fetal abdomen and pelvis at day 82 of gestation. The full length of the fetal kidney is shown (*white arrows*). The fetal renal pelvis is indicated by the *black arrow* and the fetal ribs are indicated by the *grey arrows*. (**b**) *Fetal renal transverse and anteroposterior diameters*. Sonogram demonstrating a transverse view of the ovine fetal abdomen at day 82 of gestation. The right fetal kidney is shown in transverse section at its midpoint, with the spine at 12 o'clock (*grey arrow*). In this plane, the anteroposterior and transverse fetal renal diameters (*white arrows*) can be evaluated.

3. A "side-on" i.e. longitudinal mid-section view of each placentome is obtained, giving a characteristic oval appearance (Fig. 2).

4. In this plane, the length and the width of the placentome are measured.

5. The cross-sectional area (CSA) in cm² of each placentome is calculated using the following formula for an ellipse: $CSA = (\pi \times (\text{placentome length} \times 0.5) \times (\text{placentome width} \times 0.5))/100$.

6. The "placentome index" (in cm²) is then calculated as the sum of these ten cross-sectional areas.

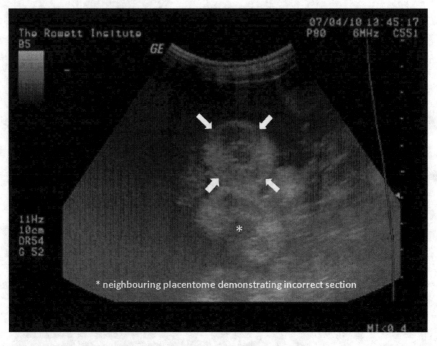

Fig. 2. *Placentome measurements*. Sonogram demonstrating the characteristic appearance of a placentome at day 82 of gestation in sheep pregnancy, viewed "side-on" in longitudinal mid-section. In this plane, the placentome width and length (*white arrows*) can be evaluated. Note the neighbouring placentome (marked with an *asterix*) insonated in the incorrect plane.

3.1.5. Amniotic Fluid Assessment

1. A general survey is made of the entire gravid uterine horn in order to identify the deepest pool of amniotic fluid—this will vary according to fetal position and movement.

2. Once located, the depth of this pool is then quantified in a vertical plane.

3. Three measurements are taken and averaged for accuracy.

3.1.6. Umbilical Artery Doppler Waveform Analysis

1. A free loop of umbilical cord is identified near its insertion into the fetal abdomen.

2. The orientation is adjusted such that the umbilical vessels are directed towards the transducer with the theta angle (θ, the angle between the direction of flow and direction of the ultrasound beam) as close to zero as possible and, importantly, below 30°.

3. Doppler waveform analysis is performed in the absence of fetal breathing and movement.

4. Once ready the colour flow mode is applied and the Doppler gate held over the umbilical artery, which can be distinguished from the umbilical vein by its characteristic pulsatile waveform (Fig. 3).

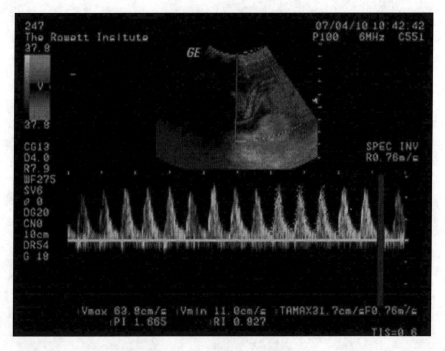

Fig. 3. *Umbilical artery Doppler waveform.* Sonogram demonstrating the characteristic umbilical artery Doppler waveform obtained at day 82 of gestation in sheep pregnancy. The artery has been insonated at an angle as close to zero as possible. The maximum (peak systolic), minimum (end-diastolic), and time-averaged mean velocities can be evaluated by tracing three consecutive waveforms.

5. The velocity setting is adjusted to ensure that the entire waveform is visible.

6. The Doppler gain is altered so that a "clean" waveform with minimum background noise is observed.

7. The outline of three consecutive uniform waveforms is manually traced in order to determine the maximum (peak systolic) velocity (V_{max}), minimum (end-diastolic) velocity (V_{min}), and time-averaged maximum velocity (TAMAX-V) over at least one cardiac cycle.

8. The above steps are repeated three times and each of the values averaged for accuracy.

9. The pulsatility index (PI) is calculated by $(V_{max} - V_{min})/$TAMAX-V.

10. The resistance index (RI) is calculated by $(V_{max} - V_{min})/V_{max}$.

11. The systolic/diastolic ratio (SDR) is calculated by (V_{max}/V_{min}).

3.1.7. Ultrasound Measurement of Uterine Artery Blood Flow

1. The anaesthetised sheep is kept under steady state conditions (maternal oxygen and carbon dioxide levels, pulse and respiratory rate, and temperature) in the supine position.

2. The main uterine artery is identified on each side of the uterus by placing the abdominal ultrasound probe parallel to the udder and turning on the colour Doppler mode. The uterine artery is seen crossing the iliac vessels and running along the uterine wall.

3. The orientation is adjusted such that the uterine artery vessel is directed towards the transducer with the insonation angle (θ, the angle between the direction of flow and direction of the ultrasound beam) as close to zero as possible and always below 30° (see Note 3).

4. Once the insonation angle is satisfactory, the pulsed wave gate is held over the uterine artery and pulsed wave Doppler is applied (see Note 3).

5. The pulsed wave Doppler pulse repetition frequency and baseline settings are adjusted to ensure that the entire waveform is visible.

6. The Doppler gain and filter are altered, so that a "clean" waveform with minimum background noise and a clear diastolic flow are observed.

7. The outline of three consecutive uniform waveforms is manually traced in order to determine the maximum (peak systolic) velocity (V_{max}), minimum (end-diastolic) velocity (V_{min}), and time-averaged peak velocity (TAMAX-V).

8. The diameter (D) of the uterine vessel is then measured using a straight portion of the uterine artery under colour or power Doppler filling. The colour or power Doppler gain should be adjusted to maximise vessel filing without overwriting the vessel walls. The callipers are then used to measure the inner–inner diameter of the vessel wall.

9. The above steps are repeated three times and each of the values averaged for accuracy.

10. The calculated uterine artery volume blood flow (cUtABF) is derived using the formula: cUtABF (ml/min) = TAMEAN-V(cm/s) × A(cm²) ×60, where the vessel area is calculated as $A = \pi(D/2)^2$ assuming the vessel to have a circular lumen. The time-averaged intensity-weighted mean velocity (TAMEAN-V) is calculated by multiplying TAMAX-V by 0.6, which is the spatial velocity distribution coefficient derived in the sheep uterine arteries (26).

3.2. Ultrasound-Guided Fetal Sampling

Ultrasound fetal sampling allows samples of fluids to be taken from the fetus without excessive disturbance and with a low miscarriage risk. Procedures can be repeated during gestation providing serial samples which can inform as to the efficacy of a prenatal gene therapy approach and the safety.

3.2.1. Amniocentesis

1. This section should be read with reference to Chapter 11, Subheading 3.4.1 for *Ultrasound-guided procedures, Intra-amniotic injection*.

2. A 22-gauge Echotip Lancet needle (Cook Medical, UK) can be used for procedures throughout gestation.

3. Up to 10 ml of amniotic fluid can be removed in the first trimester and up to 20 ml can be removed in the second trimester without pregnancy loss. In the third trimester, there is less amniotic fluid around the fetus and careful evaluation of the pockets of fluid needs to be made before the procedure is attempted.

3.2.2. Fetal Blood Sampling

1. This section should be read with reference to Chapter 11, Subheading 3.4.3 for *Ultrasound-guided procedures, Umbilical vein injection and Intracardiac injection*.

2. A 22-gauge Echotip Lancet needle (Cook Medical, UK) is used for all procedures up to 85 days of gestation. A 20-gauge echotip stylet needle (Cook Medical, UK) is used for experiments at later gestational ages.

3. The intrahepatic portion of the umbilical vein is the preferred route for fetal blood sampling in the sheep fetus, especially in fetuses before 100 days of gestation. General anaesthesia should be used for this route of blood sampling. The intracardiac route is an alternative for sampling fetal blood and can be done in awake, restrained ewes (27), but it has a 4.5% miscarriage rate and is preferably performed after 100 days of gestation.

4. A 1- to 2-ml syringe is used to confirm that the needle is in the umbilical vein within the fetal liver. Blood should be aspirated slowly into this syringe using a steady constant pressure. The syringe can be changed over to another of similar size or larger, once it is full. At early to mid-gestation however (up to 85 days), great care must be taken not to dislodge the needle from the umbilical vein during syringe changeover.

5. The small fetal blood volume must be taken into consideration when sampling in early and mid-gestation. Sampling of up to 10% of the total fetal and placental blood volume does not usually increase the risk of fetal loss during clinical fetal medicine procedures.

3.3. Invasive Maternal and Fetal Monitoring Using Telemetry or Catheters to Allow Serial Blood Sampling

Some studies require long-term monitoring of fetal or maternal condition after application of prenatal gene therapy. Implanted telemetric devices allow the data to be beamed to a nearby computer where it can be captured, stored, and then analysed off line, without disturbing the animals in their normal habitat. For long-term blood sampling, catheters can be implanted allowing serial sampling to be performed.

1. Induce general anaesthesia using intravenous thiopental sodium and maintain with 2–2.5% isoflurane in oxygen via an endotracheal tube in overnight fasted animals.

2. Place the metallic receiver RMC-1 under the operating table to catch the signal from the transmitter during surgery. This receiver is connected to the data acquisition computer via a cable.

3. In preparation for the blood pressure catheter insertion surgery, switch the transmitter "on" with the magnet and confirmed it to be "on" by bringing it close to a radio tuned into 530 KHz AM. When the transmitter is on, it is reflected on a "tuned" radio as a characteristic humming sound. Place the transmitter in normal saline solution for a minimum 10 min in order to equilibrate the catheter.

4. Clean the skin of the neck thoroughly with povidone antiseptic solution, as for the abdomen.

5. Make a 5 cm incision on the skin of the neck to the right of the trachea in the midline with a no. 11 blade, and expose the trachea and the strap muscles. The muscles are gently pushed aside to the midline to expose the carotid artery, which is then dissected free of underlying connective tissue.

6. Place two 1-0 sterile silk ties (Ethicon, New Jersey, USA) around the carotid artery, about 1 cm apart.

7. Secure the tie around the distal part of the vessel tightly to occlude it, while the lower tie is loosely tied. With a fine pair of scissors, make a small 2 mm incision through the vessel wall into the lumen, perpendicular to the length of the vessel (see Note 4).

8. Gently insert the tip of the catheter into the carotid artery lumen, and push it proximally towards the heart so that approximately 10 cm of catheter is inside the vessel when a clean blood pressure waveform can be observed on the acquisition computer.

9. The lower vessel tie is secured tightly around the vessel and catheter to keep it in place. A second tie should be secured around the vessel and catheter for safety.

10. The body of the transmitter is placed in a cavity made between the subcutaneous tissue and underlying neck muscles.

11. The subcutaneous tissue is closed with 2-0 Vicryl suture continuously.

12. The skin incision is closed with continuous 2-0 vicryl.

13. Terramycin antibiotic is sprayed over the incision.

14. The monitoring space should consist of an animal enclosure room adjacent to a computer room. The PA-D70 implanted

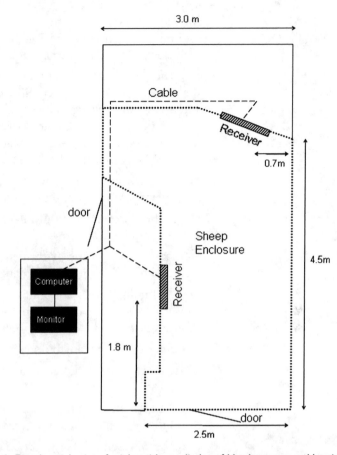

Fig. 4. *Experimental set-up for telemetric monitoring of blood pressure and heart rate in maternal or fetal sheep.* The two metallic RMC-1 receivers fixed at 53 cm above ground level on the inside of the enclosure are arranged in a way to maximise free animal movement in the space available while maintaining a good signal. The long dotted line indicates the area of the enclosure in which the sheep are free to roam. Cables run from each receiver to the computer outside the sheep enclosure and above the height of the ewe to avoid damage (*short dotted line*).

devices can transmit in a range slightly exceeding one metre and two metallic RMC-1 receivers can be arranged in a way to maximise free animal movement while maintaining a good signal to the nearby computer. The monitoring room set-up is graphically illustrated in (Fig. 4).

15. In order to monitor blood pressure and heart rate, the transmitter needs to be switched on with a magnet by holding the magnet close to the ewe's neck or ewe's abdomen (for fetal BP and HR measurement). Once "on", the transmitter telemetrically transmits the blood pressure and heart rate signal to the metallic receivers RMC-1, which further relay it to the acquisition computer located in close proximity.

1. Induce general anaesthesia using intravenous thiopental sodium and maintain with 2–2.5% isoflurane in oxygen via an endotracheal tube in overnight fasted animals.

2. Clip the fleece to the skin over the abdomen of the ewe from the suprapubic area to 3 cm above the umbilicus, and to the anterior superior iliac spine laterally on each side. On the right side of the ewe, clip further laterally around onto the flank where the skin buttons will be placed.

3. Clean the skin of the abdomen first with ethanol to remove the lanolin, then with hibiscrub to work the latter into the skin, and finally povidone antiseptic solution.

4. Drape the abdomen and right flank with sterile drapes to allow access to both sites.

5. Sterilise the flow probes with their reflector brackets disassembled by immersion in cold sterilant for minimum 15 min.

6. Make a midline lower abdominal incision in the skin from just above the pubic bone to just below the umbilicus using a no. 11 blade.

7. Carefully diathermy any blood vessels within the fat.

8. Incise the rectus sheath and open the peritoneal cavity carefully.

9. Place the metal cannula from inside the peritoneal cavity through the skin of the right flank (see Note 5).

10. Pass the disassembled flow probe and cabling into the metal cannula from the outside into the peritoneal cavity and pull them out into the abdominal incision, together with the metal cannula. Reassemble the flow probes with their reflection brackets.

11. Check the orientation of the uterus and identify the uterine arteries bilaterally. Document the number of the flow probe assigned to each uterine artery.

12. Pack back any bowel using the lap sponges. Carefully open the parietal peritoneum over the uterine arteries just proximal to the first division. Mobilise a 2 cm length of the uterine artery using dissection scissors.

13. Place the reflection bracket under the uterine artery and screw the other reflection bracket in place, so that the flow probe is around the vessel. Close the overlying peritoneum using 4-0 Prolene suture (see Note 6).

14. Remove the lap sponges and instil sodium benzylpenicillin G (3g)+gentamicin 80 mg into the peritoneal cavity before closure for antibiotic prophylaxis.

15. Close the ewe's abdomen in layers. The rectus sheath is closed with continuous 6 mm nylon tape to prevent herniation of the

abdominal contents. In adolescent ewes where the rectus sheath can be fragile, an alternative suture material is interrupted mattress sutures with 0 Prolene. The subcutaneous tissue and the skin are closed with continuous 1-0 PDS and 1-0 Vicryl, respectively.

16. Tilt the ewe to its left side to expose the right flank.

17. Carefully tunnel under the skin from the site where the skin buttons exit the abdomen, to a site a further 10 cm higher up on the ewe's right flank. Pass the skin buttons along and out onto the skin. Suture the skin buttons in place with 1-0 PDS. Close the opening into the abdomen with 1-0 PDS for the sheath and 1-0 Vicryl for the skin.

18. Terramycin is sprayed over the abdominal incision and the flank incisions. Analgesia (0.1 mg/ml IM buprenorphine) is administered to the animal.

19. Intramuscular Penstrep 3 ml (200 mg/ml procaine penicillin and 250 mg/ml dihydrostreptomycin) is given for infection prophylaxis.

20. The animal is extubated, allowed to recover and returned to the pen.

21. The day after surgery the wounds are carefully checked for soreness and inflammation.

22. Following the surgery, the battery pack and physiometer need to be connected to the flow probes via the skin buttons whenever the animal needs to be monitored. The battery pack and physiometer are secured to the animal while blood flow is being monitored (see Note 7) (Fig. 5).

23. Once connected, they telemetrically transmit the uterine blood flow signal to an acquisition computer with the Bluetooth device located in close proximity (within 2–4 m), on which the data are collected and analysed.

3.3.3. Long-Term Measurement of Fetal Blood Pressure and Heart Rate

The procedure is performed at 100–105 days of gestation.

1. Place the metallic receiver RMC-1 under the operating table to catch the signal from the transmitter during surgery. This receiver is connected to the data acquisition computer via a cable.

2. In preparation for the D70-PCTP device implantation, switch the transmitter "on" with the magnet and confirmed it to be "on" by bringing it close to a radio tuned into 530 KHz AM. When the transmitter is on, it is reflected on a "tuned" radio as a characteristic humming sound. Place the transmitter in normal saline solution for a minimum 10 min in order to equilibrate the catheter.

Fig. 5. *Pregnant sheep being monitored telemetrically for blood pressure, heart rate, and uterine artery blood flow.* The two metallic RMC-1 receivers can be seen attached to the *bars* of the stable stall. The battery pack and Physiometer are secured to the ewe by placing them in two *red* "backpacks" that are tied to the fleece.

3. Induce general anaesthesia, prepare abdomen as before for laparotomy, and open the abdomen to visualise the uterus. Pack back any bowel using lap sponges and identify the pregnant uterus (see Note 8).

4. Identify the fetal parts by gently palpating the uterus.

5. In an area of myometrium away from placentomes, place a 5 cm uterine incision just above the fetal head using a no. 21 blade (see Notes 8 and 9).

6. Exteriorise the fetal head through the uterine incision and use Babcock clamps to compress the amniotic membrane against the uterine wall in order to minimise bleeding (Fig. 6a).

7. Make a 3 cm incision on the skin of the neck to the right of the trachea in the midline with a no. 21 blade, and expose the trachea and the strap muscles (Fig. 6a).

8. The muscles are gently pushed aside to the midline to expose the common carotid artery, which is then dissected bluntly free of underlying connective tissue using a small Kelly clamp (Fig. 6b, c).

9. Place two 3-0 silk ties (Ethicon, New Jersey, USA) around the common carotid artery, about 1–2 cm apart. Throw each tie once but do not tighten the knot at this stage.

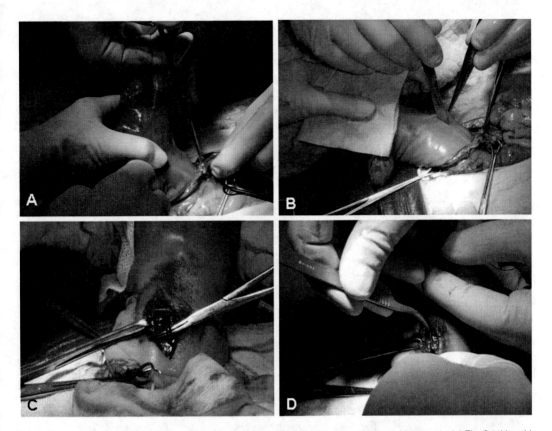

Fig. 6. *Placement of a carotid artery catheter to measure fetal sheep blood pressure and heart rate.* (**a**) The fetal head is exteriorised through the uterine incision while a Babcock clamp is applied to secure the amniotic membrane and prevent blood loss from the incision edge. (**b**) A paramedian incision is placed in the anterior fetal neck. (**c**) The right common carotid artery has been isolated and a small Kelly clamp is used to bluntly dissect the artery from the underlying connective tissue. (**d**) The common carorid artery is held between two silk ties. The vessel lumen is entered and a fine-tipped curved forceps is inserted into the lumen to maintain patency while the catheter is inserted.

10. Using the 3-0 silk ties already in place, elevate the common carotid artery from each side to prevent excessive bleeding and maintain controlled tension.

11. While maintaining controlled tension, incise the vessel wall between the two ties using fine-tipped scissors until the lumen can be identified.

12. Insert a fine-tipped curved forceps into the arterial lumen and then leave the jaws to separate and maintain luminal patency (Fig. 6d).

13. Insert one of the haemodynamic catheters of the telemetric D70-PCTP device into the arterial lumen and thread it upstream towards the aorta for a distance of around 7 cm until a clear signal to the nearby receiver could be obtained (see Note 10).

14. The silk ties are then tied, one to occlude the vessel distally and one to secure the catheter within the vessel.

15. The neck skin is closed over the catheter using 2-0 Vicryl and a few coils of the catheter are secured to the fetal skin to prevent accidental stretching of the catheter.

16. Return the fetal head into the amniotic cavity and insert the other haemodynamic catheter into the amniotic cavity to record background pressure (see Note 11).

17. Close the uterine incision in two layers using 2-0 PDS in a running fashion with the catheters passing through the incision at a distance from the angles. The first layer incorporates the whole thickness of the uterine wall along with the amniotic membrane, while the second layer buries the first to provide a water-tight seal.

18. Attach the transmitter of the D70-PCTP device to the inner surface of the abdominal wall slightly lateral to the laparotomy incision. For this, use three 3-0 Prolene interrupted sutures passed and tied through the transmitter's nylon suspensions and the abdominal wall's fascia. This step is necessary to allow easy accessibility since an external magnet is needed to turn the transmitter On and Off.

19. Remove the lap sponges and give a 3 ml intramuscular injection of procaine penicillin 200 mg/ml and 250 mg/ml dihydrostreptomycin, and a 75% intraperitoneal/25% intrauterine injection of sodium benzylpenicillin G 3g + gentamicin 80 mg for infection prophylaxis.

20. Close the ewe's abdomen in layers as described above.

21. Terramycin antibiotic is sprayed over the incision.

22. The monitoring space should be the same as that for monitoring ewe's blood pressure and heart rate.

3.3.4. Catheters to Allow Serial Fetal Sampling

1. Induce general anaesthesia using intravenous thiopental sodium and maintain with 2–2.5% isoflurane in oxygen via an endotracheal tube in overnight fasted animals.

2. Clip the fleece to the skin over the abdomen of the ewe from the suprapubic area to 3 cm above the umbilicus, and to the anterior superior iliac spine laterally on each side. On the right side of the ewe, clip further laterally around onto the flank where the skin buttons will be placed.

3. Clean the skin of the abdomen first with ethanol to remove the lanolin, then with hibiscrub to work the latter into the skin, and finally povidone antiseptic solution.

4. Drape the abdomen and right flank with sterile drapes to allow access to both sites.

5. Sterilise the polyvinyl catheters by immersion in cold sterilant for minimum 15 min.

6. Make a midline lower abdominal incision in the skin from just above the pubic bone to just below the umbilicus using a no. 11 blade.

7. Carefully diathermy any blood vessels within the fat.

8. Incise the rectus sheath and open the peritoneal cavity carefully.

9. Place the metal cannula from inside the peritoneal cavity through the skin of the right flank (see Note 5).

10. Pass the catheter (s) into the metal cannula from the outside into the peritoneal cavity and pull them out into the abdominal incision, together with the metal cannula.

11. Identify the fetal parts by gently palpating the uterus.

12. In an area of myometrium away from placentomes, place a 5 cm uterine incision just above the fetal head using a no. 21 blade (see Note 8).

13. Exteriorise the fetal head through the uterine incision and use Babcock clamps to compress the amniotic membrane against the uterine wall in order to minimise bleeding.

14. Make a 3 cm incision on the skin of the neck to the right of the trachea in the midline with a no. 21 blade, and expose the trachea and the strap muscles. The muscles are gently pushed aside to the midline to expose the common carotid artery, which is then dissected bluntly free of underlying connective tissue using a small Kelly clamp.

15. Place two 3-0 silk ties (Ethicon, New Jersey, USA) around the common carotid artery, about 1–2 cm apart. Throw each tie once but do not tighten the knot at this stage.

16. Using the 3-0 silk ties already in place, elevate the common carotid artery from each side to prevent excessive bleeding, and maintain controlled tension.

17. While maintaining controlled tension, incise the vessel wall between the two ties using a fine-tipped scissors until the lumen can be identified.

18. Insert a fine-tipped curved forceps into the arterial lumen and then leave the jaws to separate and maintain luminal patency.

19. Insert a catheter into the arterial lumen and thread it upstream towards the aorta for a distance of around 7 cm until a clear signal to the nearby receiver can be obtained (see Note 10).

20. The silk ties are then tied, one to occlude the vessel distally and one to secure the catheter within the vessel.

21. The neck skin is closed over the catheter using 2-0 Vicryl and a few coils of the catheter are secured to the fetal skin to prevent accidental stretching of the catheter.

22. Return the fetal head into the amniotic cavity

23. Close the uterine incision in two layers using 2-0 PDS in a running fashion with the catheters passing through the incision at a distance from the angles. The first layer incorporates the whole thickness of the uterine wall along with the amniotic membrane, while the second layer buries the first to provide a water-tight seal.

24. Remove the lap sponges and give a 3 ml intramuscular injection of procaine penicillin 200 mg/ml and 250 mg/ml dihydrostreptomycin, and a 75% intraperitoneal/25% intrauterine injection of sodium benzylpenicillin G 3g + gentamicin 80 mg for infection prophylaxis.

25. Close the ewe's abdomen in layers as described above.

26. Terramycin antibiotic is sprayed over the incision.

3.4. Neonatal Monitoring and Follow-up

It is important to assess the effect of prenatal gene therapy on the labour and delivery process, as well as the long-term health of the neonate. Much information can be gained by standardising the observations and samples taken at birth, so that comparison can be made between groups of treated and untreated animals. For longer term analysis, challenge tests by giving glucose or vaccination can provide useful information on the functional well-being of the neonate.

3.4.1. Neonatal Monitoring in the Immediate Postnatal Period

1. Lambing is supervised as per local husbandry procedures (see Henderson 1990 for further information (28))—the degree of intervention required is recorded in detail (see Note 12).

2. At the time of birth, any requirement for airway suctioning (routine or extensive) or use of respiratory stimulants is recorded.

3. An initial subjective assessment is made of breathing, which is documented as normal or abnormally fast, slow, or laboured.

4. An initial subjective assessment of circulation is made by observing the colour of the mucous membranes and by palpation of the left anterior chest for the cardiac impulse, which is documented either as normal or as obviously slow.

5. An initial subjective assessment of tone is made and documented as normal, decreased (floppy) or increased (stiff, neck held back in extension).

6. The lamb is dried thoroughly with a warm towel before being put onto the scales to measure birthweight.

7. The respiratory rate per minute is determined by counting the number of outward chest movements in a 15 s period and multiplication by 4.

8. The heart rate per minute is determined by palpation of the left anterior chest over a 15 s period and multiplication by 4.

9. Rectal temperature is measured using a thermometer placed 1.5–2 cm into the rectum (for as long as required to reach steady state) and recorded prior to the use of a heating lamp if required.

10. 0.5 ml oxytocin (10 IU/ml) is administered intravenously to the ewe.

11. The udder is completely stripped and the colostrum is collected in a measuring jug—the total colostrum yield (in ml) is determined and documented.

12. Colostrum is then fed back to the lamb at a dose of 50 ml/kg using a sterilised baby bottle (or sterilised lamb feeding tube if the bottle is not tolerated) (see Note 13).

13. Once delivered, the placenta is carefully examined to ensure that it is complete and to identify any abnormalities (e.g. stigmata of intrauterine infection, abnormal cotyledonary tissue or vasculature, embryo loss)—if present, these should be photographed and samples sent for histological and/or microbiological analyses.

14. The fetal cotyledons are dissected off the membranes, counted, and weighed.

15. The residual membrane weight is determined and added to the cotyledon weight to give the total fetal placental weight.

16. The lamb is given routine antibiotic prophylaxis using 0.1 mg/kg Baytril® by subcutaneous (SC) injection and vitamin E/selenium supplementation with Vitesel® by intramuscular injection.

17. The lamb is observed at frequent intervals, approximately every 10 min until it is seen to be standing independently on all four legs—time to standing is documented.

18. Routine reviews of lamb and ewe well-being and lamb weight should be checked every 4 h during the first 48 h of life then every 6 h until 72 h of age provided there are no other concerns (see Notes 14 and 15).

19. Routine antibiotics are given daily to the lamb (Baytril® 0.1 mg/kg SC) until day 5.

20. In the event of any concerns about neonatal well-being, individualised advice should be sought from the responsible veterinary surgeon.

21. After the neonatal intensive monitoring period (minimum 72 h), lambs are weighed at daily and then at weekly intervals to monitor postnatal growth velocity—umbilical girth and shoulder height (after the first 24 h) are useful additional measures for this purpose.

22. To measure the lamb girth, a length of string is wrapped around the lamb's mid-section at the level of the umbilicus and at the

right angles to the spine, then measured against a ruler—it is important to reproduce the same plane each time.

23. To measure shoulder height, the lamb is encouraged to stand still and in a neutral position without excessive extension or flexion of the back—measurements are taken at the front shoulder using the horse measuring device and are repeated to ensure reproducibility.

3.4.2. Liver Biopsy in Lambs and Sheep

1. Liver biopsy can be performed with lambs or sheep standing up without sedation by a qualified veterinary surgeon.

2. The right or left flank of the sheep is closely clipped and the site of the biopsy identified as the 11th intercostal space at a level just anterior to the tuber coxae, the most cranial part of the pelvis (see Note 16).

3. The skin is carefully cleaned with povidone iodine antiseptic cleaning solution, and infiltrated with 2.5 ml of 5% procaine hydrochloride for local anaesthesia.

4. A small nick is made in the anaesthetised skin with a no. 11 blade over the chosen site.

5. A 9 cm long 14-gauge veterinary Quick-Core Biopsy Needle (Cook, Australia) is inserted approximately 4 cm into the liver at the 11th intercostal space to obtain a liver core.

6. A small number of liver cores can be taken to ensure enough material is available for analysis.

7. To close the wound, two nylon skin sutures are placed in the skin and the wound is sprayed with Terramycin.

8. An intramuscular injection of 750 mg amoxycillin trihydrate is given for infection prophylaxis.

3.4.3. Bone Marrow Biopsy in Lambs and Sheep

1. Bone marrow can be taken from lambs at approximately 3 months of age.

2. Bone marrow may be taken from the sternum or iliac crest.

3. Clip the fleece over the chosen site and clean the skin with povidone iodine antiseptic cleaning solution.

4. Infiltrate the area with 2 ml of 2% lidocaine HCl over the periosteum.

5. Holding the biopsy needle perpendicular to the bone, advance it into the marrow space by applying light pressure and using slight rotatory motions.

6. With the needle in the marrow space, remove the stylet and connect a 10-ml syringe containing EDTA to the needle.

7. Aspirate 1 ml of bone marrow into the syringe and remove the needle.

8. Spray the wound with Terramycin.

3.4.4. Immune Response to Challenge with Transgenic Protein

1. Serum is taken from the lamb or sheep from the carotid vein using a 10-ml syringe and 19-gauge needle, just prior to immune challenge, for measurement of baseline antibody levels.

2. Four sites over the left and right buttock are chosen, and a small area of fleece (3×3 cm) is clipped closely.

3. The clipped areas are cleaned carefully with povidone iodine antiseptic cleaning solution.

4. A 1ml aqueous solution of transgenic protein, e.g. hFIX protein, is mixed with 50% Freund's complete adjuvant with great care to avoid accidental spillage (see Note 17).

5. At each site, 250 mcl of this mixture is injected subcutaneously using a 19-gauge needle (see Note 18).

6. The injection sites are carefully observed for the next week for signs of swelling and redness that could indicate a severe immune reaction.

7. Serum is collected at serial time points after protein challenge, for example 1, 2 weeks and 1 month.

3.4.5. Immune Response to Routine Vaccination

1. The lamb should be at least 4 weeks of age postnatally before this test is performed (see Note 19).

2. A baseline (day 0) blood sample is taken using the Vacutainer system into a 10-ml potassium EDTA tube.

3. The sample is centrifuged at 3,000 rpm (=2,000×g) for 20 min at 4°C and the resultant plasma is pipetted off into a storage vial.

4. 2 ml Heptavac P Plus® vaccine is administered intramuscularly.

5. Sampling and centrifugation are repeated as above after 24 h (day 1) and 48 h (day 2).

6. Serum amyloid A (SAA, an acute phase protein produced in response to inflammatory stimuli) is quantified using the Phase SAA ELISA kit—plasma should be diluted with the diluents provided (×5 for the baseline sample and ×500–2,000 for each of the post-vaccination samples).

3.4.6. Glucose Tolerance Test

1. The lamb should be at least 4 weeks of age postnatally before this test is performed.

2. The maternal udder is covered using a custom-made cloth sling in order to prevent the lamb from suckling for 3 h before the metabolic challenge begins.

3. The lamb is weighed once the udder cover is fitted.

4. The required glucose dose is calculated at 0.25 mg/kg (i.e. 0.5 ml 50% glucose solution).

5. The required volume is drawn up into one or more 10-ml sterile syringes.

6. The neck fleece is clipped both sides and the skin prepared with cleaning solution.

7. An intravenous cannula is inserted into each jugular vein and their patency checked before being sutured in place using silk. The lamb is kept still using gentle restraint by standard animal handling technique (see Note 18).

8. One side is designated for the collection of blood samples and the other side for infusion.

9. Baseline fasting blood samples are taken using a 2.5-ml syringe into pre-labelled heparinised tubes at 20 min, 10 min, and immediately prior to infusion (0 min).

10. All samples are stored on ice prior to centrifugation (see below).

11. The patency of the infusion catheter is rechecked and then the measured glucose bolus is slowly infused (e.g. 1 ml every 5 s).

12. A 5 ml normal saline flush is given at the end of the infusion to clear the catheter.

13. Post-infusion blood samples are taken as above after 5, 10, 15, 20, 25, 30, 45, 60, 90, and 120 elapsed minutes.

14. Blood samples are centrifuged and plasma harvested as before.

15. After the final blood sample is collected, the udder cover is removed and the lamb weighed.

16. After a further 15 min, the lamb is reweighed and the difference between the two weights is calculated to provide a "suckling index".

17. Plasma samples are analysed for glucose and insulin in the first instance using in-house methods (see (29) and (30) for details).

3.4.7. Blood Sampling and Analysis in Lambs and Adult Sheep (see Note 19)

1. Blood can be collected from newborn lambs as soon as they have been born, dried off, and are breathing. The best route is via the external jugular vein as it runs in the neck. Only 10 ml should be taken at this early time point, and only 5 ml in a growth restricted lamb. The lamb should be kept with the ewe during the sampling procedure to minimise animal stress. Sampling can then be done at serial time points but not more than 10% of circulating blood volume should be collected at any one withdrawal, and only up to 15% circulating blood volume collected every 4 weeks.

2. The fleece should be clipped over the external jugular vein.

3. A 19-gauge needle or butterfly is used connected to a 10-ml syringe. For growth restricted lambs, a smaller 23-gauge needle, and a 5-ml syringe should be used to avoid collapsing the vessel. The needle is inserted into the engorged vein that is

filled by compressing the vein just above the clavicle to prevent venous return to the heart.

4. Blood should be immediately collected into the specific bottle assigned for the tests required. This is particularly important when collecting blood to assess coagulation factor levels, since coagulation factors will be consumed if the blood clots.

5. Samples for blood count (haematology) should be collected into a sterile bottle containing potassium EDTA, kept at room temperature, and processed within 4 h. Samples for biochemistry, liver enzymes, and bile acids should be collected into a sterile bottle and allowed to clot at room temperature. The samples should be processed within 2 h. Samples for assessment of coagulation factors should be collected into sterile bottles containing sodium citrate to prevent blood clot formation, kept at room temperature, and processed within 4 h of collection (see Note 20).

6. A serum sample is collected for assessment of liver enzymes (see Notes 20 and 21) and bile acids (see Note 22). Normal ranges of liver function tests for lambs differ widely from adult normal ranges, for example the bone isoenzyme of alkaline phosphatase (ALP) is elevated in neonatal sheep due to their rapid skeletal growth.

4. Notes

1. Occipitosnout length is the most accurate parameter for determination of gestational age in early pregnancy. By the end of the second trimester however, the fetal head becomes progressively more difficult to access because it is buried in the ewe's pelvis. The abdominal circumference and fetal renal weight become more accurate parameters of fetal growth.

2. Fetal renal weight is very strongly correlated with fetal weight in late gestation in sheep (24). The ultrasound measurement of renal volume in turn correlates strongly with renal weight and is a good surrogate marker of fetal weight. As in the human, the abdominal circumference is also a good marker of ovine fetal growth, reflecting both abdominal fat and liver glycogen stores.

3. In measuring blood flow velocity in the uterine artery, an insonation angle above 30° is associated with a significant error in estimating the true blood flow velocity (31) and so the angle should be kept less than 30°. The characteristic uterine artery Doppler flow waveform is a low-resistance waveform composed of a systolic peak and a diastolic trough with forward

flow throughout the cardiac cycle. This is in contrast to the iliac artery Doppler waveform, which is a high-resistance waveform characterised by reversed flow during early diastole.

4. To prevent excessive blood loss from the carotid artery as the lumen is being opened, the carotid artery can be suspended and slightly stretched between the two ties by slightly lifting them. This occludes the vessel and prevents excessive blood loss when the vessel is opened. While the catheter is being inserted down the carotid artery lumen towards the heart, the proximal tie may need to be lowered slightly to allow the tip to pass.

5. The metal cannula must be placed with care in the right flank of the ewe to prevent damaging the bowel. While protecting the sharp tip of the trocar with your finger, pass the metal trocar and cannula along the underside of the right abdominal skin within the peritoneal cavity around to the right flank where the skin buttons are to be placed. Push the trocar through the skin while incising over the skin with a scalpel. Remove the trocar from the abdomen by moving your hand in reverse while protecting the sharp tip.

6. A small clot of maternal blood can be used to ensure a good signal from the flow probe within a few hours of its placement. At the start of the general anaesthesia, take a 5 ml sample of ewe's blood and allow it to clot within the 5-ml syringe. Once the flow probe is in place around the uterine artery, cut the clot to size so that it fits within the reflector bracket, and secure the parietal peritoneal around it.

7. The battery pack and Physiometer can be secured to the ewe in a number of ways. Canvas bags in which the battery pack and Physiometer are placed can be (1) sutured to the skin and fleece before the animal is recovered from general anaesthesia or (2) tied in place to the fleece if the area on the back of the ewe is not clipped.

8. If uterine artery flow probes are being placed at the same time as a fetal blood pressure and heart rate catheter, they should be placed first before hysterotomy for a better exposure and to minimise the chances of iatrogenic fetal catheter displacement.

9. Try to minimise amniotic fluid loss during the hysterotomy procedure. If amniotic fluid loss is significant, it is replaced using a continuous drip of warm Hartmann's solution or lactated Ringer's solution.

10. The threading distance should be estimated as the distance between the insertion point and the fetal heart. If the catheter is threaded for a shorter distance, the bulky tip of the haemodynamic catheter may lie in the bicarotid trunk, thus blocking the carotid circulation bilaterally (32).

11. True fetal blood pressure is calculated by subtracting the amniotic cavity pressure from the measured fetal blood pressure.

12. Term for sheep is 145 days gestation. Some models of compromised pregnancy, such as those resulting in fetal growth restriction, are also associated with preterm delivery with viable fetuses being born as early as 135 days of gestation (33). Ewes should be supervised 24 h a day 7 days a week from the earliest gestational age commensurate with live births.

13. It is well established that lambs require 50 ml/kg birthweight of good quality colostrum to allow them to obtain sufficient levels of protective antibody (34). In certain circumstances (e.g. multiple pregnancy, fetal growth restriction, impaired placental hormone production), maternal udder development and colostrum production may be insufficient. Routine determination of colostrum quantity at birth identifies lambs whose minimum requirements would otherwise not be met, who can then be supplemented with donor or formula colostrum, thereby reducing neonatal morbidity and mortality.

14. Regular weight checks are the primary surveillance tool in monitoring the well-being of lambs during the neonatal period—weight loss or failure to gain weight is an early sign of problems establishing feeding or other complications.

15. Lamb well-being can be assessed using a variety of observations. At each review, the lamb is stimulated and encouraged to stand—once up it will usually go into a big stretch, a reassuring sign of well-being. On returning to mother after being weighed, most lambs will immediately go in to suckle—this is another sign of well-being and indicates appropriate establishment and functioning of the ewe/lamb bond.

16. To aid positioning of the liver biopsy, the landmarks of the 11th, 12th, and 13th ribs can be marked on the shaved skin with a thick water-resistant pen. The area chosen for biopsy can be scanned with ultrasound to confirm the liver is adjacent to the body wall at this point. It is vital to clean the skin/fleece of large animals adequately before embarking on such minor surgery. Close clipping of the fleece or hairs from the skin helps remove dirt and a source of bacteria. Care is taken to ensure that even the finest wool is removed. We clean prior to surgery using povidone iodine antiseptic cleaning solution, which is worked into the skin/fleece using a scrubbing brush and then wiped clean. We have found judicious attention to this aspect of care reduces the rate of infection.

17. Great care must be taken during handling of the Freund's complete adjuvant to ensure that accidental inoculation does not occur since this can result in sensitisation to tuberculin as

well as chronic, local inflammation, which is poorly responsive to antibiotic therapy.

18. Proper preparation of the Freund's complete adjuvant-antigen emulsion will limit inflammation. The mycobacterial component of Freund's complete adjuvant should be resuspended before use by vortexing or shaking. An emulsion is properly prepared when it becomes thick, will not separate on standing, and will not disperse when a droplet is placed in saline. Use cold adjuvant (4°C) to improve emulsification. It is best to choose four skin sites that are widely separated to avoid coalescence. The inoculation should be kept to the minimum volume practical. Injections of Freund's complete adjuvant should be subcutaneous, because intradermal injections may cause skin ulceration and necrosis. Alternative adjuvants should be considered and utilised if and when possible.

19. Lambs can be held easily for procedures such as blood sampling and cannulation by standing over their back and gently squeezing them between your legs. This permits the operator's hands to be free to perform the procedure.

20. International standards for adult sheep for haematology, biochemistry, liver function tests, and bile acids are described (35).

21. The diagnostic usefulness of particular liver enzymes depends on the domestic animal species investigated and their age, and this must be considered when interpreting results in sheep (36). While abnormal levels of AST provide only a general indicator of hepatic injury in ruminants, the serum concentration of the liver enzyme glutamate dehydrogenase (GLDH) is the most specific for hepatocellular injury. In ruminants, GLDH is raised in hepatic necrosis and bile-duct obstruction (37). Gamma-glutamyltransferase (GGT) and alkaline phosphatase (ALP) are markers of cholestasis, GGT being associated with epithelial cells comprising the bile ductular system and ALP associated with the canalicular membrane. Serum levels of ALP fluctuate widely in normal ruminants and are less valuable than GGT in the evaluation of cholestatic disorders (37). There is only low level ALT activity in the hepatocytes of ruminants in contrast to rodent hepatocytes, and therefore the level of this liver enzyme in the serum is not diagnostically useful. The most useful liver enzymes for diagnosis of hepatic impairment are glutamate dehydrogenase (GLDH) and gamma-glutamyltransferase (GGT). The level of liver enzyme activity observed in sheep does not always correlate with the degree of functional impairment, and serum bile acid level is considered to be a better indicator.

22. The measurement of the serum total bile acid level in ruminants is difficult due to the wide range of reported normal

values, hourly fluctuations up to 60 μmol/L during feeding and occasional high levels in apparently healthy animals. Levels appear to be higher after parturition, and it may be a few weeks before they return to pre-partum values. The normal ranges quoted in the literature for studies investigating liver damage in sheep vary widely (38–40). In practice, most veterinary biochemists use the normal range that has an upper limit of 50 μmol/L (personal communication, Department of Pathology & Infectious Diseases, Royal Veterinary College, UK). Generally, there are low levels of bile acids in the fetus, and increased synthesis after delivery leads to a rise in the serum bile acids up to adult levels (41).

References

1. Tabor A, Philip J, Madsen M et al (1986) Randomised controlled trial of genetic amniocentesis in 4606 low-risk women. Lancet 1:1287–1293
2. Alfirevic Z, Sundberg K, Brigham S (2003) Amniocentesis and chorionic villus sampling for prenatal diagnosis. Cochrane Database of Systematic Reviews CD003252
3. Weiner CP, Wenstrom KD, Sipes SL, Williamson RA (1991) Risk factors for cordocentesis and fetal intravascular transfusion. Am J Obstet Gynecol 165:1020–1025
4. Peckham CS, Martin JA, Marshall WC, Dudgeon JA (1979) Congenital rubella deafness: a preventable disease. Lancet 8110: 258–261
5. Preece PM, Pearl KN, Peckham CS (1984) Congenital cytomegalovirus infection. Arch Dis Child 59:1120–1126
6. Wenstrom KD, Andrews WW, Bowles NE et al (1998) Intrauterine viral infection at the time of second trimester genetic amniocentesis. Obstet Gynaecol 92:420–424
7. Dodic M, May CN, Wintour EM, Coghlan JP (1998) An early prenatal exposure to excess glucocorticoid leads to hypertensive offspring in sheep. Clin Sci 94:149–155
8. Schellenberg JC, Liggins GC (1987) New approaches to hormonal acceleration of fetal lung maturation. J Perinat Med 15:447–452
9. Barker DJ, Osmond C, Simmonds SJ, Wield GA (1993) The relation of small head circumference and thinness at birth to death from cardiovascular disease in adult life. BMJ 306: 422–426
10. Miller SL, Loose JM, Jenkin G, Wallace EM (2008) The effects of sildenafil citrate (Viagra) on uterine blood flow and well being in the intrauterine growth-restricted fetus. Am J Obstet Gynaecol 102:e1–e7
11. David AL, Torondel B, Zachary I et al (2008) Local delivery of VEGF adenovirus to the uterine artery increases vasorelaxation and uterine blood flow in the pregnant sheep. Gene Ther 15:1344–1350
12. Porada CD, Park PJ, Tellez J et al (2005) Male germ-line cells are at risk following direct-injection retroviral-mediated gene transfer in utero. Mol Ther 12:754–762
13. Heikkilä A, Hiltunen MO, Turunen MP et al (2001) Angiographically guided utero-placental gene transfer in rabbits with adenoviruses, plasmid/liposomes and plasmid/polyethyleneimine complexes. Gene Ther 8:784–788
14. MacCalman CD, Furth EE, Omigbodun A et al (1996) Transduction of human trophoblast cells by recombinant adenoviruses is differentiation dependent. Biol Reprod 54: 682–691
15. Parry S, Holder J, Strauss JR (1997) Mechanisms of trophoblast-virus interaction. J Reprod Immunol 37:25–34
16. Koi H, Zhang J, Makrigiannakis A et al (2001) Differential expression of the coxsackievirus and adenovirus receptor regulates adenovirus infection of the placenta. Biol Reprod 64: 1001–1009
17. Bedrosian JC, Gratton MA, Brigande JV et al (2006) In vivo delivery of recombinant viruses to the fetal murine cochlea: transduction characteristics and long-term effects on auditory function. Mol Ther 14:328–335
18. Brown AS, Derkits EJ (2010) Prenatal infection and schizophrenia: a review of epidemiologic and translational studies. Am J Psychiatry 167:261–280

19. Themis M, Waddington SN, Schmidt M et al (2005) Oncogenesis following delivery of a nonprimate lentiviral gene therapy vector to fetal and neonatal mice. Mol Ther 12: 763–771

20. Wong LF, Goodhead L, Prat C et al (2006) Lentivirus-mediated gene transfer to the central nervous system: therapeutic and research applications. Hum Gene Ther 17:1–9

21. David AL, Peebles D (2007) Gene therapy for the fetus: is there a future? Best Pract Res Clin Obstet Gynaecol 22:203–218

22. Barbera A, Jones OW, Zerbe GW et al (1995) Early ultrasonographic detection of fetal growth retardation in an ovine model of placental insufficiency. Am J Obstet Gynecol 173:1071–1074

23. Kelly RW, Newnham JP (1989) Estimation of gestational age in Merino ewes by ultrasound measurement of fetal head size. Aust J Agr Res 40:1293–1299

24. Carr DJ, Aitken RP, Milne JS et al (2011) Ultrasonographic assessment of growth and estimation of birthweight in late gestation fetal sheep. Ultrasound Med Biol 37:1588–1595

25. Jeanty P, Dramaix-Wilmet M, Elkhazen N et al (1982) Measurements of fetal kidney growth on ultrasound. Radiology 144:159–162

26. Abi-Nader K, Mehta V, Wigley V et al (2009) Doppler ultrasonography for the noninvasive measurement of uterine artery volume blood flow through gestation in the pregnant sheep. Reprod Sci 17:13–19

27. Newnham JP, Kelly RW, Boyne P, Reid SE (1989) Ultrasound guided blood sampling from fetal sheep. Aust J Agr Res 40:401–407

28. Henderson DC (1990) The veterinary book for sheep farmers. Farming Press, Ipswich

29. Wallace JM, Bourke DA, Aitken RP et al (2002) Blood flows and nutrient uptakes in growth-restricted pregnancies induced by overnourishing adolescent sheep. Am J Physiol Regul Integr Comp Physiol 282:R1027–R1036

30. Macrae JC, Bruce LA, Hovell DFD et al (1991) Influence of protein nutrition on the response of growing lambs to exogenous bovine growth hormone. J Endocrinol 130:53–61

31. Dickerson KS, Newhouse VL, Tortoli P, Guidi G (1993) Comparison of conventional and transverse Doppler sonograms. J Ultrasound Med 12:497–506

32. Abi-Nader K, Mehta V, Shaw SWS et al (2010) Telemetric monitoring of fetal blood pressure and heart rate in the freely moving pregnant sheep: A feasibility study. Lab Anim Sci 45:50–54

33. Wallace JM, Luther JS, Milne JS et al (2006) Nutritional modulation of adolescent pregnancy outcome – a review. Placenta 27(Suppl A):S61–S68

34. Logan EF, Foster WH, Irwin D (1978) A note on bovine colostrum as an alternative source of immunoglobulins for lambs. Anim Prod 26:93–96

35. Jubb KVF, Kennedy PC, Palmer N (1993) Pathology of domestic animals. Academic, London

36. Meyer DJ, Harvey JW (1998) (Meyer DJ and Harvey JW, Eds.) Veterinary laboratory medicine – interpretation and diagnosis. WB Saunders Company, Philadelphia, pp 157–186

37. Kaneko JJ, Harvey JW, Bruss M (1997) Clinical biochemistry of domestic animals. Academic, San Diego

38. West HJ (1987) Changes in the concentrations of bile acids in the plasma of sheep with liver damage. Res Vet Sci 43:243–248

39. Anwer MS, Engelking LR, Gronwall R, Klentz RD (1976) Plasma bile acid elevation following carbon tetrachloride induced liver damage in dogs, sheep, calves and ponies. Res Vet Sci 20:127–130

40. Sutherland RJ, Deol HS, Hood PJ (1992) Changes in plasma bile acids, plasma amino acids, and hepatic enzyme pools as indices of functional impairment in liver-damaged sheep. Vet Clin Pathol 21:51–55

41. Hardy KJ, Hoffman NE, Mihaly G et al (1980) Bile acid metabolism in fetal sheep; perinatal changes in the bile acid pool. J Physiol 309:1–11

Chapter 15

Monitoring for Potential Adverse Effects of Prenatal Gene Therapy: Mouse Models for Developmental Aberrations and Inadvertent Germ Line Transmission

Charles Coutelle, Simon N. Waddington, and Michael Themis

Abstract

So far no systematic studies have been conducted to investigate developmental aberrations after prenatal gene transfer in mice. Here, we suggest procedures for such observations to be applied, tested and improved in further in utero gene therapy experiments. They are based on our own experience in husbandry for transgenic human diseases mouse models and breeding, rearing, and observing mice after fetal gene transfer as well as on the systematic screens for monitoring of knock-out mutant mouse phenotypes established in international mutagenesis projects (EUMORPHIA and EUMODIC and subsequently the International Mouse Phenotyping Consortium).

We also describe here the analysis procedures for detection of germ line mutations based on quantitative PCR (qPCR) by sperm-DNA analysis and breeding studies.

Key words: In utero (prenatal, fetal) gene therapy, Mouse model, Adverse effects, Developmental aberrations, Germ-line gene transfer, Real-time quantitative PCR, Spermatozoa, DNA extraction, Phenotype screens

1. Introduction

Long-term postnatal monitoring will be part of the preclinical or clinical protocols of any in utero gene therapy trial on larger animals and eventually humans. In human trials the aim will be to monitor therapeutic effects and the wellbeing of the treated patients. In experimental animals they will primarily serve to monitor long-term vector efficiency and early detection of adverse effects (see also Chapter 12). This will allow to improve vectors, modify gene therapy protocols, and devise appropriate therapeutic measures against observed adverse events. Due to limited numbers and long observation times, which may last for the life-time of the treated subjects, observation of adverse effects will, at least for a longer

Charles Coutelle and Simon N. Waddington (eds.), *Prenatal Gene Therapy: Concepts, Methods, and Protocols*,
Methods in Molecular Biology, vol. 891, DOI 10.1007/978-1-61779-873-3_15, © Springer Science+Business Media, LLC 2012

time period, be mostly anecdotal before statistically assessable patterns emerge (see Chapter 14).

To accumulate statistically validated data for testing of emerging hypotheses derived from the observation of adverse effects, to gather comparative data on vector action, and to test vectors and transgenes for unknown or potential adverse effects, fast and high-throughput screening systems are required. Due to their short generation times and the relatively low costs for breeding and maintenance small animals, and in particular mice, provide several advantages for monitoring of potential adverse effects of gene therapy following pre or postnatal applications. In addition, the mouse genome has been extensively sequenced and mapped and many natural and laboratory-created mutant strains have been characterized and linked to a large collection of phenotypes, many of them modeling known human genetic diseases (1).

The potential of adverse effects resulting from prenatal gene therapy has been reviewed in Chapters 14 and 17, Subheading 2 and monitoring for such effects on large animals is described in Chapters 12 and 14. This chapter deals with the use of the mouse model to monitor for the potential of gene therapy interventions in utero to cause developmental aberrations and inadvertent transfer of transgenic sequences to the germline of treated animals. The following Chapter 16 describes the use of cell culture as well as ex vivo and in vivo murine systems to detect gene therapy-related oncogenesis.

1.1. Developmental Aberrations

Up to now no systematic investigations to assess the risk of developmental aberrations due to in utero gene therapy have been conducted. However, as outlined in Chapter 17, Subheading 2 some observations showing that the expression of specific transgenic products can induce severe development changes have been made fairly recently (2, 3). Such effects appear to be relatively rare and will very much depend on the applied transgene, its expression level, the gestation-stage at time of application, the targeted organ system, and perhaps also the vector construct itself. They are therefore very difficult to predict and may best be discovered by implementing a set of structured procedures for systematic monitoring of the pregnant animals and their in utero treated offspring. Some basic observations will already be in place in accordance with home office regulations for good standard practice of animal care. But while these rules are primarily meant to ensure wellbeing of the experimental animals and to avoid undue suffering, in our context they should also serve to carefully follow up and document any unusual observations, including postmortem investigation on mother and fetus. Further procedures, particularly specific and longer-term postnatal observations will, however, be more extensive. Obviously, the extent of such observations will very much depend on the concrete goal of the gene therapy application. For instance, they may be minimal if already tested vectors and transgenes are used in standard applications, or more extensive if a new vector system or potentially problematic transgenes, for example ones expressing growth factors or hormones, are used; or if to follow up on unexpected observations made on larger animals or in clinical trials.

Although systematic observations have so far not been conducted in connection with prenatal gene therapy very systematic screens (EMPReSS) for monitoring experimental mice for knockout mutant mouse phenotypes on several morphological and physiological levels have been established in ongoing multicenter international projects (EUMORPHIA and EUMODIC and the International Mouse Phenotyping Consortium, IMPC) (4). They follow Standard Operational Procedures (SOPs) and cover equipment, consumables, and the applied protocols and are accessible through http://empress.har.mrc.ac.uk. Some of the screens, which are organized in Phenotype Platforms, may be suited for adaptation to the actual needs of monitoring mouse models in fetal gene therapy projects (see Subheading 3). However, so far they only cover a very limited range of phenotypes, derived from the mutant stain collection (to date 268), and several physiological phenotype systems have not yet been incorporated such as respiratory and gastroenteral. Phenotypes observed in the EUMODIC/IMPC projects are linked to genome data via http://www.europhenome.org, but the restricted number of mutant strains limits the probability to find a link between any unexpected gene therapy phenotype and genome data. While some of the principles from these screens have been incorporated into the following Subheading 3, a full application of these methods would extend beyond the aims of this chapter.

1.2. Inadvertent Germ Line Transmission

The question of germ line transmission has been a contentious issue as long as gene therapy has been attempted. Although it is generally agreed among researchers and clinicians in the field that this is not and should not be a goal of gene therapy, the possibility of inadvertent gene transfer to germ cells cannot be denied and indeed, low level germ line transduction has been found by microdissection or immuno-histochemical analysis in male and female germ cell precursors after in utero gene delivery (5, 6). However, this is thought to affect only a very small number of mature germ cells (7, 8). Unlike any other not clearly detectable iatrogenic or environmental mutagenesis, the unique power of gene therapy to transfer well-defined and easily identifiable gene sequences into the recipient organism, allows also to check treated individuals for the presence of such transgenes in any part of the body including the gonads and germ cells. An assessment of the risks of this happening and the ethical and practical implications if such event were to occur are discussed extensively in Chapter 17 of this book.

The above referenced highly sophisticated methods to investigate germ cell precursors are too specialized for routine or high-throughput application and would not necessarily give an indication of gene transfer to mature germ cells and passing on of altered genes to the following generation. This chapter therefore describes two lines of investigation (purified spermatocyte analysis and breeding studies) used for the detection of transgenic sequences in mature germ cells and as proof for trans-generational germ-line gene transfer. The specific gene sequence detection relies ultimately on the molecular biology methods of sensitive and quantitative polymerase chain reaction (qPCR) analysis on DNA isolated from gonads or sperm of the treated animals and on different tissues including gonads from their offspring (9).

It is important to note that the detection of such sequences in gonad tissue or even whole sperm samples (ejaculate) does not automatically mean that the mature germ cells themselves carry the transferred gene. The direct analysis of germ cells offers the most sensitive examination for vector sequences in a haploid pool representing genomes with the potential to give rise to genetically modified offspring following successful fertilization. Only detection of transgenic sequences in purified sperm or isolated oocytes could serve as such proof and only quantitative analyses of the degree of such gene transfer may indicate the presence of a potential risk to following generations (10).

The additional approach to detect germ line gene transfer by breeding studies with animals that have received gene therapy vectors aims to find the transferred gene sequences in tissues of their offspring (see also Chapter 17).

Real-time quantitative PCR (qPCR) can be performed using Taqman® chemistry or SYBR® Green I dye chemistry that can each provide sensitive detection of PCR amplified products from germ cells or tissue DNA. Because qPCR using SYBR® Green I can lead to the detection of false positives, Taqman® chemistry is the method of choice. The principle of qPCR is based on the selection of sequence-specific primer and probe oligonucleotide combinations designed to recognize the target DNA. In our case, a sequence of the transgene or of the vector packaging site is advised, depending on which is found most sensitive to detect vector presence. The qPCR relies on hybridization to one of the target DNA strands by a probe that is labeled at the 5′ end with a fluorescent reporter dye (6-carboxylfluorescein) and at the 3′ end with a quencher (6-carboxytetramethylrhodamine). No fluorescence will be emitted until the fluorescent reporter is cleaved from the probe by the thermostable DNA polymerase during amplification of DNA from the PCR primer. Hence, in real-time measurement of PCR, amplification can be detected and quantified by measuring fluorescence during amplification in the exponential phase of amplification where the quantity of products is directly proportional to the amount of DNA target template. Most important in this process is the generation of a standard curve produced by qPCR of vector copies in genomic

DNA "spiked" with known dilutions of the target vector. This allows determination of the assay sensitivity with regards to vector copies. The genomic DNA is also subjected to measurement of the GAPDH housekeeping gene or a Y-chromosome sequence by qPCR as a control. Ultimately, final proof can also be confirmed by sequence analysis of the PCR product.

2. Materials

2.1. Detection of Developmental Aberrations

1. Balance (up to 1 g for tissues, over 0.5 g for whole body weight).
2. Ruler.
3. Photo and/or Video camera.
4. Micrometer caliper screw gauges for skin thickness measurements.

2.2. Detection of Germ Line Transmission

2.2.1. Animals and General Materials

1. Wild-type male and female mice.
2. Surgical scissors and forceps.
3. 4-0 silk suture.
4. Phosphate-buffered saline (PBS).
5. 5-ml Syringe.
6. Inverted microscope.
7. Petri dishes.
8. Fluorescent particles (FluoSpheres, Invitrogen, Carlsbad, CA, USA) or tattoo ink.

2.2.2. DNA Extraction from Sperm

1. PBS. Store at 4°C.
2. Lysis buffer: 6 M guanidinium thiocyanate, 30 mM sodium citrate (pH 7.0), 0.5% Sarkosyl, 0.20 mg/ml proteinase K, and 0.3 M b-mercaptoethanol.
3. Isopropanol.
4. Tris/EDTA buffer (TE): 10 mM Tris/HCl (pH 7.4–8.0) 0.1 mM EDTA (store at –20°C).
5. Isopropanol (Sigma).
6. 70% Ethanol.

2.2.3. qPCR Assay for Vector Copy Number

1. Real-Time PCR instrument (e.g., the 7900HT Real-Time PCR Thermal Cycler–Applied Biosystems).
2. TaqMan® Universal PCR Master Mix with AmpErase UNG Gold polymerase.
3. dNTP (Applied Biosystems). Store at –20°C.

4. AMRA probe and unlabeled primers specific to the vector packaging signal or gene under investigation carried by the vector (Applied Biosystems) dissolved in TE buffer.

Y-specific qPCR probes and TaqMan probe:

TSPY (reverse primer) 5′-GAG AAC CAC GTT GGT TTG AGA TG-3′.

TSPY (forward primer) 5′-TCC TTG GGC TCT TCA TTA TTC TTA AC-3′.

and TaqMan probe:

6 FAM-TCC TGG ATC AGA GTG GCT TAC CCA GG TAMRA.

5. 96-Well plates (Applied Biosystems).

6. MicroAmp® Optical adhesive film (Applied Biosystems).

3. Methods

3.1. Detection of Developmental Aberrations by Observation on Pregnant, Neonatal, and Postnatal Mice

These methods concentrate on procedures during pregnancy and neonatal life based on our experience. For extended phenotype observations, we refer to the above-mentioned EMPReSS screens (see http://empress.har.mrc.ac.uk and http://www.europhenome.org).

3.1.1. Litter Size

Usual litter size is 3–4 fetuses per uterus horn for inbred strains, however, outbred strains, such as CD1, may carry 12 or more fetuses per uterus. Exposure time of the uterus should be minimized therefore, if necessary, the number of injections may require limitation to only two to three fetuses. Injections should be performed starting at the top of the horn (see Chapter 10, Subheading 3).

3.1.2. Fetal Death

Occasional early fetal death and resorption in utero is a common phenomenon. However, increased fetal death requires follow-up. Procedural stress, inadequate animal maintenance, or excessive vector dose or impurity can cause fetal death (for optimal animal care and procedures see Chapter 10, Subheading 3).

These reasons excluded specific developmental defects that can be instrumental in fetal death:

1. Keep record of number of delivered neonates vs. total number of fetuses and number of injected fetuses.

2. Timing of fetal death and sampling of fetuses by cesarean section before resorption may be required to ensure access to fetal material for morphological/molecular analysis.

3. Check reduced number of neonates for vector presence to verify connection between gene transfer and demise.

3.1.3. Increased Abortion Rate

Usually at later stages of pregnancy, this can also indicate developmental aberrations. Cesarean section to avoid cannibalism may be needed to obtain fetal material for morphological and molecular analysis or to ensure life birth and survival of neonates for fostering (procedures for fostering see Chapter 10, Subheading 3).

3.1.4. Investigation of Neonates and Maternal Mice

1. Neonates of a litter should be counted; any difference to original fetal number at injection time and any neonatal death should be recorded and survivors checked for gene transfer vector as above.

2. Behavior of mother towards neonates should be recorded—most commonly this would be a maternal tendency to cannibalize one or more pups from the litter, or abandonment. However, this may be induced by maternal illness as much as by problems with the neonatal health status. If cannibalism occurs see above (see also Note 1).

3. Behavior of neonates should be recorded, e.g., feeding, movement, activity (see Note 2).

4. Feeding can be monitored in the first few days when the milk is clearly visible in the stomach through the abdominal wall.

5. Neonates should be weighed at birth and for some outbred strains less susceptible to handling stress as frequently as every day. Other strains may be more susceptible to handling stress; so frequency of weighings must be judged on a strain-by-strain basis.

6. First morphological inspection of neonates should be performed as soon as possible. Make note of gross malformations or nutritional status (runted?).

   Whole body: length of body, thickness of skin (using micrometer screw gauge).

   Coat dorsal and ventral: variation in color, presence, patterning.

   Skin: texture, thickness, moisture, color, and patterning on the body, nose, ear, hand, foot, and tail.

   Head: anomalies in the length, width, symmetry, and curvature of the head; abnormalities in vibrissae; number color and shape of teeth; shape, position, and size of ears; opacity, size, positioning, color, and closure of eyes.

   Genitalia: size, presence, and structure. (Color may vary for female mice according to the stage of the oestrous cycle).

   Forelimb: size, flexibility, shape, structure, and number of digits present on the forepaws.

   Hindlimb: size, flexibility, shape, structure, and number of digits present on the hindlimbs.

   Tail: presence, length, width, curvature, and any subtle kinks.

7. Blood can be collected within the first 4 days of birth by puncture of the superior temporal vein, which drains from the eye to the jugular vein, with a 29-Gauge needle. Beyond 4 days of age the skin is too thick to access this vessel but puncture of the lateral tail vein becomes possible.

8. Transgene detection in fetal/neonatal tissues to make sure that any fetal aberration is associated with the gene transfer.

3.1.5. Extended Postnatal Observations

These investigations will depend on the nature of the gene therapy intervention, concentrating either very specifically on organ systems where adverse effects on development might be expected from the nature of the transgene or they could follow some of the screens complied in the Phenotype Platforms devised by EUROMORPHIA:

1. Access EMPReSS via http://empress.har.mrc.ac.uk.

2. Go to Phenotype Platform.

3. Search presently developed platforms for appropriate screens:

   Behavior and cognition.

   Clinical Chemistry and Hematology.

   Hormonal and Metabolic Systems.

   Cardiovascular.

   Allergy and infectious diseases.

   Sensory Systems.

   Bone, Cartilage, Arthritis, Osteoporosis.

   Gene Expression.

   Necropsy Exam, Pathology Histology.

   In any case they should contain continuous daily observations to check wellbeing and normal behavior; weight monitoring throughout life–weekly for the first 4 weeks of life, monthly thereafter; a second general inspection at 9 weeks according to the dysmorphology screen (can be found within the Bone, Cartilage, Arthritis, Osteoporosis platform); a thorough necropsy (see Necropsy Exam, Pathology Histology platform) at the end of the observation time or at untimely death; and perhaps behavioral analysis following the Behavior and Cognition platform if any abnormal behavior is seen.

3.2. Detection of Germ Line Transmission by Spermatozoa Analysis and Breading Studies

The easiest way to purify spermatozoa from contaminating cells of the male genitourinary tract is to perform an in vivo swim-up purification. During spermatozoa movement up the genital tract they are separated from other cellular components of the ejaculate.

3.2.1. Purification of Spermatozoa

1. Cage males overnight with two to three female mice of reproductive age.

2. Early the following morning check the females for vaginal plugs.

3. Sacrifice females with vaginal plugs and perform a midline laparotomy to expose the uterus.

4. If the female mated during the previous night, the uterus should be swollen as it contains male sperm.

5. Make a ligature around the neck of the cervix with 4-0 silk suture and excise the uterus intact by cutting distal to the ligature, cutting also the vessels supplying the uterus and ovaries (see Note 3).

6. Place the intact uterus in a petri dish and flush the contents of each horn out with phosphate buffered saline using a 5-ml syringe.

7. Inspect the contents under the microscope for the presence of spermatozoa (see Note 4).

3.2.2. DNA Isolation and RT-PCR on Spermatozoa

Genomic DNA can be isolated from sperm using a modified procedure developed by Hossain et al. (1997) (11) and quantitative PCR (qPCR) is performed with probes and primers specific to the vector packaging signal or gene carried by the vector under investigation to detect and quantify vector in sperm or other tissues of interest.

1. Sperm is centrifuged at full speed in a bench top centrifuge for 5 min and supernatant removed. The sperm is washed with PBS and resuspended in lysis buffer at 3×10^6 to 2×10^7 sperm/ml and incubated at 55°C for 2 h. An equal volume of cold isopropanol is added to precipitate the DNA (see Note 5). The DNA is washed twice in 70% ethanol then resuspended in 200 μl TE buffer (see Note 6).

2. Measure the concentration of genomic DNA by spectrophotometry at 260 and 280 nm. 260/280 nm ratios should exceed 1.80.

3. To detect the relative amount of target sequence in different samples a dilution set of vector "spiked" sample should be prepared.

4. The genomic DNA should be diluted to 30 mg/ml with nuclease-free dH_2O and DNA samples "spiked" with between 1 and 10^5 vector copies to generate a standard curve by qPCR. To test the quality of the reactions each sample should be subjected to qPCR of either the 18S or GAPDH (housekeeping control) genes (supplied by Applied Biosystems). For the analysis of sperm, qPCR should be performed on a second internal control for a male-specific endogenous sequence, the mouse testis-specific Y encoded protein (TSPY) pseudogene, using Mouse Y-specific primers (12) TSPY (reverse primer) 5′-GAG AAC CAC GTT GGT TTG AGA TG-3′ TSPY (forward primer) 5′-TCC TTG GGC TCT TCA TTA TTC TTA AC-3′ and TaqMan probe 6 FAM-TCC TGG ATC AGA GTG GCT TAC CCA GG-TAMRA.

5. Prepare a single PCR master mix for all samples to be assayed. For each sample the mix will need to be prepared with 7 µl of distilled water, 10 µl of TaqMan Universal PCR Master Mix, and 1 µl of TaqMan gene expression assay (100 µm of TAMRA probe and unlabelled primers dissolved in TE buffer). TaqMan gene expression assays used should include sequences specific to GAPDH (Y-specifc DNA) and viral vector transgene or packaging signal.

6. Add 18 µl of master mix to each sample well in a frozen (−20°C) 96-well plate. 2 µl of genomic DNA is then added to each well. The reactions should be performed in triplicates and the standard curve of the DNA templates should be performed on the same plate. The plate is sealed with 96-well MicroAmp® Optical adhesive film.

7. Amplification reactions are performed using the thermal cycler (e.g., the 7900HT Real-Time PCR Thermal Cycler–Applied Biosystems) with parameters specific for the target sequence under examination (provided by Applied Biosystems). This will produce a standard curve of log vector copy number vs. Ct value (see Note 7). Intercept, Ct, and slope values should then be used to calculate copy numbers in experimental samples.

8. When tissues are used where cells are diploid, the DNA values can be converted into cell numbers using the calculations described below. Mass of the haploid genome and the number of base pairs are related by the formula:
$n = (m \times N_a)/M$ where n is the number of base pairs (haploid human cell $= 2.9 \times 10^9$ bp), m is the mass of the DNA, N_a is Avogadro's number (6.02×10^{23} bp/mol), and M is the average molecular weight of a base pair (660 g per mol). Mass per diploid cell $= ((2 \times 2.9 \times 10^9) \times 660)/6.02 \times 10^2$ g. Therefore, 1 diploid cell $= 5.92 \times 10^{-12}$ g and 1 ng of DNA from diploid cells $= 169$ cells.

3.2.3. Breeding Studies

Although analysis of extremely high numbers of male gametes is possible using sperm swim-up, the same is not possible with female mice. Superovulation is possible but collection of oocytes usually requires sacrifice of the mother and is difficult and time-consuming. Compare Chapter 12 Section 9.3. In contrast breeding studies permit analysis of tissue of pups at birth for detection of transgene by qPCR.

1. One female mouse that has been proven positive for (somatic) gene transfer and one male should be placed together in a cage for around 10 min, for the scent of the male to be transferred to the female. The female should be the subject of the investigation and the male should be wild-type, ideally from an outbred strain such as MF1 or CD1 (in principle this procedure could also be applied so that the male is the subject of investigation) (see Note 8).

2. The male should be removed and another male swapped into its place. The male should be gently coaxed to sniff the genitalia of the female. The scent of the previous male, and the scent of the female's glands induce the male to copulate.

3. The male can remain in the same cage as the female for up to 14 days. If, after this time, the female is not pregnant (swollen sides, fetuses palpable under anesthesia) then the above procedure is repeated.

4. If the female is pregnant, the male is removed and should be recorded as an effective stud for further breeding.

5. The female should be observed carefully and on the day that the pups are born they should be removed from the cage and tissues collected from each one for qPCR.

4. Notes

1. When inspecting the neonates it is best to remove the female to another cage. Otherwise, removal of the neonates will cause the mother to disrupt the nest while searching for them. Avoid disrupting the shape of the nest while removing the neonates for inspection.

2. For unique identification of neonates it is useful to perform a subcutaneous tattoo. By tattooing the paws up to 16 neonates per litter can be identified (1 no-mark, 4 unique single marks, 6 unique double paw marks, 4 unique triple paw marks, and 1 with all four marks). The tattoo can be made with fluorescent particles (FluoSphere) or tattoo ink.

3. For sacrifice of females in swim-up experiments, sacrifice can be performed by a Schedule 1 method, however, care must be taken not to rupture the uterus by excess force.

4. For detection of sperm it is not necessary to observe motile sperm as the morphology is very distinct.

5. In cases of incomplete lysis indicated by a transparent gel matrix to which the DNA pellet remains attached, the pellet should be mechanically separated from the protein gel with the help of fine forceps and scissors. For further purification a phenol/chloroform extraction can be added.

6. When DNA is precipitated the alcohol can be decanted or drawn out using a pipette. Ideally, as much alcohol should be removed as possible but since 70% ethanol is going to be used to wash the DNA it is acceptable not to remove all the residual ethanol before washing. After washing the DNA should be dried to remove the alcohol. At this stage the pellet should not be completely dried out as residual alcohol prevents successful resuspension in TE.

7. $^{\Delta\Delta}$Ct calculations after real-time PCR are used to provide relative amplification levels between the "standard" and experimental samples. $^{\Delta\Delta}$Ct values from this standard sample should be generated from a dilution set to find the optimal DNA concentration for real-time PCR. This may then be compared with unknown amples under investigation at the DNA concentration of choice. Between experimental and control sets, experiments should be run in triplicate to generate standard deviations of the means. Confidence intervals at 95% should be set and p-values calculated (values < 0.05 should be regarded significant).

8. For breeding studies, males and females should, ideally, be no younger than 7 weeks and no older than 12 weeks.

Acknowledgements

We wish to thank Paul Potter and Andrew Blake for introducing us to the International Mouse Phenotyping project and for helpful comments.

References

1. Rosenthal N, Brown SDM (2007) The mouse ascending: perspectives for human disease models. Nat Cell Biol 9:997–999

2. Gonzaga S, Henriques-Coelho T, Davey M et al (2008) Cystic adenomatoid malformations are induced by localized FGF10 overexpression in fetal rat lung. Am J Respir Cell Mol Biol 39: 346–355

3. Tarantal AF, Chen H, Shi TT et al (2010) Overexpression of TGF-{beta}1 in foetal monkey lung results in prenatal pulmonary fibrosis. Eur Respir J 36(4):907–914

4. Gates H, Mallon AM, Brown SDM (2010) High-throughput mouse phenotyping. Methods 53:394–404

5. Porada CD, Park PJ, Tellez J et al (2005) Male germ-line cells are at risk following direct-injection retroviral-mediated gene transfer in utero. Mol Ther 12:754–762

6. Ye X, Gao GP, Pabin C, Raper SE, Wilson JM (1998) Evaluating the potential of germ line transmission after intravenous administration of recombinant adenovirus in the C3H mouse. Hum Gene Ther 9:2135–2142

7. Schuettrumpf J, Liu JH, Couto LB et al (2006) Inadvertent germline transmission of AAV2 vector: findings in a rabbit model correlate with those in a human clinical trial. Mol Ther 13: 1064–1073

8. Park PJ, Colletti E, Ozturk F et al (2009) Factors determining the risk of inadvertent retroviral transduction of male germ cells after in utero gene transfer in sheep. Hum Gene Ther 20:201–215

9. Lee CC, Jimenez DF, Kohn DB, Tarantal AF (2005) Fetal gene transfer using lentiviral vectors and the potential for germ cell transduction in rhesus monkeys (*Macaca mulatta*). Hum Gene Ther 16:417–425

10. Kazazian H (1999) An estimated frequency of endogenous insertional mutations in humans. Nat Genet 22:130

11. Hossain AM, Rizk B, Behzadian A, Thorneycroft (1997) Modified guanidinium thiocyanate method for human sperm DNA isolation. Mol Hum Reprod 3:953–956

12. Wang LJ, Chen YM, George D et al (2002) Engraftment assessment in human and mouse liver tissue after sex-mismatched liver cell transplantation by real-time quantitative PCR for Y chromosome sequences. Liver Transpl 8:822–828

Chapter 16

Monitoring for Potential Adverse Effects of Prenatal Gene Therapy: Genotoxicity Analysis In Vitro and on Small Animal Models Ex Vivo and In Vivo

Michael Themis

Abstract

Gene delivery by integrating vectors has the potential to cause genotoxicity in the host by insertional mutagenesis (IM). Previously, the risk of IM by replication incompetent retroviral vectors was believed to be small. However, the recent observation of leukaemic events due to gamma retroviral vector insertion and activation of the *LMO-2* proto-oncogene in patients enrolled in the French and British gene therapy trials for X-SCID demonstrates the need to understand vector associated genotoxicity in greater detail. These findings have led to the development of in vitro, ex vivo, and in vivo assays designed to predict genotoxic risk and to further our mechanistic understanding of this process at the molecular level. In vitro assays include transformation of murine haematopoietic stem cells by integrating retroviral (RV) or lentiviral (LV) vectors and measurement of cell survival resulting from transformation due to integration mainly into the Evi1 oncogene. Ex vivo assays involve harvesting haematopoietic stem cells from mice followed by gene transfer and re-infusion of RV or LV infected cells to reconstitute the immune system. Insertional mutagenesis is then determined by analysis of clonally dominant populations of cells. The latter model has also been made highly sensitive using cells from mice predisposed to oncogenesis by lack of the P53 and Rb pathways.

Our investigations on fetal gene therapy discovered a high incidence of liver tumour development that appears to be associated with vector insertions into cancer-related genes. Many genes involved in growth and differentiation are actively transcribed in early developmental and are therefore in an open chromatin configuration, which favours provirus insertion. Some of these genes are known oncogenes or anti-onco-genes and are not usually active during adulthood. We found that in utero injection of primate HIV-1, HR'SIN-cPPT-S-FIX-W does not result in oncogenesis as opposed to administration of non-primate equine infectious anaemia virus (EIAV), SMART 2 lentivirus vectors and, most recently, the non-primate pLIONhAATGFP (FIV) vector, which both give rise to high frequency hepatocellular carcinoma. The peculiar integration pattern into cancer-related genes observed in this model makes the fetal mouse a sensi-tive tool, not only to investigate long-term vector-mediated gene expression, but also vector safety in an in vivo system with minimal immunological interference. The identification of distinct differences in geno-toxic outcome between the applied vector systems i.e. EIAV or FIV vectors versus HIV may indicate a particular biosafety profile of the HIV-1-based vector, which renders it potentially suitable for safe prenatal gene therapy.

Key words: Genotoxicity, Retrovirus vector, Lentivirus vector, Virus integration, Gene expression

Charles Coutelle and Simon N. Waddington (eds.), *Prenatal Gene Therapy: Concepts, Methods, and Protocols*,
Methods in Molecular Biology, vol. 891, DOI 10.1007/978-1-61779-873-3_16, © Springer Science+Business Media, LLC 2012

1. Introduction

The potential for permanent gene transfer offered by retrovirus vectors (RV) presents a genotoxic risk to the host. This is partly due to the preference for retroviruses to integrate in a semi-random way into actively transcribed genes with open chromatin configuration. Target site selectivity is also believed to be the reason for the higher genotoxic risk of retroviruses compared to their lentivirus counterparts since retroviruses integrate primarily near promoter regions, whereas lentivirus insertion occurs within genes' transcriptional units (1–4). Retrovirus-mediated insertional mutagenesis (IM) has been studied for several years, and the theoretical estimates for the potential frequency of mutation of a haploid locus were supported by cell culture assays that relied on mutant selectivity. Mutagenesis involving inactivation of the *hprt* locus, as a haploid genome target in male cells, or the promotion of cells to growth factor independence showed this risk to be in the order of 10^{-5} to 10^{-7} per provirus insertion (5–8). Therefore, the likelihood of adverse events caused by RV when used to correct genetic diseases in the clinic was considered remote. Unfortunately, the unexpected development of clonal dominance attributed to RV-mediated IM in X-SCID and CGH patient trials highlights the need to understand more about the association between vector insertion and genotoxicity in the host and to develop safer integrating gene therapy vectors (9–14).

As a result of the X-SCID clinical trial in France in which the first leukaemic patients were identified, ex vivo and in vitro models have been developed to examine RV and lentivirus vector (LV) genotoxicity. These models have confirmed that insertion by γ RV and, to a lesser extent, self-inactivating (SIN) RV and LV with this design, into proto-oncogenes can contribute to leukaemic development (15–21) that depends on factors such as the integrated vector copy number, vector configuration and even the transgene carried by the vector (17, 20, 22). More recently, the host cell transcription status, in combination with the mutational potential of the vector, has been shown to play a role in the emergence of clonal dominance (17, 22–24).

In vitro and ex vivo genotoxicity models developed thus far have mainly been concerned with cells of the haematopoietic system since gene therapy using stem cells is considered a more controllable way of introducing genetic modification to the host than by vector administration directly in vivo. Models to identify RV and LV genotoxicity following gene delivery directly in vivo to a variety of tissues with complex patterns of temporal and spatial gene expression have yet to be established.

We have recently developed two genotoxicity models. One is an in vitro assay that involves insertional mutagenesis of a haploid locus and the other an in vivo assay based on vector delivery to the fetal circulation in utero. Particularly, the latter highlights the importance of a direct in vivo genotoxicity model and confirms the need to design safer vector systems. The latter model is also pertinent to the safety assessment of prenatal gene therapy.

1.1. In Vitro Assays to Measure Vector-Related Genotoxicity

Our in vitro genotoxicity assay is an adaptation of the *hprt* in vitro assay in V79 Chinese hamster male cells that previously was universally employed as a genotoxicity model of chemical carcinogenesis and mutagenesis by ionising radiation (25–30). Using this system, instead of treating cells by chemicals, RV infection is used to determine IM-mediated loss of HPRT activity by gene inactivation following selection for *hprt* mutants. The mechanisms leading to this are still unclear; however, insertion within either introns or exons may lead to inactivation by interruption of the gene's coding sequences (6, 8) or by vector splicing to create novel RNA transcripts producing truncated proteins, which no longer perform their normal function (31). It is important to note that the *hprt* assay provides an estimated measure for gene inactivation of a single allele of a cellular gene mimicking mutagenesis that could be the initiating event leading to tumour development, for example of a tumour suppressor gene. Actual, genotoxicity to the organism to be caused in this way would require the unlikely chance of insertion and disruption of both alleles of the same gene by the vector unless loss of this allele already existed in the host. Alternatively, this measure could indirectly represent the frequency of mutagenesis as a result of single gene activation by an integrating vector, where the genotoxic effect on a single allele may suffice to induce uncontrollable cellular proliferation.

The *hprt* assay is easily performed and can be applied to almost any cell line with a single *hprt* copy but should ideally have a low frequency of spontaneous *hprt* mutagenesis and a stable karyotype. One advantage provided by the *hprt* system is that existing *hprt* mutants can be purged from culture populations using HAT pretreatment. Then IM-induced *hprt* mutants can be isolated using 6-thioguanine (6TG) selection, which reduces the number of background mutants in the analysis.

Initially, replication competent RV was found to be able to induce *hprt* mutation at a frequency of between 1×10^{-6} and 1×10^{-7} (6). The theoretical estimate of virus-induced mutagenesis of this locus is between 10^{-5} and 10^{-6}. This calculation is based on the genome being approximately 3×10^{6} kb in size with a typical gene, such as the hrpt gene, being around 30 kb in size. Therefore, an insertional mutagenesis library of 3×10^{6} cells each with a single provirus insertion would be expected to yield 30 hprt mutants. If an exon sequence alone is required for mutagenesis, this would bring the expected frequency to approximately 1×10^{-5} considering the hprt cDNA to be around 2 kb in size (32). In our hands, *hprt* mutagenesis by attenuated RV was found to occur at a frequency similar to that of the replication competent vector of 3.6×10^{-6} at high multiplicity of infection (8).

Although the frequency of IM obtained by the *hprt* assay relates to gene inactivation, as already stated, this frequency can also be applied to activation of a single gene by vectors carrying an intact 3′ LTR with the potential to drive host gene expression, if located in close proximity to the integrated vector. In conclusion, the *hprt* assay provides a measure of gene inactivation and the materials to study the mechanisms associated with this mode of genotoxicity. For example, the frequency of mutagenesis caused by integration into exon sequences, or due to insertion into introns followed by aberrant splicing by the vector, can be measured. The assay does not, however, provide directly a measure of gene activation and therefore assays involving transformation in vitro and in vivo have been developed.

Based on the theoretical estimate of IM at a single gene locus, Modlich et al. (2006) (39) developed a cell culture model assay to test genotoxicity by RV- or LV-mediated gene up-regulation in haematopoietic cells. This model also shows that SIN configuration vectors have the capacity to cause mutagenesis, albeit at a lower level than vectors with conventional LTRs. Their model uses transformed primary mouse cells that are selected by limiting dilution following RV exposure. Most importantly, insertion into the *Evi1* oncogene coupled with gene up-regulation is shown to be a common event and confirms the importance of the vector configuration in the transformation process.

The advantage of using cell culture systems as a measure of genotoxicity is that a rapid estimate of the potential of a vector to cause deleterious side effects to the genome may be obtained. Also, the material obtained to analyse insertional mutagenesis at the molecular level is free from contaminating polyclonal cells that are difficult to remove from the clonally derived dominant cells found with in vivo models. The avoidance of animals also makes these assays less costly.

The disadvantage is that the models rely on selection systems to obtain mutant cells, and therefore, the frequency of mutagenesis may not be representative of RV or LV genotoxicity at every locus. These assays do, however, identify vectors that cause mutagenesis and may be useful to obtain comparative frequencies between vector systems such as RV and LV to cause genotoxicity. Whether these models can be used to provide quantification of vector genotoxicity in vivo is, however, debatable since they do not take into account temporal and spatial gene expression that could bias vector integration into genes with different propensities to influence oncogenesis.

1.2. Ex Vivo Models of Genotoxicity

An ex vivo genotoxicity assay developed by Montini et al. (17) exploits a tumour-prone mouse model, *Cdkn2a−/−*, in which loss of p16^{ink4a} and p19ink^{Arf} genes results in the deficiency of the p53 and Rb1 pathways and leads to tumour development with a predictably early onset. Because this model is predisposed to oncogenesis, it is highly sensitive to vector mutagenicity as a result of oncogene activation or tumour suppressor gene inactivation with the development of malignant infiltrates in organs of lymphoid or myeloid origin. Infection of Lin⁻ bone marrow cells from this model with RV or LV, which are then transplanted into wild-type lethally irradiated mice, results in tumour development in a dose–dependent manner and also shows dependence on LTR activity for gene activation by comparison of active LTRs from MLV with self-inactivating (SIN) LTRs (17). More recently, this model has also been used to identify the potential of the spleen focus-forming virus (SFFV) promoter to activate oncogenes depending on its position in the LV backbone (33). Hence, this model has been proposed as useful to quantify the potential of a given vector for genotoxicity.

The advantage of the ex vivo model is that it follows the clinic protocol of gene transfer to stem cells by engraftment into the recipient to test for the development of clonal dominance. In addition, the model is very sensitive since the stem cells used reproducibly develop into leukaemias in a relatively short time frame for an in vivo model. The disadvantage of this model is that oncogenesis relies on mutations already in the host cell that could contribute to genotoxicity, and therefore mutagenesis by the vector may be biased to pathways particular to loss of P53 and Rb and thus not reveal alternative routes to leukaemia caused by the vector under examination.

1.3. In Vivo Models of Genotoxicity

The in vivo genotoxicity model developed in our laboratory is particularly relevant to the in utero gene therapy approach, in that it relies on direct in vivo vector application into the fetal circulation. By administration of LV to the fetal blood circulation via yolk-sac vessel injection at E16 gestation in outbred, fully immunocompetent MF-1 mice (although theoretically any strain that tolerates the injection procedure may be used), we can achieve gene transfer to most organs, although the liver is mainly transduced (34). Introduction of a primate HIV-1-based vector HR'SIN-cPPT-S-FIX-W and non-primate EIAV SMART vectors, each providing human factor IX (hFIX) gene expression to MF-1 and hFIX knockout (KO) mice, showed only the non-primate vector to be associated with high frequency (80%) liver tumour development (35). These observations were also made with the β-galactosidase reporter as the transgene, and more recently, when we used the pLIONhAATGFP FIV-derived vector (unpublished data), and therefore we attributed genotoxicity to the non-primate vector backbones or their configurations (each shows little homology with HIV vectors). The analysis of these tumours by Southern blotting showed each tumour to be clonally derived. By using inverse (36) or ligation adaptor-mediated (37) PCR to identify provirus insertions in or close to RefSeq genes (within a 100 kb integration site window as a theoretical distance by which vector insertion is believed to influence the expression of a gene carrying the integrated vector), we were able to assign genes as candidates for association with cancer development. More than 50% of these genes were found to be registered in the Mouse Retroviral Tagged Cancer Gene Database (RTCGD) (38). Conversely, the HIV-1-based LV in this model was not associated with tumour development. These data suggest a potential of this assay for analysis of vector safety since many genes involved in cell growth, differentiation and development, known to have oncogenic potential, are actively transcribed and are therefore targets for vector integration. Given that only the EIAV and FIV vectors, but not the used HIV vector, induce tumours, the fetal model described here appears to be highly sensitive to identify vector configurations with genotoxic potential following integration into the host genome. Our data also indicate that the applied HIV vector may be safe for prenatal gene therapy. More analyses are clearly needed since the exact mechanisms underlying the described differences between the oncogenic potential of these two vectors are still unknown.

It is important to note that each of the discussed genotoxicity assays uses rodent cells as a measure of insertional mutagenesis. These cells are more predisposed to tumour development than human cells, and therefore each must be viewed with certain caution as reliable predictors for mutagenesis in the clinic. The observations of vector genotoxicity in the clinic have, however, been recapitulated using these models and therefore important information can be gained by each test system as to the mechanisms of oncogenesis caused by each vector and ways to improve vector design for safety.

Each model has advantages and disadvantages. The in vitro models are restricted to selection of mutagenesis using defined phenotypes and therefore may not identify the result of mutagenicity of genes not involved in *hprt* mutagensis or early stage transformation. These models do, however, provide genotoxicity data within a short time period of within approximately 1–2 months. The ex vivo model may also not uncover alternative pathways of oncogenesis such as those involved in solid tumours; however, the development of adverse effects is found between around 60 and 380 days as opposed to the fetal model that identifies genotoxicity leading to the development of solid tumours but takes between 154 and 666 days.

In conclusion, the importance of genotoxicity assays to understand the cause and to measure the risk of adverse effects by gene therapy vectors on the host cannot be overstated. With the currently used genotoxicity assays in cell culture and by skilful in vivo vector application to wild-type or genetically modified animal models, we are becoming more confident that gene therapy to adults or even prenatally will be possible with minimal side effects by application of integrating vectors that have been demonstrated with safe design.

Following vector administration, genotoxicity assays depend on molecular techniques that combine established protocols, which have been carried out in most laboratories for several years as well as recently devised methods more specifically designed to determine the outcome of vector/host interactions. Since genotoxicity assays are under continual research and development, the assays presented in this chapter will be the subject to constant improvement. Their use should be considered, however, before clinical vector application, in order to reveal potential differences in genotoxic outcome that may depend on the vector configuration and/or the transgene. At this time, there is no universally accepted model to predict genotoxicity, and no single assay that should be considered the most appropriate model. Anyone therefore intending to reach the clinic with his or her vector of choice may wish to use more than one of the test assays described here if lifelong gene supplementation by integrating vectors is envisaged. Consideration should also be made on the time frame for experimentation since in vivo models, unless predisposed to tumourigenesis, may take the lifetime of the host to determine safety, whereas in vitro models give result within a much shorter time frame.

For completeness, in this chapter we present some of the most current in vitro and in vivo model systems used to test RV and LV genotoxicity. These assays may be informative in efforts to improve vector biosafety and especially the in utero system, may be applied if prenatal gene therapy is under consideration

2. Materials

2.1. Cell Culture and the hprt Insertional Mutagenesis Assay

1. V79E cells (8) can be obtained from Brunel University, Uxbridge, Middlesex, UB8 3PH, UK. Available from Dr M. Themis.

2. Dulbecco's Modified Eagle's Medium GlutaMAX™ supplemented with 1% each Penicillin and Streptomycin antibiotics and 10% fetal calf serum stored at 4°C (FCS).

3. Solution of 0.25% trypsin and 1 mM ethylenediamine tetraacetic acid (EDTA) stored at 4°C.

4. DEAE dextran 0.5 mg/ml stock solution (100×). Pass through a 4-μm filter and stored at room temperature.

5. HAT Media Supplement (50×). Pass through a 4-μm filter and store at –20°C Hybri-Max® containing hypoxanthine, aminopterin, and thymidine diluted to 1× with culture medium stored at 4°C.

6. Phosphate-buffered saline (1×) stored at 4°C.

7. 10 mg/ml 6 Thioguanine (6TG, ≥98%). Pass through a 4-μm filter store at –20°C.

2.2. In Vitro Cell Transformation Assay

1. C57B16/J mice (Charles River Laboratories).

2. Enrichment of Murine Haematopoietic Progenitors kit. Store at –20°C (StemCell Technologies).

3. StemSpan SFEMexpansion medium stored at 4°C (StemCell Technologies).

4. Murine cytokine cocktail composed of 100 ng/ml stem cell factor, 100 ng/ml thrombopoietin, 100 ng/ml Flt3-ligand. Store at –20°C (StemCell Technologies).

5. 20 ng/ml Interleukin-3 Store at –20°C (PeproTech).

6. Retronectin stored at 4°C (Clonetech).

7. Iscove's Modified Dulbecco's Media (IMDM) supplemented with 10% FCS, 1% each penicillin and streptomycin antibiotics and 2 mM glutamine stored at 4°C.

8. 24-well and 96-well non-tissue culture plates.

9. Antibodies for Sca1 and c-KIT. Store at –20°C (Abcam®).

10. Jenner Blood Stain Solution (Newcomer Supply Co) made with 20 ml Jenner Stain and 20 ml dH$_2$O. Store at room temperature.

11. Giemsa Solution (Fisher). Make fresh with 1 ml Giemsa Stain and 50 ml dH$_2$O. Store at room temperature.

12. 1% Acetic Acid Store in an acids cabinet at room temperature.

13. Tissue slides and control slides for special stain requested.

14. Microwaveable tissue slides and coverslips (Abnova).

15. Synthetic mounting media (Marivac).

16. Methanol. Store in a solvent cabinet at room temperature.

2.3. Ex Vivo Mutagenesis Assay

1. *Cdkn2a–/–* mice (Charles River Laboratories).

2. Enrichment of Murine Haematopoietic Progenitor kit.

3. StemSpan SFEM expansion medium (StemCell Technologies). Store at 4°C.

4. Cytokine cocktail: 100 ng/ml Stem cell factor, 100 ng/ml thrombopoietin, 100 ng/ml Flt3 ligand, and 20 ng/ml interleukin-3 (Pepro Tech). Store at –20°C.

5. RPMI medium supplemented with 10% FCS with 1% Pen/strep and 2 mM glutamine. Store at 4°C.

6. FVB/N.129 mice (Charles River Laboratories).

7. Red blood cell lysis (StemCell Technologies) with 7% ammonium chloride Store at 4°C.

8. PBS. Store at 4°C.

9. 40 μm nylon cell strainers (StemCell Technologies).

10. 5% rat serum, store at –20°C.

11. 2% fetal calf serum, store at –20°C then at 4°C after thawing.

12. 2 μg/ml R-phycoerythrin-conjugated monoclonal antibodies (BD Biosciences Pharmigen). Store at –20°C.

13. Propidium iodide. Keep at room temperature.

14. Paraformaldehyde. Store in 5 ml aliquots at –20°C.

2.4. Fetal Genotoxicity Assay

1. Sodium citrate (1/9 dilution in dH_2O). Store at 4°C.

2. Isopentane. Store in a solvents cupboard.

3. Paraformaldehyde. Store in 5 ml aliquots at –20°C.

2.5. Southern Blotting

1. DNA extraction buffer (1×): 10 mM Tris/HCl (pH 7.5), 100 mM ethylene tetraacetic acid (EDTA), (pH 8.0), 25 μg/ml RNase H and 5% SDS. Keep at room temperature.

2. Proteinase K, 100 mg/ml stock solution in distilled water. Store in aliquots of 500 μl at –20°C.

3. Phenol equilibrated in 50 mM Tris/HCl (pH 8.0). Store at 4°C.

4. Chloroform. Store in a solvents cabinet at room temperature.

5. 100% Ethanol analytical grade. Store in a solvents cabinet at room temperature 6. 0.6% Agarose gel made with TAE running buffer: 4 mM Tris-acetate and 1 mM EDTA (pH 8.0) and 0.5 μg/ml ethidium bromide. Keep at room temperature.

6. Ethidium bromide: 10 mg/ml stock solution. Keep at room temperature wrapped in foil or suitable dark container.

7. Nylon membrane Amersham Hybond™-XL.

8. Transfer buffer sodium citrate (10× SSC). 20× SSC buffer concentrate used by diluting with dH_2O that contains 3 M sodium chloride and 0.3 M sodium citrate (pH 7.0). Store at room temperature.

9. Watman 3MM paper and paper towels.

10. Depurination solution: 0.2 N HCl. Keep in a dedicated acids cabinet.

11. Denaturation buffer (1×): 0.4 N NaCl, 1 M NaCl. Make fresh each time used.

12. Neutralisation buffer (1×): 0.5 m Tris/HCl (pH 7.2), 1 M NaCl. Store at room temperature.

2.6. Northern Blotting

1. TRIzol reagent® Store at 4°C (Invitrogen).

2. Polytron homogeniser (Qiagen).

3. Isopropanol. Store in a solvents cabinet at room temperature.

4. Chloroform. Store in a solvents cabinet at room temperature.

5. dH_2O treated with diethylpyrocarbonate (DEPC). Store at room temperature.

6. Denaturation buffer: 50% deionised formamide, 2.2 M formaldehyde. Make fresh each time used.

7. Running buffer: 400 mM 10× (N-morpholino) propanesulphonic acid (MOPS) buffer (pH 7.0), 100 mM Na Acetate, 10 mM EDTA (pH 7.0). Store at room temperature.

8. Loading dye: 50% glycerol, 5% formaldehyde, 0.25% bromophenol, 1 mM EDTA. Store at room temperature.

9. Ethidium bromide: 1 mg/ml stock solution. Keep at room temperature wrapped in foil or suitable dark container.

10. Transfer buffer sodium citrate (20× SSC) (pH 7.0). Store at room temperature.

11. RNA size marker 100 bp ladder. Store in aliquots at −20°C.

12. 5% Acetic acid. Store in an acids cabinet.

13. 0.1% Methylene Blue. Store at room temperature.

14. Hybridisation buffer: 0.4 M Na_2HPO_4, 6% SDS, 1 mM EDTA. Store at room temperature.

15. Wash buffers: (1) 40 mM Na_2HPO_4 5% SDS, 1 mM EDTA. (2) 40 mM Na_2HPO_4, 1% SDS, 1 mM EDTA. Store at room temperature.

16. Amersham Rediprime II DNA labelling system. Store in aliquots at −20°C (GE Healthcare).

17. Bio-Spin™ 6 column (Bio-rad).

18. Kodak BioMax® MR-1 autoradiography film (PerkinElmer).

2.7. Inverse PCR

1. Restriction enzyme (4 base pair cutter) with a recognition sequence close to (<100 bp) and 3′ of the RV or LV LTR.

2. Design two primers that will hybridise to each end of the vector 5′ LTR. Ensure these are 5′ to the enzyme 1 restriction site. One primer should read in a 5′ direction outwardly from the vector to the integration site and host genome and the second should read in the 3′ direction into the vector genome. Ideally, 20–23 oligomer primers allow melting temperatures high enough to avoid non-specific amplification products. Store in aliquots at −20°C.

3. Choose another restriction enzyme to cut between these primers (enzyme 2).

4. T4 ligase.

5. PCR reagents: sterile dH_2O, Taq polymerase buffer, dNTPs, $MgCl_2$, Taq polymerase. Store at −20°C.

6. TAE buffer: 4 mM Tris-acetate and 1 mM EDTA (pH 8.0). Store at room temperature.

7. TA cloning kit® (Invitrogen).

8. Agar plates with 50 μg/ml ampicillin and spread with 40 mg/ml X-gal solution. Make fresh each time used or can be stored up to 2 weeks at 4°C.

9. LB broth (1 L): 15 g Tryptone, 7.5 g yeast extract, 7.5 g NaCl, dH_2O up to 1 L. Stored at room temperature.

10. PCR-positive colonies identified and isolated from steps 9 and 10 should be sequenced to identify the inserted genomic DNA. This step is usually performed using a commercial DNA sequencing facility.

2.8. Ligation Adaptor-Mediated PCR

1. PCR reagents: Taq polymerase buffer, dNTP's, $MgCl_2$, Taq polymerase. Store at −20°C.

2. 500 mM Tris/HCl (pH 7.5). Store up to 2 weeks at 4°C.

3. 100 μM LC1-5′ GACCCGGGAGATCTGAATTCAGTGGCA CAGCATTAGG-3′.

   100μM LC2-5′ AATTCCTAACTGCTGTGCCACTGAATT CAGATC-3′. Aliquot primers and store at −20°C.

4. Dynal Kilobase binder Kit® (Invitrogen).

5. PBS and BSA (Sigma-Aldrich). Store at 4°C.

6. Klenow Polymerase 2 U/μl. Store at −20°C.

7. Enzyme reaction buffer. Store at −20°C.

8. Hexanucleotide mix. Store at −20°C.

9. 4,000 μg/ml TSP509I enzyme. Store at −20°C.

10. Micron-YM30 concentrator (Millipore). Keep at room temperature.

11. 10 mM ATP. Store at −20°C.

12. 2 Uμl T4 Ligase. Store at −20°C.

13. 0.1 M NaOH. Keep at room temperature.

14. TAE buffer: 4 mM Tris-acetate and 1 mM EDTA (pH 8.0). Keep at room temperature.

15. TA cloning kit®.

16. Agar plates with 50 μg/ml ampicillin and spread with 40 mg/ml X-gal solution. Make fresh each time used or can be stored up to 2 weeks at 4°C.

17. LB broth (1 L): 15 g Tryptone, 7.5 g yeast extract, 7.5 g NaCl dH_2O up to 1 L. Store at room temperature.

2.9. Real-Time PCR

1. Trizole reagent. Store at 4°C.

2. Reverse transcriptase. Store at –20°C.

3. Random hexamer primers. Store at –20°C.

4. High capacity cDNA synthesis kit. Store at –20°C.

5. TaqMan Gene Expression Assays specific to each gene under investigation. Store at –20°C.

6. TaqMan® Universal PCR master mix with AmpErase UNG (Applied Biosystems). Store at –20°C.

3. Methods

3.1. The hprt Insertional Mutagenesis Assay

As the spontaneous frequency of *hprt* mutagenesis in V79 cells is expected to be in the region of 10^{-5} to 10^{-6} (8, 32), 10^{7} cells should be used in the assay to obtain a sufficient number of mutant clones for molecular analyses. Typically, cells should be treated with HAT medium prior to infection by the vector of choice to remove existing *hprt* negative cells that would complicate the identification of *hprt*-mutated clones as a result of IM. After exposure to infection, negative selection for *hprt* mutants is carried out using 6TG. Clones that grow in 6TG are then isolated and expanded for molecular analyses to determine provirus copy number and vector insertion sites. As the assay selects for loss of *hprt* function, there is no need to determine protein expression.

Provirus copy number is determined using Southern analysis of DNA isolated from each clone and compared to that of spontaneously developing clones isolated without prior exposure to RV and of cells isolated directly after HAT exposure providing the wild-type pattern of the intact *hprt* gene. Southern analysis also identifies rearrangements of the *hprt* locus possibly due to chromosome breakage resulting from the introduction of fragile sites by the vector or due to vector recombination.

Provirus insertion sites can be identified using a number of methods. One of the most efficient technologies developed for this purpose is ligation adaptor-mediated PCR (LAM) (37), although inverse PCR may also be used (36). Once the genes with provirus insertions have been identified, these may be used to confirm *hprt* integration and/or insertions in alternative genes, which will be used to measure changes in gene expression levels possibly as a result of RV insertion. Loss of gene expression can also be identified by Northern analysis on total RNA isolated from each clone using the gene of interest, into which vector was inserted, as a probe. This will also provide identification of novel transcripts resulting from vector-driven gene expression or aberrant splicing between the host gene and the RV. The *hprt* assay will therefore provide information on whether the vector under examination can interfere with gene expression once inserted. Although this would be expected to occur from integration into *hprt* exon sequences by any integrating vector, a safe vector would not be expected to generate *hprt⁻* cells once inserted into intronic *hprt* DNA.

3.1.1. Infection of Cells and Selection of hprt⁻ Mutants

1. V79E cells (8) are maintained on 100 mm culture dishes in DMEM 10% FCS at 37°C in an atmosphere of 10% CO_2. Prior to IM assay 5×10^4 cells per dish are treated to remove *hprt* negative mutants in HAT medium (1×) for 72 h. Cells are then trypsinised and seeded at 1×10^6/100 mm dish for infection of a total of 1×10^7 cells (10 dishes per vector dose to be used). An additional set of cells should be seeded as controls that will not be subjected to RV infection (see Notes 1–3).

2. Virus dilutions at multiplicities of infection (MOI) of between 1 and 100 (ideally 1, 10, and 100) are complexed for 20 min with 5 μg/ml of DEAE dextran and placed into each 100 mm dish containing 10 ml medium and left overnight for infection to occur. The next day the medium is changed to remove virus and cells are re-fed with 10 ml fresh medium and allowed a 24 h recovery prior to selection for loss of the *hprt* phenotype in 6TG (1×). Clones that have lost *hprt* expression should appear as colonies that may be harvested approximately 10 days following the addition of 6TG.

3. Clones may be removed from culture dishes using Pasteur pipettes or standard cloning rings. Care should be taken when isolating clones to prevent cross-contamination of cells between clones. This may be achieved by removing only one colony per 100 mm plate if desired. Additionally, isolated clones may be seeded onto 96-well plates by diluting cells to 40 cells/10 ml (100 μl/well) in order that 30–40 colonies per 96-well plate can be expanded for isolation. Isolated clones are then grown in T25 flasks before transfer to 100 mm dishes to obtain between 10^6 and 10^7 cells for DNA and RNA isolation. Each clone should be stored in liquid nitrogen for future work.

3.2. In Vitro Cell Transformation Assay

This assay follows the protocol of Modlich et al. (2006) (39) and relies on transformation of primary mouse haematopoietic cells by RV. Cells are infected then plated using limited dilutions followed by growth and analysis of re-plated cells stimulated to have growth advantage as a result of IM by RV integration. Using this assay, vector insertion is commonly found in known oncogenes and vector effects on these genes can be observed as a result of cellular transformation. Hence, this assay can be used as an indicator of LTR activity or internal promoter read-through and for the generation of aberrant RNA transcription.

1. Lin⁻ purified murine stem cells of haematopoietic origin isolated from bone marrow are obtained from C57B16/J mice and

enriched using the Enrichment of Murine Hematopoietic Progenitors kit. Cells are then plated at a density of 1×10^6 cells/ ml in StemSpan SFEM expansion medium on non-tissue culture dishes. Two days later cells are used for infection by the vector chosen for genotoxicity assessment.

2. Six-well non-tissue culture dishes are coated with Retronectin at 1.5 µg/cm² as per manufacturer's instructions.

3. Each culture well is then pre-loaded with the vector of choice by addition of 1 ml of vector stock to the wells followed by centrifugation for 10 min at $1,000 \times g$ at 10°C. This is repeated four times leaving 1 ml of medium in each well. The MOIs usually used to infect cells are between 1 and 10 (typically 2×10^6 cells are to be infected/well).

4. A 1 ml aliquot containing 2×10^5 cells is added to each well, and the dish is centrifuged at $1,500 \times g$ for 1 h at 30°C. Cells are then left for infection overnight.

5. Transfer the cells to a fresh pre-loaded plate for a second transduction.

6. Transduction levels and vector copy number/cell should be monitored. The measurement of the levels of transduction will depend on the transgene. An identical vector carrying a reporter gene may be used for this in parallel, where flow sorting is used or a haemocytometer to count visibly transduced cells. Additionally, real-time PCR can be used with primer/probe sets designed to recognise the virus packaging signal for average copy number analysis.

7. Following infection cells are expanded in culture medium with a murine cytokine cocktail composed of 100 ng/ml stem cell factor, 100 ng/ml thrombopoietin, 100 ng/ml Flt3-ligand and 20 ng/ml interleukin-3 whilst adjusting the cell density to 5×10^5 cells/ml and gradually changing the medium to IMDM with 10% FCS, 1% Pen/strep and 2 mM glutamine plus cytokines. After 2 weeks cells are transferred to 96-well culture plates at 100 cells/well and the number of positive wells with cell growth is counted as the re-plating cell frequency. The re-plating frequency is the measure of the transformation potential of the vector. Calculations should be made based on Poisson statistics using L-Calc software (Stem Cell Technologies, Vancouver, BC, Canada) compared to control mock-infected cells (39).

8. Cell markers should be identified using FACS analysis for Gr1, CD11b, CD45R/B220, CD3e, and TER119, Sca1 and c-KIT. Cells should be treated as in Section 3.3.5. Cell morphology should be analysed using May-Grunwald/Giemsa staining.

9. May-Grunwald/Giemsa staining: Add methanol and leave for 6 min. Add Jenner Solution and leave for 6 min. Add Giemsa Solution and leave for 45 min. Rinse in dH$_2$O. Add 1% Acetic

Acid and dip four times. Rinse again, dyhydrate clear and cover with a cover slip. Positive cells will appear with a pink cytoplasm and blue nuclei.

10. Positive clones isolated should be subjected to Southern analysis and inverse or LAM PCR to identify vector insertion sites followed by BLAST (http://www.ncbi.nlm.nih.gov/genome/seq/MmBlast.html) and BLAT (http://genome.ucsc.edu) searches of the mouse genome. On Southern analysis distinct provirus bands should appear in contrast to a smear from the control DNA without provirus insertions (Fig. 2).

11. Changes in gene expression for each gene identified should be analysed by Northern blotting or real-time PCR.

3.3. Ex Vivo Mutagenesis Assay

1. Lin⁻ cells are isolated from *Cdkn2a−/−* mice following sacrifice by flushing femurs with 2% PBS using a 21-gauge needle.

2. Cells are then purified using the Enrichment of Murine Haematopoietic Progenitor kit as per the manufacturer's instructions and grown in StemSpan SFEM expansion medium with stem cell factor, thrombopoietin, Flt3 ligand, and interleukin-3 at 1×10^6 cells/ml.

3. After pre-stimulation, cells are infected (Subheading 7.4.2, steps 2–5) for 12 h up to three times and then re-suspended in RPMI medium (supplemented with 10% FCS with 1% Pen/strep and 2 mM glutamine plus a murine cocktail composed of cytokines-100 ng/ml) stem cell factor, 100 ng/ml, thrombopoietin 100 ng/ml Flt3-ligand and 20 ng/ml interleukin-3- whilst keeping the cell density to 5×10^5 cells/ml.

4. FVB/N.129 mice (Charles River Laboratories) are transplanted at 6 weeks of age. Mice are lethally irradiated with 11.5 Gy and injected with 7.5×10^5 cells/animal (include a cohort of mock-infected cells for control mice) via the tail vein.

5. FACS analysis: Six weeks later, mice should be bled and blood cells lysed with 7% ammonium chloride. Cells of the bone marrow, spleen, thymus, and lymph nodes should also be harvested by washing each tissue in PBS then physically disrupted and passed through a 40 μm nylon cell strainer after blood cell lysis to obtain single cells. 1×10^6 cells are blocked in 5% rat serum with 2% fetal calf serum in PBS for 15 min. Aliquots of cells (150 μl) should be mixed with R-phycoerythrin-conjugated monoclonal antibodies at 2 μg/ml. For each cell type, lineage-specific antibodies should then be used. Cells should be stained with 2.5 g/ml propidium iodide (PI) after a 30 min incubation at 4°C and washing with PBS containing 2% fetal bovine serum. The software of choice (such as WinMDI 2.8) to analyse the cells should be set to exclude at least 98% of PI-labelled non-viable cells, and after removal of these, at least 5,000

antibody-positive/transgene-positive PI clear cells may be scored. Each lineage of cells will then be available for histological examination.

6. For histological examination, tissues should be fixed in 4% paraformaldehyde and sectioned (4 μm thick) then stained with haematoxylin and eosin. The tissue will then be ready for analysis and will require specialist histopathological expertise. From this, for each animal scoring may be performed for the following: 0 = no pathological infiltration; 1 = mild infiltrate; 2 = moderate; 3 = heavy.

7. To perform secondary transplants of vector treated cells, bone marrow is collected from each treated mouse, and 2×10^5 to 2×10^6 cells are used for transplantation to sub-lethally irradiated FVB/N.129 mice (5.75 Gy). Mice are treated once again as described above to identify leukaemic cells. Malignant infiltrates will appear in primary and secondary transplanted mice.

3.4. Fetal Genotoxicity Assay

1. Fetal administration of RV is described in Chapter 10. Typically, a vector dose of 1×10^7 IU is used for injection of each mouse fetus at E16 gestation (MF-1 mice or the strain of choice). Retrovirus vectors will have been assayed for titre as described in Chapter 5. Animals should be maintained in a healthy manner under the appropriate husbandry conditions for the strain in use in accordance with Home Office guidelines and with the appropriate Home Office Licenses. Four fetuses may be injected per mouse to achieve optimum survival. To serve as control injections three mice should be injected with vector buffer only, three with the vector less transgene, three with integrase mutations to prevent virus integration, three without tansgene and mutated integrase and six mice should not be injected. In our hands, experimental animal mortality following LV injection was initially 20% (40). This has been reduced by careful husbandry to below 5%. Based on this frequency, we performed a power calculation using the software, "PS Power and Sample Size Calculations" (41), which is based on a chi-square test, using alpha = 0.05, power = 0.8 and a spontaneous tumour incidence (p0) of 5%. With an expected tumour incidence (p1) of 80%, which is based on our published observations (40), the required sample size per group (assuming equal number in each group) to test each vector for genotoxicity is 8. This should yield between six and eight liver-treated and mock-treated animals.

2. To test each vector prior to the establishment of mouse cohorts, a single pregnant female should be injected with the vector under examination to obtain up to four fetuses for sacrifice. Forty-eight hours after injection, liver samples should be taken

for molecular analyses from this early time-point cohort for analysis of successful vector infection. If the transgene on the vector permits the evaluation of positive expression, this should also be monitored. Infection may also be determined using real-time PCR for the vector packaging signal.

After successful gene transfer by a vector, mouse cohorts for genotoxicity analysis can be studied. All animals from these experimental and control groups are allowed to come to birth. The resulting offspring animals should be weighed monthly and monitored for signs of sickness according to Home Office guidelines and palpated at this time under anaesthesia (see Note 4).

3. Mice that develop any signs of abdominal growths should undergo laparotomy to determine whether tumours are developing. If possible, no tumour should be allowed to reach more than 1.3 mm^2 in size (Fig. 1).

4. Tumour-bearing mice, to be sacrificed (using a schedule 1 procedure), should be bled before sacrifice. Blood (100 μl) should be prevented from coagulation by mixing with 12 μl sodium citrate solution. The sample is centrifuged in a bench top centrifuge at $6,000 \times g$ for 5 min and the clear surface layer removed

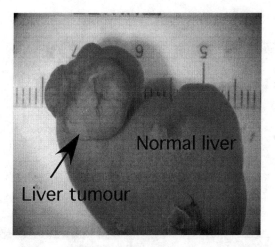

Fig. 1. Tumour development in a fetally treated adult mouse with EIAV lentivirus. Fetal mice were injected at E16 gestation with 1×10^7 EIAV SMART 2 LV. Tumours develop from about 4 months after birth in the treated animals. Mice are palpated each month and those suspected of tumour development are subjected to laparotomy. Tumours appear as large, highly vascular growths. Histological analysis of each liver tumour shows them to be hepatocellular carcinomas surrounded by normal liver architecture. Tumour tissue is subjected to DNA extraction followed by Southern analysis that indicates clonal development in most cases (40). LAM PCR of tumour DNA reveals the position of the vector in genes following BLAST and BLAT analysis. Tumour RNA is then subjected to gene expression analysis by real-time PCR to determine the involvement of the vector on the inserted gene.

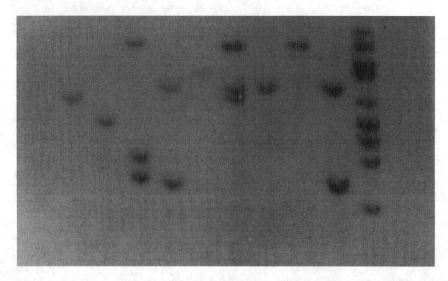

Fig. 2. Southern analysis for provirus copy number in infected V79 clones. V79E cells infected with MLV RV at an MOI of 10. Cells were exposed to RV for 12 h followed by removal of virus and the addition of fresh medium. Cells were allowed to expand in culture, then individual clones isolated and grown for DNA analysis. 10^7 V79E cells were used to isolate genomic DNA, which was restricted with the enzyme Hind III, which cuts the vector once leaving the transgene sequence and 3' LTR intact for hybridisation with a transgene probe. DNA was transferred to nylon membrane and the transgene probe labelled with a-^{32}P-CTP using a random prime labelling kit (Amersham Rediprime II DNA labelling sytem, GE Healthcare) was hybridised to the membrane. Positive hybridisations show provirus integrations vary in number from 1 to 9. Lane 1; uninfected V79E cells, lane 2–11 infected V79Es.

from the cells below. The cells may be re-suspended in 0.5 ml cell freezing medium and each sample stored frozen for subsequent analysis if required. Isolated white blood cells and plasma will also be useful to identify oncogenesis in blood-borne cells. The plasma also serves to identify replication competent vector in the blood. All tissues should be harvested from these mice to determine the bio-distribution of gene transfer vectors.

5. Tissues for molecular assays should be collected in cryovials and frozen in isopentane chilled by liquid nitrogen (see Note 5). For histological analyses, tissues should be placed in 7 ml bijou bottles immersed in 4% paraformaldehyde to be fixed over night followed by three washes in PBS then stored in 70% ethanol. These can then be paraffin embedded as described in Chapter 10. Sample tissues including the liver, heart, lung, kidney, brain, gonads, muscle, spleen, and thymus should be collected from all experimental and control animals.

6. PCR should be performed on maternal tissues (Fig. 2) to determine vector spread during the in utero gene transfer procedure as described in Chapter 13 using primers designed for the vector of choice.

3.5.1. Genomic DNA Extraction and Southern Analysis

1. Mouse tissue (100–500 mg) or 1×10^7 V79E cells are used to isolate genomic DNA for vector copy number analysis or *hprt* gene profile. The method used follows that described by Sambrook et al. (42) with minor modifications (see Notes 6–9). Tissues are placed in mortar and a pestle used in the presence of liquid nitrogen to grind tissues into a fine dust before the addition of 1 ml extraction buffer using a 1-ml tip with the end cut to create a wide aperture followed by the addition of 2 ml additional extraction buffer. The mix is then transferred to a 50-ml Falcon tube. The mixture should appear viscous (see Note 6).

   Alternatively, if cells in culture are used, they are trypsinised (1×) for 2 min and the trypsin is inactivated with DMEM containing 10% FCS. Cell pellets are then re-suspended in 1 ml followed by a further 2 ml extraction buffer before the mix is transferred to a 50-ml Falcon tube.

2. To the viscous mix 5 µl of RNase H is added followed by incubation for 1 h at 37°C. Proteinase K 5 µl is then added followed by incubation for 2 h at 55°C with occasional gentle mixing. Ensure to mix the enzyme/buffers into the DNA gently with a sterile tip to aid comprehensive digestion.

3. An equal volume of phenol is added to the solution, which is then gently inverted four to five times before centrifugation at 9,000 rpm to separate the organic from the aqueous component containing the DNA (represented by the top layer). A cut 1 ml pipette tip is used to remove the required aqueous solution in aliquots into a fresh 50-ml falcon tube. The procedure is repeated using a 1:1 Phenol/chloroform mix and finally using a chloroform only extraction. Three volumes of ethanol are added to the final aqueous solution to precipitate the DNA, which will appear as a white "cotton-like" mass after gentle inversion. No salt is needed in the precipitation step. The unwanted ethanol is gently poured away leaving the DNA precipitate in the tube. Distilled water is added briefly to the DNA to wash away residual salt followed by rapid removal (~2–10 s). Alternatively, 70% ethanol can be used. The DNA is then dried almost to completeness then 1 ml of distilled water is used to re-suspend the DNA in on ice overnight.

4. 10 µg of genomic DNA should be digested with the appropriate restriction enzyme (see Note 7) to obtain the desired fragments to be separated on agarose gels. Gels are usually prepared at 0.6% with 0.05 µg/ml of ethidium bromide and run over a 4 h period to achieve clear separation of bands.

5. Gels are then transferred to nylon membranes for hybridisation with virus probes or the cDNA of the gene of interest. Firstly, cut upper left corner of the gel for orientation, then wash in distilled water for 30 min, and soak in 10× SSC for 5 min.

Build a platform by using a "bridge" of 3MM Whatman paper over an inverted gel former placed in a plastic tray filled with 20× SSC. Place the gel on top of the "bridge". Remove air bubbles between the gel and 3MM paper (rolling with a 10-ml tissue culture pipette over the gel can perform this). Pre-wet the Nylon membrane (cut to the exact size of the gel) in 10× SSC and place over the gel. Remove air bubbles between the gel and the membrane. Cover the nylon membrane with three sheets of 3MM paper (cut to the exact size of the gel) and pre-wet in 20× SSC. Place tissues papers on top of the 3MM to draw the 20× SSC from the plastic tray through the gel to the nylon membrane. A weight of approximately 500 g should be placed on the tissues and left for nucleic acid transfer overnight (see Note 8). The following day dismantle the assembly and mark the nylon membrane with a pencil to identify the side exposed to the gel. Place the membrane between two sheets of 3MM paper and bake in an oven at 80°C for 2 h to fix the nucleic acid permanently to the membrane.

6. DNA or RNA may be used as a probe. As DNA is most routinely used in our laboratory, discussion of DNA probes will be focussed upon. Many RV and most LV vectors use the WPRE sequence in the vector backbone. This can serve as an ideal probe to identify the vector genome. Alternatively, LTR-specific probes can be generated. In the case of the *hprt* gene, the cDNA of this gene should be used. (Exon 1 appears not be identifiable using this cDNA and hence an exon 1 PCR probe must be generated for hybridisation).

For *hprt* analysis, an *hprt* ORF may be obtained from GeneCopoeia™ to generate a radioactive probe for hybridisation. To prepare an *hprt* exon 1 PCR amplified product, exon 1 DNA *hprt* primers must be used (Table 1). The probe is then purified using Bio-Spin™ 6 columns prior to hybridisation on nylon membranes. The PCR mix consists of 0.25 U Taq polymerase (Promega), 0.25 mM dNTPs (Promega), 1.5 mM $MgCl_2$ 5 mM of d(AGT)TPs plus 50 mCi of 3,000 Ci/mM alpha ^{32P}-CTP.

PCR of DNA sequences, up to 6 kb in length, can be achieved using the method described by Ponce and Micol (43).

7. DNA probes are easily prepared after restriction from plasmid vectors followed by isolation from agarose gel and purification using a gel purification kit. Alternatively, the probe may be generated by PCR. The DNA is labelled highly efficiently using a random prime labelling kit. Typically, 100 ng of cDNA is used with the Amersham Rediprime II DNA labelling system to generate a probe using $[\alpha\text{-}^{32}P]dCTP$ that provides specific activity of approximately 10^9 dpm/μg DNA. Unincorporated nucleotides are removed using Bio-Spin™ 6 column (Bio-rad)

Table 1
Primers for hprt amplification

Primer	Product size (bp)	MgCl$_2$ (mM)	Sequence
Exon 1	321	1.5	F, 5′ GTA CCT GGC CCC AGG AGC CAC C 3′ R, 5′ TCC GCT CTG CTG AAG AGT CCC G 3′
Exon 2	166	1.5	F, 5′ AGC TTA TGC TCT GAT TTG AAA TCA GCT G 3′ R, 5′ ATT AAG ATC TTA CTT ACC TGT CCA TAA TC 3′
Exon 3	220	1.33	F, 5′ CCG TGA TTT TAT TTT TGT AGG ACT GAA AG 3′ R, 5′ AAT GAA TTA TAC TTA CAC AGT AGC TCT TC 3′
Exon 4	191	1.33	F, 5′ GTG TAT TCA AGA ATA TGC ATG TAA ATG ATG 3′ R, 5′ CAA GTG AGT GAT TGA AAG CAC AGT TAC 3′
Exon 5	247	1.33	F, 5′ AAC ATA TGG GTC AAA TAT TCT TTC TAA TAG 3′ R, 5′ GGC TTA CCT ATA GTA TAC ACT AAG CTG 3′
Exon 6	145	1.33	F, 5′ TTA CCA CTT ACC ATT AAA TAC CTC TTT TC 3′ R, 5′ CTA CTT TAA AAT GGC ATA CAT ACC TTG C 3′
Exons 7 and 8	243	1.33	F, 5′ GTA ATA TTT TGT AAT TAA CAG CTT GCT GG 3′ R, 5′ TCA GTC TGG TCA AAT GAC GAG GTG C 3′
Exon 9	734	1.67	F, 5′ CAA TTC TCT AAT GTT GCT CTT ACC TCT C 3′ R, 5′ CAT GCA GAG TTC TAT AAG AGA CAG TCC 3′

to remove unincorporated nucleotides as per manufacturer's instructions.

8. To hybridise to the membrane with the probe of choice, pre-hybridise the membrane for 1 h at 65°C with hybridisation buffer, then replace with fresh hybridisation buffer preheated to 65°C. The labelled probe is heated to 95°C to separate the DNA strands and then added directly. Hybridisation should be left overnight at 65°C. The following day, after pouring away the labelled buffer containing the probe, wash the membrane twice at 65°C for 10 min with wash 1. Monitor the radioactivity carefully with a Geiger counter to ensure signal is present.

If a strong reading exists, then wash membrane twice for 10 min at 65°C with wash 2. As soon as the radioactive reading falls to below between 5 and 10 counts per second, expose the membrane to an autoradiogram film (see Note 9). The length of exposure should be monitored by film development initially after 1 h and, depending on the presence or absence of bands, repeated for longer of shorter periods to obtain optimal results.

Table 1 Primers for hprt amplification

Cycling parameters:

Exon 1: 94°C/5 min, 70°C/19 min; 1 cycle, 94°C/1 min, 60°C/1 min, 70°C/45 s; 30 cycles.

Exons 2–9: 94°C/5 min, 70°C/13 min; 1 cycle, 94°C/1 min, 58°C/1 min, 70°C/45 s; 30 cycles.

3.5.2. RNA Extraction Assay and Northern Blotting

1. Transfer frozen tissue or cells (50–100 mg) to a 15-ml tube with 1 ml TRIzol® and homogenise for 1 min in a polytron followed by the addition of 200 μl chloroform and gentle mixing. Incubate for 3 min at room temperature and then centrifuge at $12,000 \times g$ for 15 min and transfer the aqueous phase into a fresh Eppendorf tube (pre-washed in DEPC water and autoclaved). Add 500 μl isopropanol and centrifuge at $12,000 \times g$ for 10 min at 4°C, then 500 μl of 70% ethanol is added to the pellet to wash and centrifuge at $7,500 \times g$ for 5 min at 4°C. The supernatant should be discarded and the pellet dried by air for 10 min. The pellet should then be re-suspended in 100 μl DEPC-H_2O water by incubation for 10 min at 60°C.

2. To analyse the RNA 2 μg should be dissolved in 11 μl denaturation buffer. Add 1 μl of ethidium bromide and denature at 65°C for 15 min. Load sample onto a 1% agarose gel in MOPS buffer plus 5% formaldehyde (100–120 ml gel) and set the gel to run at 25 V to separate RNA using size marker of choice (usually 100 bp ladder). Run gel in 1×MOPS running buffer until marker is ~2/3 down the length of the gel.

   The RNA is then transferred to a nylon membrane (as described in Subheading 3.5.1, step 5). The nylon membrane with RNA fixed can be stained by washing in 5% acetic acid for 5 min. Stain with 0.1% Methylene Blue in 5% Acetic acid for 5 min, then de-stain with dH$_2$O until bands appear and mark on the blot the positions of the 28S, 18S and the different bands of the size marker.

   Hybridisation of DNA probes should be performed as described in Subheading 3.5.1, step 8.

Inverse PCR is carried out according to the protocol described by
Silver and Keerikatte (1989) (36) with minor modifications.

1. Choose a restriction enzyme (enzyme 1) that cuts frequently
   within the mammalian genome (4 base pair) and that has a
   recognition sequence close to <100 bp and locates 3′ of the RV
   or LV LTR.

2. Design two primers that will hybridise to each end of the vector 5′ LTR, one to read outwardly in the 5′ direction from the
   5′ LTR and the other to read in a 3′ direction from the 5′ LTR.
   Ensure these are both 5′ to the enzyme 1 restriction site.
   Hence, one primer will read in a 5′ direction outwardly from
   the vector to the integration site and host genome and the
   second should read in the 3′ direction into the vector genome.
   Ideally, 20–23 oligomer primers allow melting temperatures
   high enough to avoid non-specific amplification products.

3. Choose another restriction enzyme to cut between these primers (enzyme 2).

4. Digest 1 μg of Genomic DNA with enzyme 1 according to
   manufacturer's instructions in a total of 40 μl in an Eppendorf
   tube. Heat inactivate the enzyme for 10 min at 90°C. Make
   sure to spin down the contents of the tube before opening
   (10 s at maximum speed).

5. Ligate the restricted DNA with T4 ligase at 16°C for 4 h to
   generate circles containing the vector LTR bordered by integrated genomic DNA.

6. Take 10 μl of the ligation mix and digest with the second
   enzyme (enzyme 2), heat inactivate and spin down as before to
   linearise the DNA circles, thus producing integrated genomic
   DNA bordered by known LTR sequences that will be recogn-
   ised by the primers.

7. Using 2 μl of this sample, amplify the DNA using the following PCR mix: 12.4 μl dH_2O, 2 μl of 10× buffer Taq buffer,
   0.5 μl of 10 mM dNTPs, 0.6 μl of 50 mM $MgCl_2$, 1 μl of
   10 μM primer 1, 1 μl of 10 μM primer 2. The reaction volume
   is 20 μl in total. Set PCR conditions according to TM values
   5°C below primer melting temperatures, typically, 94°C for
   60 s, 60°C for 45 s, and 72°C for 60 s for 30 cycles.

8. To check for amplification, run a 5 μl aliquot of the PCR products using bromophenol blue dye and 100 bp ladder marker in
   a 2% agarose gel made with TAE buffer and electrophorese in
   TAE.

9. The PCR should then be cloned using a TA cloning kit. Briefly,
   4 μl of fresh PCR product containing 50 ng of DNA is mixed
   with 1 μl of the linear pCRII®-TOPO® plasmid and 1 μl of salt
   solution each supplied by the manufacturer. This is placed on

ice for 5 min. It is important to transform pUC19, provided with the kit, in parallel to test the transformation efficiency of the DH5α competent cells supplied.

10. Transform the PCR (and controls pUC19 and dH$_2$O only) into DH5α competent cells by adding 5 μl of DNA to the cells on ice for 30 min. Heat shock the cells for 42 s at 42°C, then return on ice for 2 min. Add 800 μl of SOC medium and shake the cells at 37°C for an hour. Pre-prepared agar plates containing 50 μg/ml ampicillin and spread with 40 mg/ml X-gal solution are then used to spread a dilution of one in ten of the cells. The plates are then incubated at 37°C overnight.

11. The following day white colonies are picked (using a sterile tooth pick) from agar plates and "dipped" into a PCR tube containing an identical reaction mix used for PCR in step 7 to check for the presence of the desired product. This is run on an agarose electrophoresis gel with an agarose percentage dependent on the size of the PCR product expected.

12. In parallel, the colony should be "dipped" into LB broth and grown overnight to purify the plasmid containing the vector LTR/inserted genomic DNA for sequencing.

13. PCR-positive colonies identified and isolated from steps 11 and 12 should be sequenced to identify the inserted genomic DNA. This step is usually performed using a commercial DNA sequencing facility.

14. Sequence data should be used for BLAST (http://www.ncbi.nlm.nih.gov/genome/seq/MmBlast.html) and BLAT searching (http://genome.ucsc.edu) of the mouse genome. During this search, vector sequences should be identified and removed from the genomic DNA sequence in order to prevent alignment to the vector during the search. The resulting sequence is then available to identify the vector insertion site(s). Searches should be made within a window of 100 kbp near to RefSeq genes. Molecular functions and gene ontologies can be identified during this search to find candidate genes for consideration as potentially involved in tumourigenesis. Genes identified in this way are also checked for listing in the Mouse Retroviral Tagged Cancer Gene Database (RTCGD) http://rtcgd.ncifcrf.gov.

3.5.4. Ligation-Mediated PCR

Ligation-mediated PCR (LAM PCR) is carried out according to the protocol described by Schmidt et al. (2002) (44) (see Note 10).

1. 100 ng of genomic DNA purified from infected tissue or cells is pre-amplified in a linear PCR using 0.25 pmol of vector-specific 5′-biotinylated LTR primer. Ideally, a 23–25 oligomer primer is designed for this, and conditions are set appropriate for the primer. Typically, samples are subjected to the following

conditions: denaturation at 95°C for 5 min, annealing for 1 min (at the appropriate primer TM less 5°C), extension at 72°C for 1 min for 50 cycles is followed by a further 50 cycles with the addition of fresh Taq polymerase. Finally, an extension at 72°C for 10 min is used.

2. Biotinylated DNA is captured and enriched using Dynal® beads. Re-suspend the beads by vortexing in an Eppendorf tube, then place 20 μl of the beads into a fresh tube. Place the Eppendorf in a magnetic carrier and remove the supernatant from the beads. Remove the tube from the rack and re-suspend in 20 μl of wash solution. Return to the rack, remove supernatant again, and re-suspend the beads in 50 μl of binding solution taking care not to cause foaming. Add 50 μl of linear PCR product to the magnetic beads, mix and incubate for 1 h (at room temperature) with occasional mixing. The sample is then returned to the magnet. Remove and discard the supernatant and then re-suspended the beads in 100 μl of dH_2O.

3. The sample is removed from the magnet and a Klenow mix is prepared consisting of 14.5 μl dH_2O, 0.5 μl 100 μm dNTP 2 μl 10× React2Buffer, 2 μl 10× hexanucleotide mix, and 1 μl of 2 U/μl Klenow; 20 μl of this mix is used to re-suspend the sample. The mixture is then incubated at 37°C for 1 h.

4. Prepare a DNA linker cassette. This is designed to ligate to DNA, which has been restricted with Tsp509I (or any other regular four base pair cutter of the mammalian genome not present in between the primer used in the first PCR step and upstream 5′ LTR sequences) as per the manufacturer's instructions. Linker sequences are LC1: (5′-GACCCGGGAGA TCTGAATTCAGTGGCACAGCATTAGG-3′) and LC2: (5′-AATTCCTAACTGCTGTGCCACTGAATTCAG ATC-3′), respectively (40). A mix of 40 μl of each linker LC1 and LC2 are mixed with 55 μl 500 mM Tris/HCl (pH 7.5) and 20 μl 500 mM $MgCl_2$. The linker cassette master mix is heated to 95°C for 5 min. The mix is allowed to cool at room temperature for 1 h generating the linker cassette.

5. To the sample, 80 μl of dH_2O is added and once again immobilised using the magnet and the supernatant is replaced with a 20 μl restriction digest containing Tsp509I according to the manufacturer's instructions and incubated at 65°C for 1 h.

6. During this incubation, the linker cassette is concentrated from 200 to 80 μl using a Micron-YM30 concentrator. Briefly, 200 μl of the linker cassette is added to the membrane of the concentrator and centrifuged at 13,000 rpm for 8 min; 20 μl of dH_2O is added to the column, then centrifuged for 1 min at 13,000 rpm. After 1 min, the column is inverted and re-centrifuged for 2 min. dH_2O is then added to reach a final volume of 80 μl.

7. The Tsp509I restricted sample is placed in the magnet and the sample washed in 80 μl of dH$_2$O. This is repeated with 100 μl of dH$_2$O. To the washed sample is added 10 μl of a ligation mix consisting of 5 μl dH$_2$O, 1 μl of the linker cassette, 1 μl of 10 mM ATP, 2 μl ligation buffer, and 1 μl of 2 U/μl T4 ligase. This is left overnight at room temperature.

8. The sample is then immobilised on the magnet and washed twice in 100 μl dH$_2$O; 5 μl of freshly made NaOH is added to the sample for 10 min to denature the sample.

9. PCR primers designed to recognise the viral LTR and LC1 (5′-GATCTGAATTCAGTGGCACAG-3′) are used to amplify the linker, the captured genomic sequence and part of the vector LTR. PCR amplification follows the conditions described in step 1. The PCR products are then used for a second round PCR using nested primers for the vector LTR and the LC1 linker (5′AGTGGCACAGCAGTTAGG3′).

10. PCR amplicons may be checked by adding to a 3% agarose gel. The PCR products may either be cloned using the TA cloning Kit as described (Subheading 7.4.1.5, step 9) followed by conventional Sanger sequencing or is directly sequenced using 454 pyrosequencing. The latter method is usually beyond the ability of most laboratories due to the cost of equipment and should be sought commercially.

3.5.5. Real-Time PCR

1. Genomic DNA obtained from cells or tissues is to be used for virus copy number investigation. For gene expression analysis, RNA should be isolated from cells or tissues as described (Subheading 3.5.2) using TRIzol® reagent.

2. cDNA preparation is performed using Multiscribe™ Reverse transcriptase and random hexamer primers as per manufacturer's instructions on 125 ng RNA.

3. 2 μl cDNA from the reaction is used as template for TaqMan Gene Expression Assays specific to each gene.

4. For genomic or cDNAs, a reaction (per sample) should be prepared on ice using the pre-prepared TaqMan® Universal PCR master mix containing No AmpErase UNG (2×) with AmpliTaq Gold® DNA polymerase, dNTPs and buffers with 1 μl of the primer/probe set for the gene under examination and dH$_2$O to give a total of 10 μl. Triplicate reactions are then transferred to a 96-well plate (depending on the thermal cycler of choice).

5. Amplification reactions are performed using the thermal cycler of choice, an example being the 7900HT Real-Time PCR Thermal Cycler (Applied Biosystems) with parameters specific for the gene or virus packaging signal under examination. This provides CT values for amplifications (see Notes 11 and 12).

6. The relative expression level of each gene can be calculated by the ΔΔCt method (45) normalised to *18s* or *GAPDH RNA* expression (housekeeping gene control), and represented as fold change relative to mock-transduced samples (calibrator).

4. Notes

Infection of V79 cells

1. When infecting V79 cells with virus stocks, the vector may be filtered through a 0.2-μm or 0.4-μm filters before addition to the target cells. Care should be taken when filtering the vector since they easily block. This is particularly true for 0.2 μm filters. As a result, virus titre will drop significantly. Unless bacterial contamination is an issue, the 0.4-μm filter is the best option. Even so, care should be taken to renew the filter if vector passage through the filter is slow, indicating blockage.

2. When isolating infected *hprt⁻* V79 clones be sure not to contaminate the clones with each other as this leads to spurious insertion data. Ideally, remove clones from separate plates or use the 96-well method to isolate pure clones resistant to 6TG. Care should also be taken before isolating clones when handling culture plates or re-feeding plates, so that cross-contamination of clones is avoided.

3. Never allow cells to become confluent in culture as this alters phenotype, i.e. growth characteristics and cells begin to undergo senescence. Cells also become difficult to remove from cell culture dishes/plates by trypsin if allowed to overgrow.

Monitoring mice following vector administration in utero

4. Treated mice expected to develop tumours should be monitored daily to prevent animal suffering and sudden death, which results in loss of valuable material that can be used for genotoxicity analysis.

5. All tissue samples should always be subjected to snap freezing as described in Chapter 10.

Preparation of genomic DNA and Southern analysis

6. When preparing genomic DNA from tissues using liquid nitrogen, care should be taken to avoid spillage and to work in an area with plenty of ventilation. Between samples the use of a pestle and mortar should involve thorough washing of this equipment carefully between tissues to avoid contamination that would be picked up during PCR. Samples should be thoroughly re-suspended in extraction buffer either slowly or using a pipette tip that is cut at the end to allow viscous DNA to pass through.

7. Enzyme restriction of genomic DNA is commonly found to be problematic. This is easily overcome by addition of the restriction enzyme reagents to the DNA followed by gentle swirling of the mix for 2–3 min using the end of a 200 µl pipette tip to thoroughly mix the sample. Occasional mixing can be used during restriction. Make sure that the restriction period does not exceed the manufacturer's stated time since many enzymes have "star" activity that leaves DNA far too digested to identify virus bands after hybridisation with radiolabelled probes.

8. During Southern transfer do not use a weight greater than 500 g as this will compress the paper towels on top of the nylon membrane and prevent even transfer of DNA to the nylon membrane.

9. When washing off probes from DNA that are non-specific, make sure to use short washes with least stringency to start with to prevent loss of signal. Monitor radioactivity after each wash carefully and at any point place membranes under film. Too much signal is better than no signal at all!

LAM PCR

10. During LAM PCR, it is important not to produce bubbles when pipetting samples up and down. This reduces washing of the sample and results in loss of DNA for PCR.

Real-time PCR controls

11. $\Delta\Delta$Ct calculations after real-time PCR are used to provide relative amplification levels between the mock-treated and experimental samples. Where vector copy analysis is being performed, a tumour or cell clone should be used with a known virus copy number obtained using Southern analysis. $\Delta\Delta$Ct values from this standard sample should be generated from a dilution set to find the optimal DNA concentration for real-time PCR. This may then be compared with unknown samples under investigation at the DNA concentration of choice.

12. Between experimental and control sets, experiments should be run in triplicate or quadruplicates to generate standard deviations of the means. Confidence intervals at 95% should be set and p-values calculated (values <0.05 should be regarded significant).

Acknowledgements

The author would like to thank Professors Charles Coutelle and Dr. Manfred Schmidt for advice and LAM PCR methodology, respectively, and Dr Brian Bigger and Emma Osenjindu for real-time PCR advice. This work was supported by the Faculty of Medicine of Imperial College London and by a Brunel University BRIEF award.

References

1. Rohdewohld H, Weiher H, Reik W et al (1987) Retrovirus integration and chromatin structure: moloney murine leukemia proviral integration sites map near DNase I-hypersensitive sites. J Virol 61:336–343

2. Wu X, Li Y, Crise B et al (2003) Transcription start regions in the human genome are favored targets for MLV integration. Science 300: 1749–1751

3. Varmus HE, Padgett T, Heasley S et al (1977) Cellular functions are required for the synthesis and integration of avian sarcoma virus-specific DNA. Cell 11:307–319

4. Mitchell RS, Beitzel BF, Schroder AR et al (2004) Retroviral DNA integration: ASLV, HIV, and MLV show distinct target site preferences. PLoS Biol 2:E234

5. Stocking C, Bergholz U, Friel J et al (1993) Distinct classes of factor-independent mutants can be isolated after retroviral mutagenesis of a human myeloid stem cell line. Growth Factors 8:197–209

6. King WMDP, Lobel LI, Goff SP et al (1985) Insertional mutagenesis of embryonal carcinoma cells by retroviruses. Science 228: 554–558

7. Grosovsky AJ, Skandalis A, Hasegawa L et al (1993) Insertional inactivation of the tk locus in a human B lymphoblastoid cell line by a retroviral shuttle vector. Mutat Res 289:297–308

8. Themis M, May D, Coutelle C et al (2003) Mutational effects of retrovirus insertion on the genome of V79 cells by an attenuated retrovirus vector: implications for gene therapy. Gene Ther 10:1703–1711

9. Hacein-Bey-Abina S, Le Deist F, Carlier F et al (2002) Sustained correction of X-linked severe combined immunodeficiency by ex vivo gene therapy. N Engl J Med 346:1185–1193

10. Cavazzana-Calvo M, Hacein-Bey S, de Saint Basile G, Gross F, Yvon E, Nusbaum P, Selz F, Hue C et al (2000) Gene therapy of human severe combined immunodeficiency (SCID)-X1 disease. Science 288:669–672

11. Hacein-Bey-Abina S, Von Kalle C, Schmidt M et al (2003) LMO2-associated clonal T cell proliferation in two patients after gene therapy for SCID-X1. Science 302:415–419

12. Williams DA, Baum C (2003) Medicine. Gene therapy – new challenges ahead. Science 302:400–401

13. Howe SJ, Mansour MR, Schwarzwaelder K et al (2008) Insertional mutagenesis combined with acquired somatic mutations causes leukemogenesis following gene therapy of SCID-X1 patients. J Clin Invest 118:3143–3150

14. Ott MG, Schmidt M, Schwarzwaelder K et al (2006) Correction of X-linked chronic granulomatous disease by gene therapy, augmented by insertional activation of MDS1-EVI1, PRDM16 or SETBP1. Nat Med 12:401–409

15. Li Z, Dullmann J, Schiedlmeier B et al (2002) Murine leukemia induced by retroviral gene marking. Science 296:497

16. Baum C, von Kalle C, Staal FJ et al (2004) Chance or necessity? Insertional mutagenesis in gene therapy and its consequences. Mol Ther 9:5–13

17. Montini E, Cesana D, Schmidt M et al (2006) Hematopoietic stem cell gene transfer in a tumor-prone mouse model uncovers low genotoxicity of lentiviral vector integration. Nat Biotechnol 24:687–696

18. Nienhuis AW, Dunbar CE, Sorrentino BP (2006) Genotoxicity of retroviral integration in hematopoietic cells. Mol Ther 13:1031–1049

19. Baum C, Dullmann J, Li Z et al (2003) Side effects of retroviral gene transfer into hematopoietic stem cells. Blood 101:2099–2114

20. Modlich U, Kustikova OS, Schmidt M et al (2005) Leukemias following retroviral transfer of multidrug resistance 1 (MDR1) are driven by combinatorial insertional mutagenesis. Blood 11:4235–4246

21. Seggewiss R, Pittaluga S, Adler RL et al (2006) Acute myeloid leukemia is associated with retroviral gene transfer to hematopoietic progenitor cells in a rhesus macaque. Blood 107:3865–3867

22. Zychlinski D, Schambach A, Modlich U et al (2008) Physiological promoters reduce the genotoxic risk of integrating gene vectors. Mol Ther 16:718–725

23. Mazurier F, Gan OI, McKenzie JL et al (2004) Lentivector-mediated clonal tracking reveals intrinsic heterogeneity in the human hematopoietic stem cell compartment and culture-induced stem cell impairment. Blood 103:545–552

24. Kustikova OS, Schiedlmeier B, Brugman et al (2009) Cell-intrinsic and vector-related properties cooperate to determine the incidence and consequences of insertional mutagenesis. Mol Ther 17:1537–1547

25. Zhang LH, Jenssen D (1991) Characterization of HAT- and HAsT-resistant HPRT mutant clones of V79 Chinese hamster cells. Mutat Res 263:151–158

26. Zhang LH, Vrieling H, van Zeeland AA et al (1992) Spectrum of spontaneously occurring mutations in the hprt gene of V79 Chinese hamster cells. J Mol Biol 223:627–635

27. Zhang LH, Jenssen D (1994) Studies on intra-chromosomal recombination in SP5/V79 Chinese hamster cells upon exposure to different agents related to carcinogenesis. Carcinogenesis 15:2303–2310

28. Dahle J, Kvam E (2003) Induction of delayed mutations and chromosomal instability in fibroblasts after UVA-, UVB-, and X-radiation. Cancer Res 63:1464–1469

29. Dahle J, Noordhuis P, Stokke T et al (2005) Multiplex polymerase chain reaction analysis of UV-A- and UV-B-induced delayed and early mutations in V79 Chinese hamster cells. Photochem Photobiol 81:114–119

30. Jianhua Z, Lian X, Shuanlai Z et al (2006) DNA lesion and Hprt mutant frequency in rat lymphocytes and V79 Chinese hamster lung cells exposed to cadmium. J Occup Health 48:93–99

31. Bokhoven M, Stephen SL, Knight S et al (2009) Insertional gene activation by lentiviral and gammaretroviral vectors. J Virol 83:283–294

32. Goff SP (1987) Insertional mutagenesis to isolate genes. Methods Enzymol 151:489–502

33. Montini E, Cesana D, Schmidt M et al (2009) The genotoxic potential of retroviral vectors is strongly modulated by vector design and integration site selection in a mouse model of HSC gene therapy. J Clin Invest 119:964–975

34. Waddington SN, Mitrophanous KA, Ellard FM, Buckley SM, Nivsarkar M, Lawrence L, Cook HT, Al-Allaf F, Bigger B, Kingsman SM, Coutelle C, Themis M (2003) Long-term transgene expression by administration of a lentivirus-based vector to the fetal circulation of immuno-competent mice. Gene Ther 10:1234–1240

35. Themis M, Waddington SN, Schmidt M et al (2005) Oncogenesis following delivery of a nonprimate lentiviral gene therapy vector to fetal and neonatal mice. Mol Ther 12:763–771

36. Silver J, Keerikatte V (1989) Novel use of polymerase chain reaction to amplify cellular DNA adjacent to an integrated provirus. J Virol 63:1924–1928

37. Schmidt M, Carbonaro DA, Speckmann C et al (2003) Clonality analysis after retroviral-mediated gene transfer to CD34+ cells from the cord blood of ADA-deficient SCID neonates. Nat Med 9:463–468

38. Akagi K, Suzuki T, Stephens RM et al (2004) RTCGD: retroviral tagged cancer gene database. Nucleic Acids Res 32:D523–D527

39. Modlich U, Bohne J, Schmidt M et al (2006) Cell-culture assays reveal the importance of retroviral vector design for insertional genotoxicity. Blood 108:2545–2553

40. Themis M, Waddington SN, Schmidt M et al (2005) Oncogenesis following delivery of a nonprimate lentiviral gene therapy vector to fetal and neonatal mice. Mol Ther 12:763–771

41. Dupont WD, Plummer WD Jr (1998) Power and sample size calculations for studies involving linear regression. Control Clin Trials 19:589–601

42. Sambrook J, Fritsch FE, Maniatis T (2001) Molecular Cloning: A Laboratory manual, second edition. Cold Spring Harbour Laboratory Press (1) Chapter 6. 6.23

43. Ponce MR, Micol JL (1992) PCR amplification of long DNA fragments. Nucleic Acids Res 20:623

44. Schmidt M, Zickler P, Hoffmann G et al (2002) Polyclonal long-term repopulating stem cell clones in a primate model. Blood 100:2737–2743

45. Pfaffl MW (2001) A new mathematical model for relative quantification in real-time RT-PCR. Nucleic Acids Res 29:e45

Chapter 17

Risks, Benefits and Ethical, Legal, and Societal Considerations for Translation of Prenatal Gene Therapy to Human Application

Charles Coutelle and Richard Ashcroft

Abstract

The still experimental nature of prenatal gene therapy carries a certain degree of risk, both for the pregnant mother as well as for the fetus. Some of the risks are procedural hazards already known from more conventional fetal medicine interventions. Others are more specific to gene therapy such as the potential for interference with normal fetal development, the possibility of inadvertent germ line gene transfer, and the danger of oncogenesis. This chapter reviews the potential risks in relation to the expected benefits of prenatal gene therapy. It discusses the scientific, ethical, legal, and social implications of this novel preventive approach to genetic disease and outlines preconditions to be met in preparation for a potential future clinical application.

Key words: Prenatal (fetal, in utero) gene therapy, Genetic disease, Disease prevention, Risk/benefit, Fetal development, Germline gene transfer, Oncogenesis, Ethical principles, Legal, Societal, Adverse effects, Disability, Preimplantation selection, Genetic screening, Abortion

1. Introduction

The great advances in human genome sequencing have given us the tools for detection of almost any mutation causing monogenetic disease as well as for devising novel approaches to therapy, among them gene therapy. However, besides many problems in achieving effective long-term postnatal gene therapy without side effects, even successful therapeutic gene expression would often come too late for many of these diseases with early onset manifestations.

As described in detail in the previous chapters of this book (see Chapters 9–12), animal research on in utero gene therapy has demonstrated successful gene transfer to virtually all disease-relevant fetal organs and tissues. It has shown that expression in utero can provide tolerance to the (therapeutic) foreign protein in postnatal

Charles Coutelle and Simon N. Waddington (eds.), *Prenatal Gene Therapy: Concepts, Methods, and Protocols*,
Methods in Molecular Biology, vol. 891, DOI 10.1007/978-1-61779-873-3_17, © Springer Science+Business Media, LLC 2012

life, thus avoiding the problem of immune reactions, common to present postnatal gene therapy. Gene transfer to stem cells and propagation of the transferred genes in their progeny by clonal cell-expansion has also provided prolonged expression of the foreign proteins in fetal and postnatal life. All these advances have culminated in several independent first proofs of principle for therapeutic and, in some cases even for life-long curative effects of intrauterine gene therapy in rodent models of serious human genetic disorders. Moreover, the use of the sheep and nonhuman primate models has demonstrated that minimally invasive technologies as applied in human fetal medicine could potentially be adapted for successful intrauterine gene application to the human fetus (see Chapters 11 and 12).

These achievements confirm the conceptual prediction, that prenatal gene therapy could prevent early onset manifestation of serious genetic diseases by effectively accessing affected tissues including their stem cell compartment, that would be more difficult to reach later in life. This, together with the induction of immune tolerance to the therapeutic protein would provide the basis for a life-long disease correction (1).

It therefore appears that technically all prerequisites for a clinical application of intrauterine gene therapy of selected human genetic diseases have been met, with a good chance of therapeutic success. This seems particularly true for hemophilia B where lasting therapeutic levels of human FIX without development of neutralizing antibodies have been achieved following prenatal human FIX gene delivery to mice (2) and non-human primates (3).

However, as of now and in contrast to postnatal gene therapy, no such trials are planned or have been undertaken. This restraint is governed mainly by considerations concerning both the perceived and established risks of in utero gene therapy as well as the uniquely different medical, ethical, and legal perception of prenatal vs. postnatal gene therapy by individuals and society (4).

In 1998, the US NIH Recombinant Advisory Committee (RAC) responded to a preproposal to conduct human in utero gene therapy trials for hemoglobinopathies and adenosine deaminase deficiency (ADA) (5–7) by requesting more experimental investigations and assessment of known and potential risks (also compare GTAC in UK (8)). Since then many experimental studies and theoretical considerations have increased our insight into the risks of gene therapy including those specific to in utero application. This chapter reviews the experimental facts relating to the main potential risks of prenatal gene delivery and discusses their medical, ethical, and legal consequences for the future of human in utero gene therapy. We also attempt to recommend research approaches in order to study the societal and personal perceptions of the new opportunities presented by the progress in prenatal gene therapy research.

Like all Fetal Medicine interventions, human prenatal gene therapy involves both the mother and the fetus. While centered on the health of the fetus, its delivery would have to be an entirely maternal decision, and maternal wellbeing should always take precedence in case of complications (9–13). This principle, which is expressed in English law, will also govern the legal and ethical consideration of this chapter. Similar considerations have previously been made with respect to fetal interventional therapies such as fetal surgery and placental gene therapy (14). Risks of the procedure such as injuries, infections, premature labor, and fetal death, common to Fetal Medicine interventions, have been dealt with elsewhere (10). An overview of the various potential adverse effects of in utero gene therapy and approaches to monitor for such events is given in Chapters 14 and 15.

2. Specific Risk Factors of Prenatal Gene Therapy Causing Ethical Concern

The main identified or presumed specific risks of prenatal gene therapy that cause ethical concern are the potential to cause disruption of normal fetal development, to carry an increased risk for germ line transmission of the genetic modification, and to cause genotoxicity and/or oncogenesis. Due to the novelty and high degree of uncertainty of outcome the unifying concern here is the balance between the ethical principles of beneficence (acting to the benefit of the patient) and nonmaleficence (avoiding harming the patient).

2.1. Risk of Disruption of Normal Fetal Development (See also Chapter 15)

Some examples of how viruses and drugs can cause disturbances of normal fetal development are presented in Chapter 14. Given these examples it is difficult to exclude that a particular transgenic protein or RNA with a known postnatal therapeutic function may have an unknown (harmful) effect when expressed prenatally.

A recent investigation on fetal mice by Flake's group (15) demonstrated the temporal dependence of transgene expression after prenatal gene delivery, due to changing accessibility of the vector to tissues and stem cell compartments and the variation of expression of cellular receptors during early fetal development. The same group has previously also observed the unexpected development of adenomatous malformations in the lung of rat fetuses after prenatal over-expression of fibroblast growth factor 10 (FGF 10) (16). More recently Tarantal's group reported that adenovirus-mediated transient over-expression of transforming growth factor (TGF) β1 in the fetal monkey lung during the second or third trimester led to severe pulmonary and pleural fibrosis (17). These observations highlight how the very rapid changes in cellular composition and environment during the dynamic processes of fetal cell differentiation and organogenesis, as well as the level of transgenic expression, may compromise the intended effects of our therapeutic vector systems. Such effects would presumably be specific for each gene product.

The great diversity of potentially therapeutic gene sequences excludes general predictions, although particular care may be warranted when applying certain transgenes, such as those encoding growth factors or hormones. In order to avoid such problem it may be possible to target the cell type, expression level, and time of expression by regulated vector design (reviewed in ref. 18). Some information towards the safety of potentially therapeutic transgene expression in utero may be gained by investigation of the natural expression of the particular protein during normal fetal development. For instance CFTR, the protein which is absent or mutated in cystic fibrosis (CF), is expressed in normal human fetuses in all major organs, which become affected in a CF patient; lung expression even at higher levels than postnatally (19–21). The blood coagulation factor IX (FIX) is already expressed physiologically in 5–10 weeks old human fetuses at a plasma level of 10% of normal adult values (22) and remains at this relatively low level until birth (23). However, AAV-mediated expression of hFIX up to ninefold the normal level has caused no adverse effects prenatally or postnatally in macaques illustrating the need for case-by-case investigations and the value of primate studies with the respective therapeutic gene in preparation for human application (3). This may still not eliminate all potential risks to the human fetus and close monitoring throughout the pregnancy and following birth will be mandatory (see also Chapter 14).

2.2. Risk of Germline Transfer (See also Chapter 15)

The risk of germline transmission of the gene therapy construct is often discussed with the misconception of it being an aim of fetal gene therapy. It is therefore important to stress that fetal somatic gene therapy, like postnatal gene therapy, aims only at the treatment of individual patients, not of their offspring! Any gene therapy study will aim to prevent as far as possible the accidental transfer of the therapeutic gene to germ cells of the treated patient (24). However, there is a valid concern that gene transfer in utero could pose an increased risk of inadvertent germline gene transfer. Prenatal gene therapy in humans is not expected to be conducted before 7 weeks of gestation. At this time the germ cells are compartmentalized in the genital ridge (25). The risk of accidental blood-borne or topical gene transfer should therefore not be greater than in postnatal gene therapy. Vector DNA sequences have been found by sensitive PCR analysis in testes and ovaries of experimental mice after vector administration in utero (26), but the transduced cells were not exactly identified in this study. By elaborate microdissection or immunohistochemical analysis low numbers of genetically modified germ cells were found in testes and ovaries of sheep and monkeys, respectively, after intraperitoneal injection in utero (24, 25). The transient detection of AAV vector sequences in human sperm samples during a postnatal clinical gene therapy study for hemophilia (27) led to discontinuation of this trial although it was never investigated whether gene transfer to spermatocytes had actually occurred. No vector sequences were found in spermatocytes in specific studies carried out in rabbits, rats, and dogs (28) after postnatal AAV-injection, in mice after pre- or postnatal AAV administration (29), and in mice after lentiviral gene delivery in utero (2). Equally no vector sequences could be found in oocytes from maternal macaques after FIX-AAV in utero gene delivery to their fetuses (3). Vector transmission to subsequent generations could not be detected after in utero application of AAV in mice (29) or of retrovirus vectors in sheep (30), and extensive breeding experiments on mice after postnatal adenovirus gene therapy also showed no transmission to 578 offspring of crosses in which either one or both parents had received recombinant adenovirus (31) (see also Chapter 15).

In contrast to the nonspecific and therefore not detectable changes in the germ line, which arise naturally, due to environmental toxins, or medical procedures such as X-ray or chemotherapy, the detection of the well-defined sequence-specific gene therapy constructs is possible and should therefore be carried out as part of any gene therapy study. But even if a transmission is found in germ cells, any risk assessment should consider the endogenous insertion rate (32) and take a balanced view on the benefits and risks the transmission of a therapeutic gene sequence may have for the treated patients and for future generations. It should also be noted that even if proof for transfer of a therapeutic transgene to germ cells of a treated patients is found, this does not mean that all germ cells are affected or that procreation has to be avoided. If germ cell gene transfer does occur during in utero gene therapy, it is likely to affect only a very small percentage of mature germ cells (33, 34) making the risk of therapeutic gene transfer sequences to the offspring very low to start with. And even then it is possible to prevent further transmission reliably by in vitro fertilization and preimplantation genetic diagnosis; or prenatal diagnosis and, where appropriate, termination of pregnancy.

Currently the main reasons why the scientific community opposes deliberate germline manipulation of the human germ line are firstly that based on present knowledge, there is no sound medical need for such manipulation, and secondly that the techniques which would have to be used to achieve this efficiently (i.e., transfer of manipulated nuclei, oocyte pronuclear gene injection, spermatogonia gene transfer), carry a far too high risk of severe adverse effects, making them ethically unacceptable for human application (4, 24).

2.3. Risk of Oncogenesis (See Also Chapter 16)

That integrating gene therapy vectors can lead to tumor development has been shown by the occurrence of lymphoproliferative disease in 5 of the 21 children treated by gene therapy with retroviral vectors for the deadly immune deficiency X-SCID. This potential for oncogenesis is a result of the relatively random insertion of the vector into active genes in the host genome, which can lead to activation of oncogenes or inactivation of tumor-suppressor genes. Most of these genes have a physiological function in the regulation of growth and differentiation processes and are therefore particularly active during fetal development. They would thus be preferred integration sites for prenatally delivered gene transfer vectors and in utero gene therapy could therefore be particularly susceptible to oncogenesis by integrating vectors. Such a risk has indeed been demonstrated in fetal mice with vectors derived from the horse anemia lentivirus (EIAV). Approximately 100 days after systemic application of this vector system in utero or neonatally, almost all treated animals developed liver tumors (35). Similar results were also obtained when vectors derived from the Feline Immune deficiency virus (FIV) were used (Themis unpublished). In contrast, no tumors were found over the lifetime of treated mice, when vectors based on the human immunodeficiency virus (HIV) were applied, indicating that most likely some virus component in the EIAV and FIV vectors, respectively, plays a decisive role in triggering the oncogenic process. This may give us clues to the mechanism of the induction of oncogenesis and for the construction of safer vectors. It is worth noting that in postnatal stem cell gene therapy, such as the X-SCID trials we are also dealing with actively dividing and differentiating cells. In both cases safer integrating vectors or the development of long-term expressing nonintegrating vector constructs are urgently required. Concerning the oncogenesis risk it is also relevant to note that the occurrence of leukemogenesis in the SCID trials was seen several months after gene therapy and the liver tumors observed in mice developed in approximately mid-life of the in utero treated animals. This long-term postnatal risk, which will require monitoring of fetally treated individuals for a lifetime may add to the ethical difficulties in choosing an in utero gene intervention using integrating vectors.

3. Risk/Benefit Assessment and Related Ethical and Legal Considerations

It is important to emphasize that "a previable fetus is totally dependent on a pregnant women's autonomous decision for its status in medicine" (36). This includes any decision on the fate of the pregnancy, including therapeutic interventions or termination within legal and medical time limits. This would also imply that the mother could not be sued by her offspring for decisions to reject or accept fetal gene therapy (13). Thus, for practical legal purposes we are mainly concerned with ensuring that fetal gene therapy poses no or little risk to the mother. But for ethical purposes we are also concerned with the much larger question of how far fetal gene therapy can be of benefit to the fetus and future child, and what risks it may pose to its health and development, against the background of its underlying condition.

Following prenatal diagnosis of many severe and early onset genetic diseases for which no satisfactory therapies are available (see Chapter 2), the present alternatives are acceptance of the predictable serious problems of a severely disabled child and its care or the termination of pregnancy. Prenatal gene therapy would add a third option to deal with such conditions. Choosing between these three options is significantly different to decision-making in postnatal gene therapy, which is presently predominantly seen as a last attempt to treat or prolong life in severe disease, after all conventional treatment options have been exhausted. In these cases therapeutic failure will be a disappointing but accepted outcome, which importantly, will however not worsen the given situation.

In contrast, effective prevention of disease without adverse effects will usually be hoped for when choosing in utero gene therapy over termination of pregnancy. It has therefore been argued, that given the present risks of failure or even harm of in utero gene therapy, its preference over termination would be very unlikely; and furthermore, that in vitro fertilizations followed by preimplantation diagnosis/embryo selection would anyhow be an alternative reproductive choice for a family with a known genetic risk (37, 38). Although preimplantation genetic diagnosis has, no doubt, been proven to work clinically, this procedure based on hormone-induced ovulation, is very demanding, expensive, and not without risks (39). Most importantly, however, this procedure is only applicable with prior knowledge of the genetic risk in this family, but not once pregnancy is under way. Indeed the most effective application of prenatal gene therapy, once proven to be reliable and safe, would be achieved by combination with carrier screening and early prenatal diagnosis that is now becoming established for a range of severe genetic diseases (40).

Clearly more work to generate safer vector systems and testing for therapeutic and adverse effects by long-term animal studies, preferably on primate models, is required to assure benefit and reduce potential harm before clinical application of in utero gene therapy can be considered. Even so, it will not be possible to completely exclude risks, and clinical applications will require particular care to assure that very high ethical standards are applied to safeguard autonomy in decision-making. This should include comprehensive and timely information given to the pregnant women (and her partner). All three options of dealing with an affected fetus must remain available for autonomous decision. The intended benefits and potential risks of each option and in particular those of prenatal gene therapy must be comprehensively presented and explained, making sure that this information is really fully understood and no coercion in either directions is exerted. The pregnant women should also be guaranteed the freedom to withdraw from such a clinical trial and to have a termination of pregnancy at any time within medically and legally accepted time limits, even after prenatal gene therapy has been delivered.

There are few circumstances in medicine more susceptible to investigator bias and patient misinformation than fetal therapy, making counseling the pregnant mother a particularly difficult task (for example, see ref. 41). Fetal therapy and gene therapy have been sensationalized in the public sector and expectations and hopes are often magnified. The parents are often unusually susceptible to investigator bias or optimism because they have just received the information that their fetus is effected and emotional detachment and rational objectivism are difficult when considering treatment of a child, even if unborn. Many of the risks and potential benefits are in reality undefined and therefore susceptible to investigator and patient interpretation. These considerations emphasize the requirement for institutional oversight of fetal therapy, and a structured and well executed informed consent process, supported by clear ethical guidelines, to provide an appropriate context for counseling and care (42 and Flake, personal communication).

Given the particular situation of in utero gene therapy it would be ethically questionable to conduct pure phase 1 clinical trials, which would study safety aspects but not aim at delivering therapeutic benefits (4, 43). It may, thus, be ethically much more acceptable to first drive preclinical studies, preferable on non-human primates, to a level of safety and efficiency, which would provide sufficient confidence to allow human trials to be started at the phase 2/3 level; with very close monitoring of safety aspects and, if possible, prenatal analysis of therapeutic success.

The most likely first applications of fetal gene therapy would probably be conducted for one of the serious life-threatening monogenic diseases for which no curative postnatal therapy is available (see Chapter 2) in families who decline a termination of pregnancy, but may place their last hope of having a healthy child on prenatal gene therapy. The genetic status of the fetus would have been established by DNA diagnosis on parents and the fetus. Noninvasive fetal DNA analysis can now be achieved for some single gene disorders (44), although for the majority invasive fetal testing is required. Nevertheless, it is likely that a certain prenatal diagnosis would be possible by 12 weeks of gestation, allowing sufficient time for prenatal gene therapy to be applied. Preferably initially a disease would be chosen that does not require fine tuning of gene expression and in which the success of therapy can be assessed already in utero by biochemical tests, so that in case of failure, a termination of pregnancy could still be performed.

It is of course not yet known if the immune tolerance acquired by in utero gene therapy may be beneficial to a repeat of gene therapy after birth in case the therapeutic effect of prenatal gene delivery is lost. Equally it is unclear if successful prenatal gene therapy can be repeated on a further affected child. Most likely these outcomes will very much depend on the actual transgene and vector in question. However, to avoid this uncertainty and since the genetic risk to the family will be known from the first affected pregnancy it may be preferable to use preimplantation genetic diagnosis and embryo selection to avoid the particular genetic disease in future offspring.

Once established, prenatal gene therapy is likely to be much easier to perform and to be more cost effective than ex vivo preimplantation genetic diagnosis and embryo selection. Depending on its safety record and the acceptance by society it may also change views on prenatal genetic screening and could find application for less severe diseases such as coagulopathies or even hyperlipidaemia as well as prenatal gestation conditions like preeclampsia. This would extend, what presently appears to be a very sophisticated, highly specialized potential therapeutic application to a much broader use with equal accessibility within the heath care system, and thereby also justify the present costs of its development.

4. Beyond Risk–Benefit Analysis

The approach described so far in this chapter follows the standard analysis in clinical trial ethics of trying to analyze the risks and benefits of the procedure under trial in order to ensure that the trial, and the intervention under trial, are beneficial to the trial subject (or at least not harmful), and that the risks and benefits to trial participants are fair, both in terms of the offer of participation being fair to the participant, and fair in terms of the distribution of risk across the populations of actual and potential trial participants (45–47). However, this approach has been criticized for being too "formal" and overlooking some of the critical issues in the ethics of human genetic modification (48).

Four issues are central here. The first is whether it is right, in principle, to modify the human genome (48, 49). While this raises important questions of ethics, philosophy, and human rights, to some extent it is missing the point of contemporary human gene therapy, including fetal gene therapy, which is conceived as an intervention to modify gene expression and function in individuals suffering specific gene defects or clinical illness, rather than as "human genetic engineering" focussing on improving or altering human individuals or humanity as such.

There remains considerable ethical controversy regarding human germline modification. As pointed out above, this is not an aim of current individual-focused adult or fetal gene therapy (24). However, it might be argued that in utero therapy may strengthen the case for deliberate germline modification on the basis that germline gene transfer may be both technically simpler and more effective in achieving the preventive goal of somatic in utero gene therapy and that it would be sustained in following generations.

This argument is, however, seriously flawed since the preventive goal, is already achievable even more effectively by in vitro fertilization/preimplantation selection, based on prior genetic knowledge for a given family; knowledge, which, of course, would also be required for germline modification. Unlike fetal gene therapy, germline modification is therefore not applicable in the case of an already established affected pregnancy. Furthermore, and most importantly, no effective and safe in vitro or ex vivo method is available at present or in the foreseeable future, which would be acceptable for gene transfer into human germ cells and as already mentioned, there is presently no realistic medical need for such intervention.

One could argue that successful treatment of genetic disease in the absence of germ line correction is unethical because it will alter the human genome by increasing the frequency of the defect in the population as an effect of procreation by the treated individuals. Given a broad based successful fetal (or even postnatal) gene therapy program (or for that matter pharmaceutical treatment) for a particular early onset genetic disease one could envisage an increase in the heterozygote frequency of the particular mutant gene in the population after many generations. However, in practical terms of disease manifestation and the need for therapy in the offspring of a treated individual it is, worth remembering that heterozygocity will lead only in dominant conditions to disease manifestation and furthermore that any such person will be carefully monitored long-term as part of any clinical gene therapy program. The "prior knowledge" of the individual gene mutation in a treated individual will trigger preconception DNA analysis of the partner for this gene. Based on this information, for dominant diseases or, in the rare case of the partner carrying a defect in the same gene in recessive disease, preimplantation selection of unaffected embryos can be carried out. This will already counteract the propagation of potentially harmful mutations. Probably only very rarely would fetal gene therapy on an affected fetus be required in such families.

It is, however worth remembering, that we are just at the dawn of molecular therapy and that we can expect techniques of safe mutation-specific correction to be developed long before any population effect of present therapy related gene pool alteration could emerge. Such developments could also make safe germ line alteration possible. If this happens it is for future generation to decide on the need and the ethical implications of its practical application.

A second issue concerns how far in utero therapy is problematic from the point of view of the rights of disabled people. A very important argument has been mounted, to the effect that choosing to treat or prevent certain kinds of disabilities, or to consider certain disabilities as essentially bad, or to overlook the dimensions of disability, which are social rather than biological in nature, is both conceptually incoherent and disrespectful of disabled people (50, 51). This is a complex, controversial argument, which is not uncommon among disability activists. However, it can hardly be denied that disability impairs people, and that being disabled is not something inherently beneficial. This does not mean that we do not recognize that dealing with disability can bring out the best in affected people as well as in their caretakers. We also point out, that the social dimensions of disability are a consequence rather than the cause of the underlying genetic problem and that by offering means to prevent the biological problem we are in no way denying the need to address the medical and social problems of already existing disability or to enable affected people to live a fulfilled life! Furthermore, for present purposes we would also note, first that in utero gene therapy expands the range of options open to parents, rather than reducing them; and second that it offers ways to prevent or treat a genetic disorder, i.e., goals which are consistent with established good medical practice. Fetal gene therapy is therefore, with regards to its aims certainly not *more* problematic than any other modern medical approach to treating and managing disability.

This takes us to the third problem, which is that in utero gene therapy could promote the image of the fetus or embryo as independent "prenatal life," thereby weakening women's power to determine whether or not to continue their pregnancies, and undermining the priority of their own care and interests during pregnancy over the interests (or future interests) of the child-to-be (10, 11). As noted in Subheading 1 and Subheading 3, we maintain in accordance with current law that the previable embryo is *not* an independent legal entity. However, women may come to feel obliged to continue a pregnancy and undergo fetal gene therapy in utero, rather than terminate the pregnancy. This raises the debate of whether it is ethical to counsel a mother about an unproven but potentially beneficial fetal therapy prior to her decision to continue her pregnancy. Depending upon the nature of the counseling and the bias or equipoise of the investigators, such counseling could induce considerable pressure. Such psychological pressure is less probable in the stage of clinical trials of in utero gene therapy, where the risks of failure and the burdens of research participation are likely to be significantly salient to women's decision-making, and take-up is expected to be restricted to women who decline termination of pregnancy as a point of principle. But it may become relevant once fetal gene therapy becomes an established clinical option in some conditions. Such pressure is, however, not entirely new and actually spans a wide spectrum of perinatal conditions. It applies to various nongene-therapy operative and conventional medical interventions on the fetus, which may be required to ensure or improve postnatal life. It will influence decision making on procedures that may be urgently required to improve maternal survival chances such as an unexpected need for chemo- or radio-therapy during pregnancy. It also applies to severe lifestyle changes such as the requirement of a special maternal diet to avoid the adverse consequences of toxic maternal metabolites on fetal brain development in cases of maternal phenylketonuria and extends to "simple things" such as giving up smoking and alcohol intake during or even before pregnancy or ultimately, even deciding whether to have the baby at all.

In short, these examples illustrate how important it is to safe-guard that deciding to continue pregnancy is a conscious and autonomous maternal decision. Once made, this involves obliga-tions towards the wellbeing of the future child, which may restrict some of the mother's freedoms. However, she should be able to reverse her decision within medical and legal time frames.

The social and medical environment including gene therapy should *expand* women's options and enable free and informed decision making during pregnancy. This is an area where detailed social science investigations into women's attitudes, rights and responsibilities during pregnancy, are necessary and these investi-gations should include the possibilities of fetal gene therapy.

A fourth and final problem is that of values. This goes beyond the purely rational analysis of the complex trade-offs of risk and benefit, often under substantial uncertainty as to the nature, proba-bility, and short and long-term outcomes of particular decisions. In the situation of a woman trying to choose whether to terminate her pregnancy because of the disability she fears her child may have if it is born, or whether to expose that child to the risks of fetal gene therapy, part of this choice can be managed through reflection on risk and benefit. Unfortunately, in many circumstances, the true risks and benefits are unknown and in that circumstance decision-making must be predominantly based on a patient's values; for instance the value of having or not having a child; the value of having an affected child versus not having a child at all; the value of personal health/freedom versus the wish for a healthy child; and many more.

Not only is this intellectually very challenging, it is emotionally very fraught. It is arguable that counseling may be of benefit, but only up to a point; and that making this decision, which the preg-nant woman (with or without her partner's participation and engagement) must make, involves placing a perhaps intolerable burden on her shoulders. However, it is equally the case that deny-ing her that option and more to the point, denying her the possi-bility of a therapy, which could offer significant benefits to her child would be both paternalistic and harmful. This is where the process of non-directive counseling becomes paramount. The ethical cor-nerstones of fetal intervention are beneficence-based obligation to the mother and fetus, the primacy of maternal autonomy in the decision making process, the dependent moral status of the fetus, and the prioritization of the accuracy and integrity of the informed consent process.

5. Research Agenda

Fetal gene therapy raises important, though we believe tractable, moral issues. For this preventive approach to human genetic disease to become therapeutically viable, not only technical success is required, but also its transition into clinical trials that lead to broader clinical application. The major obstacles to these transitions are regulatory and social, as much as scientific and clinical. As outlined above, on the one hand we need clearer and better evidence to permit robust risk assessment, which would warrant transition into clinical trials (4). On the other hand, we also need greater clarity about the social and ethical constraints and trade-offs, which would morally underwrite such progression to clinical use. Some of these issues are to do with "public engagement" with the technology to increase knowledge, awareness and understanding of the goals and technical possibilities. Others are to do with "acceptability": under what conditions, if any, would this technology become acceptable to the public, and with what degree of caution or enthusiasm. And finally, we have to understand better what the fundamental public moral attitudes are here—how do we grasp what the moral principles animating public debate in this area are, beyond those entrenched in regulatory bodies and academic bioethics (48, 52). Some of these issues may be susceptible to large-scale social surveys; others to laboratory and field psychological research; others may require qualitative social research (53, 54). In order to develop this knowledge, it will be important to develop a "roadmap" of the transition points from bench to bedside, in order to understand which decisions lie where, who makes them, and how they make them.

Developing such a roadmap is a task for both medical scientists and experts in social research. On the one hand it, should aim to inform and educate the public and in particular those sections of it to whom prenatal gene therapy would be most relevant, about the medical benefits as well as about the possible risks of prenatal gene therapy. On the other hand, it should serve to gather information and organize feedback on public perception to this new technology. Without wanting to restrict the scope of this undertaking we suggest to include the following survey groups: individuals and families affected by genetic disease, support groups for such families, disabled patients organizations, randomly chosen pregnant women, and representative groups of different strata of the general population, as well as social and health care professionals with day to day experience of the effects of genetic disease. In a first attempt we would like to itemize some of the principal questions, which should be addressed in preparation for translation of in utero gene therapy into human trials as follows:

- What is known about genetic diseases in general and perhaps also about a specific genetic disease, before and after being exposed to targeted information?

- What is known about the benefits and hazards of gene therapy and in particular prenatal gene therapy before and after being exposed to targeted information?

- When is it acceptable to expose a woman to medical or surgical interventions for the sake of her fetus during pregnancy?

- What kinds of disease or disorder are considered serious enough to warrant in utero intervention?

- How effective must an in utero intervention be for it to be worth considering?

- How likely must it be to succeed, at any given level of effectiveness, for it to be worth considering?

- How much risk to the woman of the procedure is tolerable?

- How much risk to the fetus (in terms of potential miscarriage) of the procedure is tolerable?

- How much risk, and at what severity, to the fetus in terms of lifetime illness or disability (over and above that caused by the pre-existing genetic disorder) is tolerable?

- What risk of inadvertent germline modification in the fetus is tolerable?

- Are there ever reasons, other than those related to risk and benefit, which would make in utero therapy unacceptable?

- Does the potential, or actual, effectiveness of fetal gene therapy mean that a mother (or the parents) are morally obliged to try this therapy, either for the good of their unborn child, or for the benefit of society?

- Does the potential of fully or partly curing genetic disorders in utero mean that parents are responsible if their child develops a genetic disorder because they *did not* use the therapy?

- Does the potential of curing, relieving, or preventing genetic disorder alter how you feel about people with disability, now or in future?

- Do you think fetal gene therapy is "genetic engineering"? Or "just like any other medicine"? Or something else?

- Who is the main beneficiary of fetal gene therapy: The mother? The fetus? Society? Someone else?

Acknowledgements

The authors wish to thank Alan Flake, Anna David, and Donald Peebles for valuable commments and suggestions.

References

1. Coutelle C, Douar A-M, College WH, Froster U (1995) The challenge of fetal gene therapy. Nat Med 1:864–866

2. Waddington S, Nivsarkar M, Mistry A, Buckley SMK, Al-Allaf F, Bigger B, Holder M, Kemball-Cook G, Mosley KL, Brittan M, Ali R, Gregory L, Cook HT, Thrasher A, Tuddenham EGD, Themis M, Coutelle C (2004) Permanent phenotypic correction of Haemophilia B in immunocompetent mice by prenatal gene therapy. Blood 104:2714–2721

3. Mattar CN, Nathwani AC, Waddington SN, Dighe N, Kaeppel C, Nowrouzi A, McIntosh J, Johana NB, Ogden B, Fisk NM, Davidoff AM, David A, Peebles D, Valentine MB, Appelt JU, von Kalle C, Schmidt M, Biswas A, Choolani M, Chan JK (2011) Stable human FIX expression after 0.9G intrauterine gene transfer of self-complementary adeno-associated viral vector 5 and 8 in Macaques. Mol Ther 9:1950–1960

4. Kimmelman J (2010) Gene transfer and the ethics of first-in-human research. Cambridge University Press, Cambridge

5. RAC (2000) Prenatal gene transfer: scientific medical and ethical issues, a report of the Recombinant DNA Advisory Committee. Hum Gene Ther 11:1211–1229

6. Couzin J (1998) RAC confronts in utero gene therapy proposal. Science 282:27

7. Zanjani ED, Anderson WF (1999) Prospects for in utero human gene therapy. Science 285:2084–2088

8. Committee, GTA (1998) Report on the potential use of gene therapy in utero. Department of Health, London. Available at: http://www.advisorybodies.doh.gov.uk/genetics/gtac/inutero.htm. Accessed 2 Feb 2011

9. Roberts MA (1998) Child versus childmaker: future persons and present duties in ethics and the law New York: Rowman and Littlefield. Rowman and Littlefield, New York

10. Scott R (2002) Rights, duties and the body: law and ethics of the maternal-fetal conflict. Hart, Oxford

11. Dickenson DLe (2002) Ethical issues in maternal–fetal medicine. Cambridge University Press, Cambridge

12. Peters PG Jr (2004) How safe is safe enough? Obligations to the children of reproductive technology. Oxford University Press, Oxford

13. Todd S (2010) Actions arising from birth. In: Grubb A, Laing J, McHale J (eds) Principles of medical law, 3rd edn. Oxford University Press, Oxford, pp 265–323

14. David A, Ashcroft R (2009) Placental gene therapy. Obstet Gynaecol Reprod Med 19: 296–298

15. Endo M, Henriques-Coelho T, Zoltick PW, Stitelman DH, Peranteau WH, Radu A, Flake AW (2010) The developmental stage determines the distribution and duration of gene expression after early intra-amniotic gene transfer using lentiviral vectors. Gene Ther 17: 61–71

16. Gonzaga S, Henriques-Coelho T, Davey M, Zoltick PW, Leite-Moreira AF, Correia-Pinto J, Flake AW (2008) Cystic adenomatoid malformations are induced by localized FGF10 overexpression in fetal rat lung. Am J Respir Cell Mol Biol 39:346–355

17. Tarantal AF, Chen H, Shi TT, Lu CH, Fang AB, Buckley S, Kolb M, Gauldie J, Warburton D, Shi W (2010) Overexpression of TGF-{beta}1 in foetal monkey lung results in prenatal pulmonary fibrosis. Eur Respir J 36:907–914

18. Toscano MG, Romero Z, Munoz P, Cobo M, Benabdellah K, Martin F (2011) Physiological and tissue-specific vectors for treatment of inherited diseases. Gene Ther 18:117–127

19. Tizzano EF, Chitayat D, Buchwald M (1993) Cell-specific localization of CFTR mRNA shows developmentally regulated expression on human fetal tissues. Hum Mol Genet 2: 219–224

20. Tizzano E, O'Brodovich H, Chitayat D, Benichou J, Buchwald M (1994) Regional expression of CFTR in developing human respiratory tissues. Am J Respir Cell Mol Biol 10:355–362

21. Broackes-Carter FC, Mouchel N, Gill D, Hyde S, Bassett J, Harris A (2002) Temporal regulation of CFTR expression during ovine lung development: implications for CF gene therapy. Hum Mol Genet 11:125–131

22. Hassan HJ, Leonardi A, Chelucci C, Mattia G, Macioce G, Guerriero R, Russo G, Mannucci PM, Peschle C (1990) Blood coagulation factors in human embryonic-fetal development: preferential expression of the FVII/tissue factor pathway. Blood 76:1158–1164

23. Mibashan RS, Rodeck CH, Thumpston JK, Edwards RJ, Singer JD, White JM, Campbell S (1979) Plasma assay of fetal factors VIIIC and IX for prenatal diagnosis of haemophilia. Lancet 1:1309–1311

24. Rasko JEJ, O'Sullivan GM, Ankeny RA (eds) (2006) The ethics of inheritable genetic modification: a dividing line? Cambridge University Press, Cambridge

25. Gillman J (1948) The development of the gonads in man, with a consideration of the role of fetal endocrines and histogenesis of ovarian tumours. Contrib Embryol 32:81–92

26. Seppen J, van der Rijt R, Looije N, van Til NP, Lamers WH, Oude Elferink RP (2003) Long-term correction of bilirubin UDPglucuronyltransferase deficiency in rats by in utero lentiviral gene transfer. Mol Ther 8: 593–599

27. Manno CS, Arruda VR, Pierce GF, Glader B, Ragni M, Rasko J, Ozelo MC, Hoots K, Blatt P, Konkle B, Dake M, Kaye R, Razavi M, Zajko A, Zehnder J, Nakai H, Chew A, Leonard D, Wright JF, Lessard RR, Sommer JM, Tigges M, Sabatino D, Luk A, Jiang H, Mingozzi F, Couto L, Ertl HC, High KA, Kay MA (2006) Successful transduction of liver in hemophilia by AAV-Factor IX and limitations imposed by the host immune response. Nat Med 12:342–347

28. Arruda VR, Fields PA, Milner R, Wainwright L, De Miguel MP, Donovan PJ, Herzog RW, Nichols TC, Biegel JA, Razavi M, Dake M, Huff D, Flake AW, Couto L, Kay MA, High KA (2001) Lack of germline transmission of vector sequences following systemic administration of recombinant AAV-2 vector in males. Mol Ther 4:586–592

29. Jakob M, Muhle C, Park J, Weiss S, Waddington S, Schneider H (2005) No evidence for germline transmission following prenatal and early postnatal AAV-mediated gene delivery. J Gene Med 7:630–637

30. Porada C, Tran N, Eglitis M, Moen RC, Troutman L, Flake AW, Zhao Y, Anderson WF, Zanjani ED (1998) In utero gene therapy: transfer and long-term expression of the bacterial neo(r) gene in sheep after direct injection of retroviral vectors into preimmune fetuses. Hum Gene Ther 9:1571–1585

31. Ye X, Gao GP, Pabin C, Raper SE, Wilson JM (1998) Evaluating the potential of germ line transmission after intravenous administration of recombinant adenovirus in the C3H mouse. Hum Gene Ther 9:2135–2142

32. Kazazian H (1999) An estimated frequency of endogenous insertional mutations in humans. Nat Genet 22:130

33. Schuettrumpf J, Liu JH, Couto LB, Addya K, Leonard DG, Zhen Z, Summer J, Arruda VR (2006) Inadvertent germline transmission of AAV2 vector: findings in a rabbit model correlate with those in a human clinical trial. Mol Ther 13:1064–1073

34. Park PJ, Colletti E, Ozturk F, Wood JA, Tellez J, Almeida-Porada G, Porada C (2009) Factors determining the risk of inadvertent retroviral transduction of male germ cells after in utero gene transfer in sheep. Hum Gene Ther 20: 201–215

35. Themis M, Waddington SN, Schmidt M, von Kalle C, Wang Y, Al-Allaf F, Gregory L, Nivsarkar M, Themis M, Holder M, Buckley SMK, Dighe N, Ruthe A, Mistry A, Bigger B, Rahim A, Nguyen TH, Trono D, Thrasher AJ, Coutelle C (2005) Oncogenesis following delivery of a non-primate lentiviral gene therapy vector to fetal mice Molecular Therapy. Mol Ther 12: 763–771

36. Fletcher J, Richter G (1996) Human fetal gene therapy: moral and ethical questions. Hum Gene Ther 7:1605–1614

37. Scott R (2007) Choosing between possible lives: law and ethics of prenatal and preimplantation genetic diagnosis. Hart, Oxford

38. Wilkinson S (2010) Choosing tomorrow's children: the ethics of selective reproduction. Oxford University Press, Oxford

39. Gelbaya TA (2010) Short and long-term risks to women who conceive through in vitro fertilization. Hum Fertil (Camb) 13:19–27

40. Scott SA, Edelmann L, Liu L, Luo M, Desnick RJ, Kornreich R (2010) Experience with carrier screening and prenatal diagnosis for 16 Ashkenazi Jewish genetic diseases. Hum Mutat 31:1240–1250

41. Adzick N, Thom EA, Spong CY, Brock JW, Burrows PK, Johnson MP, Howell LJ, Farrell JA, Dabrowiak ME, Sutton LN, Gupta N, Tulipan NB, D'Alton ME, Farmer DL, the MOMS Investigators, N. E. J. M. F. E. a. o (2011) A randomized trial of prenatal versus postnatal repair of myelomeningocele. N Engl J Med 364:994–1004

42. Flake A (2001) Prenatal intervention: ethical considerations for life-threatening and non-life-threatening anomalies. Semin Pediatr Surg 10:212–221

43. Strong C (2011) Regulatory and ethical issues for phase I in utero gene transfer studies. Hum Gene Ther 22:1323–1330

44. Tsui NB, Kadir RA, Chan KC, Chi C, Mellars G, Tuddenham EG, Leung TY, Lau TK, Chiu RW, Lo YM (2011) Noninvasive prenatal diagnosis of hemophilia by microfluidics digital PCR analysis of maternal plasma DNA. Blood 117:3684–3691

45. Ashcroft RE, Viens AM (2008) Clinical trials. In: The Cambridge textbook of bioethics. Singer PA, Viens AM (eds) Cambridge University Press, Cambridge, pp 201–206

46. Green RM (2008) Embryo and fetal research. In: The Cambridge textbook of bioethics. Singer PA, Viens AM (eds) Cambridge University Press, Cambridge, pp 231–237

47. Meslin EM, Dickens BM (2008) Research ethics. In: The Cambridge textbook of bioethics. Singer PA, Viens AM (eds) Cambridge University Press, Cambridge, pp 187–193

48. Evans JH (2002) Playing God? Human genetic engineering and the rationalization of public bioethical debate 1959–1995. University of Chicago Press, Chicago

49. Wright S (1994) Molecular politics: developing American and British regulatory policy for genetic engineering, 1972–1982. University of Chicago Press, Chicago

50. Shakespeare TW (2006) Disability rights and wrongs. Routledge, London

51. Scully JL (2008) Disablity bioethics: moral bodies, moral difference New York Rowman and Littlefield. Rowman and Littlefield, New York

52. Jasanoff S (2005) Designs on nature: science and democracy in Europe and the United States. Princeton University Press, Princeton

53. Bauer M, Gaskell G (eds) (2002) Biotechnology: the making of a global controversy. Cambridge University Press, Cambridge

54. Sugarman J, Sulmasy DF (eds) (2001) Methods in medical ethics. Georgetown University Press, Washington, DC

INDEX

Charles Coutelle and Simon N. Waddington (eds.), *Prenatal Gene Therapy: Concepts, Methods, and Protocols*,
Methods in Molecular Biology, vol. 891, DOI 10.1007/978-1-61779-873-3, © Springer Science+Business Media, LLC 2012